# LONDON MATHEMATICAL SOCIETY STUDE[

Managing Editor: C. M. SERIES, Mathematics Institute, University of Warwick, Coventry CV4 7AL, United Kingdom

London Mathematical Society Student Texts 35

# Young Tableaux

## With Applications to Representation Theory and Geometry

William Fulton
Department of Mathematics
University of Michigan

CAMBRIDGE
UNIVERSITY PRESS

PUBLISHED BY THE PRESS SYNDICATE OF THE UNIVERSITY OF CAMBRIDGE
The Pitt Building, Trumpington Street, Cambridge, United Kingdom

CAMBRIDGE UNIVERSITY PRESS
The Edinburgh Building, Cambridge CB2 2RU, UK       http: //www.cup.cam.ac.uk
40 West 20th Street, New York, NY 10011-4211, USA   http: //www.cup.org
10 Stamford Road, Oakleigh, Melbourne 3166, Australia

© Cambridge University Press 1997

This book is in copyright. Subject to statutory exception
and to the provisions of relevant collective licensing agreements,
no reproduction of any part may take place without
the written permission of Cambridge University Press.

First published 1997
Reprinted 1999 (with corrections)

Typeset in Times Roman 11/14 pt. LaTeX $2_\varepsilon$ [TB]

A catalog record for this book is available from the British Library.

Library of Congress Cataloging-in-Publication Data

Fulton, William, 1939–
Young tableaux : with applications to representation theory
and geometry / William Fulton.
p.   cm. – (London Mathematical Society student texts : 35)
Includes bibliographical references and indexes.
1. Young tableaux.   I. Title.   II. Series.         95-47484
QA165.F83      1997                                  CIP
512′.2 – dc20

ISBN  0 521 56144 2 hardback
ISBN  0 521 56724 6 paperback

Transferred to digital printing 2003

# Contents

# Preface

The aim of this book is to develop the combinatorics of Young tableaux, and to see them in action in the algebra of symmetric functions, representations of the symmetric and general linear groups, and the geometry of flag varieties. There are three parts: Part I develops the basic combinatorics of Young tableaux; Part II applies this to representation theory; and Part III applies the first two parts to geometry.

Part I is a combinatorial study of some remarkable constructions one can make on Young tableaux, each of which can be used to make the set of tableaux into a monoid: the Schensted "bumping" algorithm and the Schützenberger "sliding" algorithm; the relations with words developed by Knuth and Lascoux and Schützenberger, and the Robinson–Schensted–Knuth correspondence between matrices with nonnegative integer entries and pairs of tableaux on the same shape. These constructions are used for the combinatorial version of the Littlewood–Richardson rule, and for counting the numbers of tableaux of various types.

One theme of Parts II and III is the ubiquity of certain basic quadratic equations that appear in constructions of representations of $S_n$ and $GL_m\mathbb{C}$, as well as defining equations for Grassmannians and flag varieties. The basic linear algebra behind this, which is valid over any commutative ring, is explained in Chapter 8. Part III contains, in addition to the basic Schubert calculus on a Grassmannian, a last chapter working out the Schubert calculus on a flag manifold; here the geometry of flag varieties is used to construct the Schubert polynomials of Lascoux and Schützenberger.

There are two appendices. Appendix A contains some of the many variations that are possible on the combinatorics of Part I, but which are not needed in the rest of the text. Appendix B contains the topology needed to assign a cohomology class to a subvariety of a nonsingular projective variety, and prove the basic properties of this construction that are needed for a study of flag manifolds and their Schubert subvarieties.

There are numerous exercises throughout. Except for a few numerical verifications, we have included answers, or at least hints or references, at the end of the text.

Our main aim has been to make the various main approaches to the calculus of tableaux, and their applications, accessible to nonexperts. In particular, an important goal has been to make the important work of Lascoux and Schützenberger more accessible. We have tried to present the combinatorial ideas in an intuitive and geometric language, avoiding the often formal language or the language of computer programs common in the literature. Although there are some references to the literature (most in *Answers and References* at the end), there is no attempt at a survey.[1]

Although most of the material presented here is known in some form, there are a few novelties. One is our "matrix-ball" algorithm for the Robinson–Schensted–Knuth correspondence, which seems clearer and more efficient than others we have seen, and for which the basic symmetry property is evident. This generalizes an algorithm of Viennot for permutations; a similar algorithm has been developed independently by Stanley, Fomin, and Roby. Appendix A contains similar "matrix-ball" algorithms for variations of this correspondence. In addition, this appendix contains correspondences between skew tableaux, related to the Littlewood–Richardson rule, which generalize recent work of Haiman.

Our proof of the Littlewood–Richardson rule seems simpler than other published versions. The construction given in Chapter 8 of a "Schur" or "Weyl" module $E^\lambda$ from a module $E$ over a commutative ring, and a partition $\lambda$, should be better known than it is. Unlike most combinatorics texts, we have preferred to develop the whole theory for general Young tableaux, rather than just tableaux with distinct entries, even where results for the general case can be deduced from those for the special tableaux. In Appendix B we show how to construct the homology class of an algebraic variety, using only properties of singular homology and cohomology found in standard graduate topology texts.

In the combinatorial chapters, it is possible to keep the presentation essen-

---

[1] In this subject, where many ideas have been rediscovered or reinterpreted over and over, such a survey would be a formidable task. In the same spirit, the references to the literature are intended to point the reader to a place to learn about a topic; there is little attempt to assign credit or find original sources.

tially self-contained. The other chapters, however, are aimed at the relations of these combinatorial ideas to representation theory and geometry. Here the goal of self-containment works against that of building bridges, and we have frankly chosen the latter goal. Although we have tried to make these chapters accessible to those with only a rudimentary knowledge of representation theory, algebraic geometry, and topology, the success of some sections may depend on the reader's background, or at least on a willingness to accept some basic facts from these subjects.

For more about what is in the book, as well as the background assumed, see the section on notation, and the introductions to the three parts. A few words should perhaps be added about what is not in this book. For much of the story presented here there are analogues for general semisimple groups, or at least the other classical groups, with a continually growing literature. We have not discussed any of this here; however, we have tried to develop the subject for the general linear group so that a reader encountering the general case will have some preparation. We have not discussed other notions about tableaux, such as bitableaux or shifted tableaux, that are used primarily for representations of other groups. Although some constructions are made over the integers or arbitrary base rings – and we have preferred such constructions whenever possible – we do not enter into the distinctive features of positive characteristic. Finally, we have not presented the combinatorial algorithms in the format of computer programs; this was done, not to discourage the reader from writing such programs, but in the hope that a more intuitive and geometric discussion will make this subject attractive to those besides combinatorialists.

This text grew out of courses taught at the University of Chicago in 1990, 1991, and 1993. I am grateful to to J. Alperin, S. Billey, M. Brion, A. Buch (1999), P. Diaconis, K. Driessel (1999), N. Fakhruddin, F. Fung, M. Haiman, J. Harris, R. Kottwitz, D. Laksov, A. Lascoux, P. Murthy, U. Persson, P. Pragacz, B. Sagan (1999), M. Shimozono, F. Sottile, T. Steger, J. Stembridge (1999), B. Totaro, and J. Towber for inspiration, advice, corrections, and suggestions.

William Fulton
February 1996

Dedicated to the memory of Birger Iversen.

# Notation

A *Young diagram* is a collection of boxes, or cells, arranged in left-justified rows, with a (weakly) decreasing number of boxes in each row. Listing the number of boxes in each row gives a partition of the integer $n$ that is the total number of boxes. Conversely, every partition of $n$ corresponds to a Young diagram. For example, the partition of $16$ into $6 + 4 + 4 + 2$ corresponds to the Young diagram

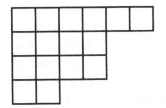

We usually denote a partition by a lowercase Greek letter, such as $\lambda$. It is given by a sequence of weakly decreasing positive integers, sometimes written $\lambda = (\lambda_1, \lambda_2, \ldots, \lambda_m)$; it is often convenient to allow one or more zeros to occur at the end, and to identify sequences that differ only by such zeros. One sometimes writes $\lambda = (d_1{}^{a_1} \ldots d_s{}^{a_s})$ to denote the partition that has $a_i$ copies of the integer $d_i$, $1 \leq i \leq s$. The notation $\lambda \vdash n$ is used to say that $\lambda$ is a partition of $n$, and $|\lambda|$ is used for the number partitioned by $\lambda$. We usually identify $\lambda$ with the corresponding diagram, so we speak of the second row, or the third column, of $\lambda$.

The purpose of writing a Young diagram instead of just the partition, of course, is to put something in the boxes. Any way of putting a positive integer in each box of a Young diagram will be called a *numbering* or *filling* of the diagram; usually we use the word *numbering* when the entries are distinct, and *filling* when there is no such requirement. A *Young tableau*, or simply

1

**tableau**, is a filling that is

(1) weakly increasing across each row

(2) strictly increasing down each column

We say that the tableau is a tableau *on* the diagram $\lambda$, or that $\lambda$ is the *shape* of the tableau. A *standard tableau* is a tableau in which the entries are the numbers from 1 to $n$, each occurring once. Examples, for the partition $(6,4,4,2)$ of 16, are

| 1 | 2 | 2 | 3 | 3 | 5 |
|---|---|---|---|---|---|
| 2 | 3 | 5 | 5 |   |   |
| 4 | 4 | 6 | 6 |   |   |
| 5 | 6 |   |   |   |   |

| 1 | 3 | 7 | 12 | 13 | 15 |
|---|---|---|----|----|----|
| 2 | 5 | 10 | 14 |   |   |
| 4 | 8 | 11 | 16 |   |   |
| 6 | 9 |   |    |   |   |

       Tableau               Standard tableau

Entries of tableaux can just as well be taken from any *alphabet* (totally ordered set), but we usually take positive integers.

Describing such combinatorial data in the plane also suggests simple but useful geometric constructions. For example, flipping a diagram over its main diagonal (from upper left to lower right) gives the *conjugate* diagram; the conjugate of $\lambda$ will be denoted here by $\tilde{\lambda}$. As a partition, it describes the lengths of the columns in the diagram. The conjugate of the above partition is $(4,4,3,3,1,1)$:

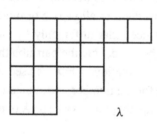         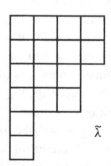

                      $\lambda$                     $\tilde{\lambda}$

Any numbering $T$ of a diagram determines a numbering of the conjugate, called the *transpose*, and denoted $T^{\tau}$. The transpose of a standard tableau is a standard tableau, but the transpose of a tableau need not be a tableau.

It may be time already to mention the morass of conflicting notation one will find in the literature. Young diagrams are also known as *Ferrers diagrams* or *frames*; sometimes dots are used instead of boxes, and sometimes, particularly in France, they are written upside down, in order not to offend Descartes. What we call tableaux are known variously as *semistandard* tableaux, or *column-strict* tableaux, or *generalized Young tableaux* (in which case our standard tableaux are just called *tableaux*). Combinatorialists also know them as *column-strict reversed plane partitions*; the "reversed" is in opposition to the case of decreasing rows and columns, which was studied first (and allows zero entries); cf. Stanley (1971).

Associated to each partition $\lambda$ and integer $m$ such that $\lambda$ has at most $m$ parts (rows), there is an important symmetric polynomial $s_\lambda(x_1, \ldots, x_m)$ called a **Schur polynomial**. These polynomials can be defined quickly using tableaux. To any numbering $T$ of a Young diagram we have a monomial, denoted $x^T$, which is the product of the variables $x_i$ corresponding to the $i$'s that occur in $T$. For the tableau in the first diagram, this monomial is $x_1 x_2{}^3 x_3{}^3 x_4{}^2 x_5{}^4 x_6{}^3$. Formally,

$$x^T = \prod_{i=1}^{m} (x_i)^{\text{number of times } i \text{ occurs in } T}.$$

The Schur polynomial $s_\lambda(x_1, \ldots, x_m)$ is the sum

$$s_\lambda(x_1, \ldots, x_m) = \sum x^T$$

of all monomials coming from tableaux $T$ of shape $\lambda$ using the numbers from 1 to $m$. Although it is not obvious from this definition, these polynomials are symmetric in the variables $x_1, \ldots, x_m$, and they form an additive basis for the ring of symmetric polynomials. We will prove these facts later.

The Young diagram of $\lambda = (n)$ has $n$ boxes in a row

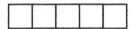

The Schur polynomial for this partition is the $n^{\text{th}}$ **complete symmetric polynomial**, which is the sum of all distinct monomials of degree $n$ in the variables $x_1, \ldots, x_m$; this is usually denoted $h_n(x_1, \ldots, x_m)$. For the other extreme $n = 1 + \ldots + 1$, i.e., $\lambda = (1^n)$, the Young diagram is

The corresponding Schur polynomial is the $n^{\text{th}}$ ***elementary symmetric polynomial***, which is the sum of all monomials $x_{i_1} \cdot \ldots \cdot x_{i_n}$ for all strictly increasing sequences $1 \le i_1 < \ldots < i_n \le m$, and is denoted $e_n(x_1, \ldots, x_m)$.

A ***skew diagram*** or ***skew shape*** is the diagram obtained by removing a smaller Young diagram from a larger one that contains it.[1] If two diagrams correspond to partitions $\lambda = (\lambda_1, \lambda_2, \ldots)$ and $\mu = (\mu_1, \mu_2, \ldots)$, we write $\mu \subset \lambda$ if the Young diagram of $\mu$ is contained in that of $\lambda$; equivalently, $\mu_i \le \lambda_i$ for all $i$. The resulting skew shape is denoted $\lambda/\mu$. A ***skew tableau*** is a filling of the boxes of a skew diagram with positive integers, weakly increasing in rows and strictly increasing in columns. The diagram is called its ***shape***. For example, if $\lambda = (5,5,4,3,2)$ and $\mu = (4,4,1)$, the following shows the skew diagram $\lambda/\mu$ and a skew tableau on $\lambda/\mu$:

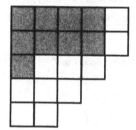

The set $\{1, \ldots, m\}$ of the first $m$ positive integers is denoted $[m]$.

---

[1] Algebraically, a collection of boxes is a skew shape if they satisfy the condition that when boxes in the $(i, j)$ and $(i', j')$ position are included, and $i \le i'$ and $j \le j'$, then all boxes in the $(i'', j'')$ positions are included for $i \le i'' \le i'$ and $j \le j'' \le j'$.

# Part I

## Calculus of tableaux

There are two fundamental operations on tableaux from which most of their combinatorial properties can be deduced: the Schensted "bumping" algorithm, and the Schützenberger "sliding" algorithm. When repeated, the first leads to the Robinson–Schensted–Knuth correspondence, and the second to the "jeu de taquin." They are in fact closely related, and either can be used to define a product on the set of tableaux, making them into an associative monoid. This product is the basis of our approach to the Littlewood–Richardson rule.

In Chapter 1 we describe these notions and state some of the main facts about them. The proofs involve relations among words which are associated to tableaux, and are given in the following two chapters. Chapters 4 and 5 have the applications to the Robinson–Schensted–Knuth correspondence and the Littlewood–Richardson rule. See Appendix A for some of the many possible variations on these themes.

# 1

# Bumping and sliding

## 1.1 Row-insertion

The first algorithm, called *row-insertion* or *row bumping,* takes a tableau $T$, and a positive integer $x$, and constructs a new tableau, denoted $T \leftarrow x$. This tableau will have one more box than $T$, and its entries will be those of $T$, together with one more entry labelled $x$, but there is some moving around. The recipe is as follows: if $x$ is at least as large as all the entries in the first row of $T$, simply add $x$ in a new box to the end of the first row. If not, find the left-most entry in the first row that is strictly larger than $x$. Put $x$ in the box of this entry, and remove ("bump") the entry. Take this entry that was bumped from the first row, and repeat the process on the second row. Keep going until the bumped entry can be put at the end of the row it is bumped into, or until it is bumped out the bottom, in which case it forms a new row with one entry.

For example, to row-insert 2 in the tableau

| 1 | 2 | 2 | 3 |
|---|---|---|---|
| 2 | 3 | 5 | 5 |
| 4 | 4 | 6 | |
| 5 | 6 | | |

the 2 bumps the 3 from the first row, which then bumps the first 5 from the second row, which bumps the 6 from the third row, which can be put at the end of the fourth row:

| 1 | 2 | 2 | **3** |
|---|---|---|---|
| 2 | 3 | 5 | 5 |
| 4 | 4 | 6 |   |
| 5 | 6 |   |   |

← 2

| 1 | 2 | 2 | 2 |
|---|---|---|---|
| 2 | 3 | **5** | 5 |
| 4 | 4 | 6 |   |
| 5 | 6 |   |   |

← 3

| 1 | 2 | 2 | 2 |
|---|---|---|---|
| 2 | 3 | 3 | 5 |
| 4 | 4 | **6** |   |
| 5 | 6 |   |   |

← 5

| 1 | 2 | 2 | 2 |
|---|---|---|---|
| 2 | 3 | 3 | 5 |
| 4 | 4 | 5 |   |
| 5 | 6 | 6 |   |

It is clear from the construction that the result of this process is always a tableau. Indeed, each row is successively constructed to be weakly increasing, and, when an entry $y$ bumps an entry $z$ from a box in a given row, the entry below it, if there is one, is strictly larger than $z$ (by the definition of a tableau), so $z$ either stays in the same column or moves to the left, and the entry lying above its new position is no larger than $y$, so is strictly smaller than $z$.

There is an important sense in which this operation is invertible. If we are given the resulting tableau, *together with the location of the box that has been added to the diagram,* we can recover the original tableau $T$ and the element $x$. The algorithm is simply run backwards. If $y$ is the entry in the added box, it looks for its position in the row above the location of the box, finding the entry farthest to the right which is strictly less than $y$. It bumps this entry up to the next row, and the process continues until an entry is bumped out of the top row. This reverse bumping can be carried out for any tableau and any box in it that is an outside corner, i.e., a box in the Young diagram such that the boxes directly below and to the right are not in the diagram. For example, starting with the tableau and the shaded box

| 1 | 2 | 2 | 2 |
|---|---|---|---|
| 2 | 3 | 3 | 5 |
| 4 | 4 | 5 |   |
| 5 | 6 | **6** |   |

the 6 bumps the 5 in the third row, which bumps the right 3 in the second row, which bumps the right 2 from the first row – exactly reversing steps in the preceding example.

There is a simple lemma about the bumping algorithm which tells about the results of two successive bumpings, allowing one to relate the size of the elements inserted with the positions of the new boxes. A row-insertion $T \leftarrow x$ determines a collection $R$ of boxes, which are those where an element is bumped from a row, together with the box where the last bumped element lands. Let us call this the **bumping route** of the row-insertion, and call the box added to the diagram for the last element the **new box** of the row-insertion. In the example, the bumping route consists of the shaded boxes, with the new box containing the 6:

| 1 | 2 | 2 | 2 |
|---|---|---|---|
| 2 | 3 | 3 | 5 |
| 4 | 4 | 5 |   |
| 5 | 6 | 6 |   |

A bumping route has at most one box in each of several successive rows, starting at the top. We say that a route $R$ is *strictly left* (resp. *weakly left*) of a route $R'$ if in each row which contains a box of $R'$, $R$ has a box which is left of (resp. left of or equal to) the box in $R'$. We use corresponding strict and weak terminology for positions above or below a given box or row.

**Row Bumping Lemma**  *Consider two successive row-insertions, first row-inserting $x$ in a tableau $T$ and then row-inserting $x'$ in the resulting tableau $T \leftarrow x$, giving rise to two routes $R$ and $R'$, and two new boxes $B$ and $B'$.*

(1) *If $x \leq x'$, then $R$ is strictly left of $R'$, and $B$ is strictly left of and weakly below $B'$.*

(2) *If $x > x'$, then $R'$ is weakly left of $R$ and $B'$ is weakly left of and strictly below $B$.*

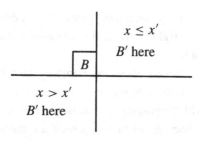

**Proof** This is a question of keeping track of what happens as the elements bump through a given row. Suppose $x \leq x'$, and $x$ bumps an element $y$ from the first row. The element $y'$ bumped by $x'$ from the first row must lie strictly to the right of the box where $x$ bumped, since the elements in that box or to the left are no larger than $x$. In particular, $y \leq y'$, and the same argument continues from row to row. Note that the route for $R$ cannot stop above that of $R'$, and if $R'$ stops first, the route for $R$ never moves to the right, so the box $B$ must be strictly left of and weakly below $B'$.

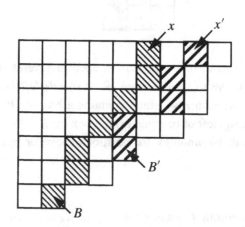

On the other hand, if $x > x'$, and $x$ and $x'$ bump elements $y$ and $y'$, respectively, the box in the first row where the bumping occurs for $x'$ must be at or to the left of the box where $x$ bumped, and in either case, we must have $y > y'$, so the argument can be repeated on successive rows. This time the route $R'$ must continue at least one row below that of $R$.  □

This lemma has the following important consequence.

**Proposition** *Let  $T$  be a tableau of shape  $\lambda$ ,  and let*

$$U = ((T \; \leftarrow \; x_1) \; \leftarrow \; x_2) \; \leftarrow \; \ldots \; \leftarrow \; x_p,$$

*for some  $x_1, \ldots, x_p$ . Let  $\mu$  be the shape of  $U$ . If  $x_1 \le x_2 \le \ldots \le x_p$
(resp.  $x_1 > x_2 > \ldots > x_p$ ), then no two of the boxes in  $\mu/\lambda$  are in the
same column (resp. row). Conversely, suppose  $U$  is a tableau on a shape
 $\mu$ ,  and  $\lambda$  a Young diagram contained in  $\mu$ , with  $p$  boxes in  $\mu/\lambda$ . If
no two boxes in  $\mu/\lambda$  are in the same column (resp. row), then there is a
unique tableau  $T$  of shape  $\lambda$ ,  and unique  $x_1 \le x_2 \le \ldots \le x_p$  (resp.
 $x_1 > x_2 > \ldots > x_p$ ) such that  $U = ((T \leftarrow x_1) \leftarrow x_2) \leftarrow \ldots \leftarrow x_p$ .*

**Proof**  The first assertion is a direct consequence of the lemma. For the converse, in the case where  $\mu/\lambda$  has no two boxes in the same column, do reverse
row bumping on  $U$ , using the boxes in  $\mu/\lambda$ , starting from the right-most
box and working to the left. The tableau  $T$  is the tableau obtained after
these operations are carried out, and  $x_p, \ldots, x_1$  are the elements bumped
out. The Row Bumping Lemma guarantees that the resulting sequence satis-
fies  $x_1 \le \ldots \le x_p$ . Similarly, if  $\mu/\lambda$  has no two boxes in the same row,
do  $p$  reverse bumpings, starting from the lowest box in  $\mu/\lambda$ ,  and working
up; again, the Row Bumping Lemma implies that the elements  $x_p, \ldots, x_1$
bumped out satisfy  $x_1 > x_2 > \ldots > x_p$ .   □

This Schensted operation has many remarkable properties. It can be used to
form a ***product tableau***  $T \cdot U$  from any two tableaux  $T$  and  $U$ . The number
of boxes in the product will be the sum of the number of boxes in each, and
its entries will be the entries of  $T$  and  $U$ . If  $U$  consists of one box with
entry  $x$ , the product  $T \cdot U$  is the result  $T \leftarrow x$  of row-inserting  $x$  in  $T$ .
To construct it in general, start with  $T$ , and row-insert the left-most entry in
the bottom row of  $U$  into  $T$ . Row-insert into the result the next entry of the
bottom row of  $U$ , and continue until all entries in the bottom row of  $U$  have
been inserted. Then insert in order the entries of the next to the last row, left to
right, and continue with the other rows, until all the entries of  $U$  have been
inserted. In other words, if we list the entries of  $U$  in order from left to right,
and from bottom to top, getting a sequence  $x_1, x_2, \ldots, x_s$ , then

$$T \cdot U = ((\ldots ((T \; \leftarrow \; x_1) \; \leftarrow \; x_2) \; \leftarrow \; \ldots) \; \leftarrow \; x_{s-1}) \; \leftarrow \; x_s.$$

For example,

$$
\begin{array}{|c|c|c|c|}
\hline 1 & 2 & 2 & 3 \\
\hline 2 & 3 & 5 & 5 \\
\hline 4 & 4 & 6 \\
\cline{1-3}
5 & 6 \\
\cline{1-2}
\end{array}
\cdot
\begin{array}{|c|c|}
\hline 1 & 3 \\
\hline 2 \\
\cline{1-1}
\end{array}
=
\begin{array}{|c|c|c|c|}
\hline 1 & 2 & 2 & 2 \\
\hline 2 & 3 & 3 & 5 \\
\hline 4 & 4 & 5 \\
\cline{1-3}
5 & 6 & 6 \\
\cline{1-3}
\end{array}
\cdot
\begin{array}{|c|c|}
\hline 1 & 3 \\
\hline
\end{array}
\cdot
$$

$$
=
\begin{array}{|c|c|c|c|}
\hline 1 & 1 & 2 & 2 \\
\hline 2 & 2 & 3 & 5 \\
\cline{1-4}
3 & 4 & 5 \\
\cline{1-3}
4 & 6 & 6 \\
\cline{1-3}
5 \\
\cline{1-1}
\end{array}
\cdot
\begin{array}{|c|}
\hline 3 \\
\hline
\end{array}
=
\begin{array}{|c|c|c|c|c|}
\hline 1 & 1 & 2 & 2 & 3 \\
\hline 2 & 2 & 3 & 5 \\
\cline{1-4}
3 & 4 & 5 \\
\cline{1-3}
4 & 6 & 6 \\
\cline{1-3}
5 \\
\cline{1-1}
\end{array}
$$

One property that is not at all obvious from the definition is the associativity of the product:

**Claim 1** *The product operation makes the set of tableaux into an associative monoid. The empty tableau is a unit in this monoid: $\emptyset \cdot T = T \cdot \emptyset = T$.*

## 1.2 Sliding; jeu de taquin

There is another remarkable way to construct the product, using skew tableaux. A skew diagram $\lambda/\mu$ which is not a tableau has one or more inside corners. An *inside corner* is a box in the smaller (deleted) diagram $\mu$ such that the boxes below and to the right are not in $\mu$. In the example

they are the fourth box in the second row and the first box in the third row. An *outside corner* is a box in $\lambda$ such that neither box below or to the right is in $\lambda$; in the example, the last boxes in the second, third, fourth, and fifth rows are outside corners. Note that it is possible for a skew diagram to arise as $\lambda/\mu$ for more than one choice of $\lambda$ and $\mu$; in this case there can be inside corners that are also outside corners:

$(3,1,1)/(2,1)$        $(3,2,1)/(2,2)$

The basic operation defined by Schützenberger is sometimes known as *sliding*, or "digging a hole." It takes a skew tableau $S$ and an inside corner, which can be thought of as a hole, or an empty box, and slides the smaller of its two neighbors to the right or below into the empty box; if only one of these two neighbors is in the skew diagram, that is chosen, and if the two neighbors have the same entry, the one *below* is chosen.[1] This creates a new hole or empty box in the skew diagram. The process is repeated with this box, sliding one of its two neighbors into the hole according to the same prescription. It continues until the hole has been dug through to an outside corner, i.e., there are no neighbors to slide into the empty box, in which case the empty box is removed from the diagram.

For example, if one carries this out for the inside corner in the third row of the above skew tableau, one gets:

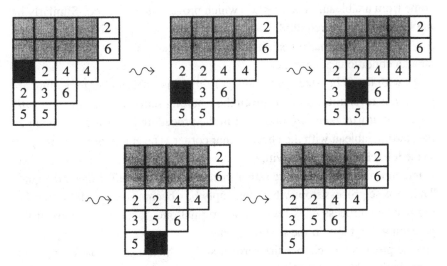

---

[1] A useful general rule, when entries in a tableau are the same, is to regard those to the left as smaller than those to the right.

It is not difficult to see that the result of this operation is always a skew tableau. Indeed, since the box which is added is an inside corner, and that which is removed is an outside corner, the shape remains a skew diagram. To see that the result is a tableau, it suffices to check that in each step of the procedure, whether the slide is horizontal or vertical, the entries in all rows remain increasing, and those in columns remain strictly increasing. The relevant boxes in a step are:

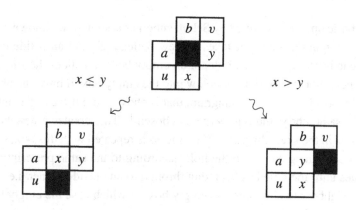

where some of the labelled boxes could be empty. In the first case, we need to know that $a \leq x \leq y$. We are given that $x \leq y$, and, since the entries come from a tableau, $a < u \leq x$, which proves what we need. Similarly in the second case, the required $b < y < x$ follows from the given $y < x$ and $b \leq v < y$. Note that the sliding rule is designed precisely to maintain the tableau conditions at each step.

As with the Schensted bumping algorithm, this Schützenberger sliding algorithm is reversible: if one is given the resulting skew tableau, *together with the box that was removed,* one can run the procedure backwards, arriving at the starting tableau with the chosen inner corner. The empty box moves up or to the left, changing places with the larger of the two entries – if there are two – and choosing the one above rather than the one to the left if they are equal. This reverse process stops when the empty box becomes an inside corner. To see that this process does reverse the original one, it suffices to look at the preceding diagram: in the first case when $x \leq y$, we have $u \leq x$, so the reverse process will choose the vertical slide, which takes us back where we were; in the other case we have $v < y$, so the reverse process is the horizontal slide, as it should be. Such a move is called a ***reverse slide.***

Given a skew tableau $S$, this procedure can be carried out from any inside corner. An inside corner can be chosen for the resulting tableau, and the procedure repeated, until there are no more inside corners, i.e., the result is a tableau. We will call this tableau a **rectification** (*redressement*) of $S$. The whole process is called the *jeu de taquin*. It can be regarded as a game[2] where a player's move is to choose an inside corner. As in many other mathematical games, the final position is independent of how the game is played:

**Claim 2** *Starting with a given skew tableau, all choices of inner corners lead to the same rectified tableau.*

We will denote the rectification of $S$ by $\mathrm{Rect}(S)$. Surprisingly, this jeu de taquin is closely related to Schensted's procedure. In fact, it can also be used to give another construction of the product of two Young tableaux. Given two tableaux $T$ and $U$, form a skew tableau denoted $T*U$ by taking (for the smaller Young diagram) a rectangle of empty squares with the same number of columns as $T$ and the same number of rows as $U$, and putting $T$ below and $U$ to the right of this rectangle:

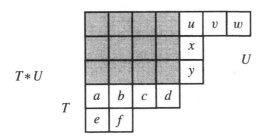

Our second construction of the product $T{\cdot}U$ is the rectification of this skew tableau: $T{\cdot}U = \mathrm{Rect}(T*U)$, which is unique by Claim 2.

**Claim 3** *This product agrees with the first definition.*

**Exercise 1** Compute the product of $T = \begin{array}{|c|c|c|c|} \hline 1 & 2 & 2 & 3 \\ \hline 2 & 3 & 5 & 5 \\ \hline 4 & 4 & 6 \\ \cline{1-3} 5 & 6 \\ \cline{1-2} \end{array}$ and $U = \begin{array}{|c|c|} \hline 1 & 3 \\ \hline 2 \\ \cline{1-1} \end{array}$

---

[2] The name "jeu de taquin" refers to the French version of the "15 puzzle," in which one tries to rearrange the numbers by sliding neighboring squares into the empty box.

by rectifying the skew tableau

|   |   |   |   | 1 | 3 |
|---|---|---|---|---|---|
|   |   |   |   | 2 |   |
| 1 | 2 | 2 | 3 |   |   |
| 2 | 3 | 5 | 5 |   |   |
| 4 | 4 | 6 |   |   |   |
| 5 | 6 |   |   |   |   |

**Exercise 2** Show that the associativity of the product (Claim 1) follows from Claims 2 and 3, by forming an appropriate skew tableau "$T*U*V$" from three given tableaux $T$, $U$, and $V$.

The proofs of the three claims will be given in the next two chapters.

# 2

# Words; the plactic monoid

In this chapter we study the word of a tableau, which encodes it by a sequence of integers. This is less visual than the tableau itself, but will be crucial to the proofs of the fundamental facts. Historically, however, the story goes the other way: the Schensted operations were invented to study sequences of integers. In this chapter we analyze what the bumping and sliding operations do to the associated words.

## 2.1 Words and elementary transformations

We write words as a sequence of letters (positive integers, with our conventions), and write $w \cdot w'$ or $ww'$ for the word which is the juxtaposition of the two words $w$ and $w'$.

Given a tableau or skew tableau $T$, we define the **word** (or **row word**) of $T$, denoted $w(T)$ or $w_{\text{row}}(T)$, by reading the entries of $T$ "from left to right and bottom to top," i.e., starting with the bottom row, writing down its entries from left to right, then listing the entries from left to right in the next to bottom row and working up to the top. A tableau $T$ can be recovered from its word: simply break the word wherever one entry is strictly greater than the next, and the pieces are the rows of $T$, read from bottom to top. For example, the word 5 6 4 4 6 2 3 5 5 1 2 2 3 breaks into 5 6 | 4 4 6 | 2 3 5 5 | 1 2 2 3, which is the word of a tableau used in examples in the preceding chapter. Of course, not every word comes from a tableau; the pieces must have weakly increasing length for the result to form a Young diagram, and, when stacked up, the columns must have strictly increasing entries. Many different skew tableaux may determine the same word. Every word arises from some skew tableau, for example by breaking the word into increasing pieces, and putting

17

the pieces in rows, each row placed above and entirely to the right of the row below.

Our first task is simply to see what the bumping process does to the word of a tableau. This will eventually tell us how the word of a product of two tableaux is related to the words of its factors. Suppose an element $x$ is row-inserted into a row. In "word" language, the Schensted algorithm says to factor the word of the row into $u \cdot x' \cdot v$, where $u$ and $v$ are words, $x'$ is a letter, and each letter in $u$ is no larger than $x$, and $x'$ is strictly larger than $x$. The letter $x'$ is to be replaced by $x$, so the row with word $u \cdot x' \cdot v$ becomes $u \cdot x \cdot v$, and $x'$ is bumped to the next row. The resulting tableau has word $x' \cdot u \cdot x \cdot v$. So in the word code, the basic algorithm is

(1)        $(u \cdot x' \cdot v) \cdot x \rightsquigarrow x' \cdot u \cdot x \cdot v$   if   $u \le x < x' \le v$.

Here $u$ and $v$ are weakly increasing, and an inequality such as $u \le v$ means that every letter in $u$ is smaller than or equal to every letter in $v$. In this code, the row-insertion of $2$ in the tableau with word $5\,6\,4\,4\,6\,2\,3\,5\,5\,1\,2\,2\,3$ can be written

$$(5\ 6)(4\ 4\ 6)(2\ 3\ 5\ 5)(1\ 2\ 2\ 3) \cdot 2 \mapsto (5\ 6)(4\ 4\ 6)(2\ 3\ 5\ 5) \cdot 3 \cdot (1\ 2\ 2\ 2)$$

$$\mapsto (5\ 6)(4\ 4\ 6) \cdot 5 \cdot (2\ 3\ 3\ 5)(1\ 2\ 2\ 2)$$

$$\mapsto (5\ 6) \cdot 6 \cdot (4\ 4\ 5)(2\ 3\ 3\ 5)(1\ 2\ 2\ 2)$$

$$\mapsto (5\ 6\ 6)(4\ 4\ 5)(2\ 3\ 3\ 5)(1\ 2\ 2\ 2).$$

Knuth described the Schensted algorithm in the language of a computer program, breaking it down into its atomic pieces. This reveals its inner structure, and is key to the proofs we will give of the claims made in Chapter 1. When we row-insert an element $x$ in a tableau $T$, we start by trying to put $x$ at the end of the first row, testing $x$ against the last entry of the row to see if that entry is larger. If it is not, we put $x$ at the end. If the last entry $z$ of the row is larger, and the entry $y$ before it also is larger than $x$, we move $x$ one step to the left and repeat the process. The steps can be listed, with the rules that govern them, as

$$
\begin{aligned}
u\,x'\,v_1 \ldots v_{q-1}\,v_q\,x &\mapsto u\,x'\,v_1 \ldots v_{q-1}\,x\,v_q && (x < v_{q-1} \le v_q)\\
&\mapsto u\,x'\,v_1 \ldots v_{q-2}\,x\,v_{q-1}\,v_q && (x < v_{q-2} \le v_{q-1})\\
\ldots &\mapsto u\,x'\,v_1\,x\,v_2 \ldots v_{q-1}\,v_q && (x < v_1 \le v_2)\\
&\mapsto u\,x'\,x\,v_1 \ldots v_{q-1}\,v_q && (x < x' \le v_1).
\end{aligned}
$$

Each of these transformations involves three consecutive letters, the last two of which are interchanged, provided the first is strictly greater than the third and no larger than the second. In other words, the basic transformation for each of these steps is:

$(K')$ $\qquad\qquad y\,z\,x \;\mapsto\; y\,x\,z$ if $\;x < y \le z.$

Let us continue, with $x$ bumping $x'$ and the $x'$ moving successively to the left:

$$
\begin{aligned}
u_1 \ldots u_{p-1} u_p\, x'\, x\, v &\;\mapsto\; u_1 \ldots u_{p-1}\, x'\, u_p\, x\, v && (u_p \le x < x')\\
&\;\mapsto\; u_1 \ldots x'\, u_{p-1}\, u_p\, x\, v && (u_{p-1} \le u_p < x')\\
\ldots &\;\mapsto\; u_1\, x'\, u_2\, u_3 \ldots u_p\, x\, v && (u_2 \le u_3 < x')\\
&\;\mapsto\; x'\, u_1\, u_2 \ldots u_p\, x\, v && (u_1 \le u_2 < x').
\end{aligned}
$$

Each of these transformations is governed by the rule

$(K'')$ $\qquad\qquad x\,z\,y \;\mapsto\; z\,x\,y$ if $\;x \le y < z.$

These two elementary transformations can be illustrated (and remembered) by the simple products or row-bumpings:

$$y\,z\,x \mapsto y\,x\,z \;\; (x < y \le z)$$

$$x\,z\,y \mapsto z\,x\,y \;\; (x \le y < z)$$

Both rules, with their inverses, allow one to interchange the two neighbors on one side of a letter $y$ if one is smaller than $y$ and the other larger than $y$; the same can be done if one is equal to $y$ provided $y$ is on the appropriate side (which as before is governed by the rule that the left of two equal letters should be regarded as smaller than the right).

An ***elementary Knuth transformation*** on a word applies one of the transformations $(K')$ or $(K'')$, or their inverses, to three consecutive letters in the word. We call two words ***Knuth equivalent*** if they can be changed into each other by a sequence of elementary Knuth transformations, and we write $w \equiv w'$ to denote that words $w$ and $w'$ are Knuth equivalent. What we have just seen amounts to proving

**Proposition 1** *For any tableau $T$ and positive integer $x$,*

$$w(T \leftarrow x) \;\equiv\; w(T)\cdot x.$$

Since the first construction of the product $T \cdot U$ of two tableaux was by successively row-inserting the letters of the word of $U$ into $T$, we have:

**Corollary** *If* $T \cdot U$ *is the product of two tableaux* $T$ *and* $U$, *constructed by row-inserting the word of* $U$ *into* $T$, *then*

$$w(T \cdot U) \equiv w(T) \cdot w(U).$$

It is a little less obvious that the Schützenberger sliding procedure preserves the Knuth equivalence of the words of the skew tableau. In the simplest cases, however, one sees again the elementary Knuth transformations:

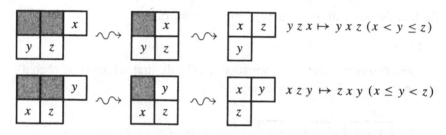

$$y z x \mapsto y x z \ (x < y \leq z)$$

$$x z y \mapsto z x y \ (x \leq y < z)$$

We claim that in fact the Knuth equivalence class of a word is unchanged by each step in a slide. Note that in such a step, the configuration may not be a skew tableau, but rather a skew tableau with a hole in it (an empty box). The word of such a configuration is defined as usual by reading the entries that occur from left to right and bottom to top. In the case of a horizontal slide the claim is obvious, since the word itself is unchanged. For a vertical slide we will have to examine the process carefully. To warm up, consider the example

$$\begin{array}{|c|c|c|}\hline u & & y \\ \hline v & x & z \\ \hline\end{array} \quad \rightsquigarrow \quad \begin{array}{|c|c|c|}\hline u & x & y \\ \hline v & & z \\ \hline\end{array}$$

Here $u < v \leq x \leq y < z$. The word changes from $v x z u y$ to $v z u x y$. It is not hard to realize this as a sequence of elementary Knuth transformations:

$$\begin{aligned} v x z u y &\equiv v x u z y &&(u < y < z) \\ &\equiv v u x z y &&(u < v \leq x) \\ &\equiv v u z x y &&(x \leq y < z) \\ &\equiv v z u x y &&(u < x < z). \end{aligned}$$

The crucial case is the generalization of this where the four corners are replaced by rows:

The assumptions are that the $u_i$'s, $v_i$'s, $y_j$'s, and $z_j$'s are (weakly) increasing sequences; that $u_i < v_i$ and $y_j < z_j$ for all $i$ and $j$; and that $v_p \leq x \leq y_1$. Let

$$u = u_1 \ldots u_p, \quad v = v_1 \ldots v_p, \quad y = y_1 \ldots y_q, \quad z = z_1 \ldots z_q.$$

We must show that

(2)      $v\, x\, z\, u\, y \equiv v\, z\, u\, x\, y.$

We argue by induction on $p$. When $p = 0$, (2) reads $x\, z\, y \equiv z\, x\, y$, or

(3)      $x\, z_1 \ldots z_q\, y_1 \ldots y_q \equiv z_1 \ldots z_q\, x\, y_1 \ldots y_q.$

If $y_1$ is inserted in a row with entries $x, z_1, \ldots, z_q$, the entry $z_1$ is bumped. By Proposition 1 we know that row-insertion respects Knuth equivalence, so $x\, z_1 \ldots z_q\, y_1 \equiv z_1\, x\, y_1\, z_2 \ldots z_q$, yielding

$$(x\, z_1 \ldots z_q\, y_1)(y_2 \ldots y_q) \equiv (z_1\, x\, y_1\, z_2 \ldots z_q)(y_2 \ldots y_q).$$

Now row-insertion of $y_2$ in the row with entries $x, y_1, z_2, \ldots, z_q$ bumps $z_2$, giving $x\, y_1\, z_2 \ldots z_q\, y_2 \equiv z_2\, x\, y_1\, y_2\, z_3 \ldots z_q$, and hence

$$(z_1\, x\, y_1\, z_2 \ldots z_q)(y_2 \ldots y_q) \equiv (z_1\, z_2\, x\, y_1\, y_2\, z_3 \ldots z_q)(y_3 \ldots y_q).$$

Continuing in this way for $k = 3, \ldots, q$, applying row-insertion of $y_k$ in the row with entries $x, y_1, \ldots, y_{k-1}, z_k, \ldots, z_q$ bumps $z_k$ to the position to the left of $x$, leading when $k = q$ to the required (2).

Now let $p \geq 1$, and assume (2) is known for smaller $p$. Set

$$u' = u_2 \ldots u_p, \quad v' = v_2 \ldots v_p.$$

We start with $v\, x\, z\, u\, y = v_1\, v'\, x\, z\, u_1\, u'\, y$. Row-inserting $u_1$ in the row with word $v_1\, v'\, x\, z$ bumps $v_1$, giving $v_1\, v'\, x\, z\, u_1 \equiv v_1\, u_1\, v'\, x\, z$ by Proposition 1, so

$$v\, x\, z\, u\, y = v_1\, v'\, x\, z\, u_1\, u'\, y \equiv v_1\, u_1\, v'\, x\, z\, u'\, y.$$

The assumed equation for $p-1$ gives $v'\, x\, z\, u'\, y \equiv v'\, z\, u'\, x\, y$, so

$$v_1\, u_1\, v'\, x\, z\, u'\, y \equiv v_1\, u_1\, v'\, z\, u'\, x\, y.$$

Finally, row-inserting $u_1$ in the row with word $v_1 v' z$ bumps $v_1$, giving the equivalence $v_1 v' z u_1 \equiv v_1 u_1 v' z$. From this we have

$$v_1 u_1 v' z u' x y \equiv v_1 v' z u_1 u' x y = v z u x y.$$

The succession of the last three displayed congruences yields (2).

The case for any vertical slide follows immediately from the case just considered, as indicated in the following diagram:

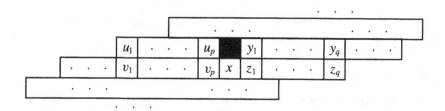

With the preceding notation, the case just considered verifies that $v x z u y \equiv v z u x y$. The required identity is obtained by preceding and following both sides by appropriate words, coming from the lower left and upper right of the diagram. This completes the proof of:

**Proposition 2** *If one skew tableau can be obtained from another by a sequence of slides, then their words are Knuth equivalent.*

Now we can state the result from which the assertions of the preceding chapter will follow:

**Theorem** *Every word is Knuth equivalent to the word of a unique tableau.*

The assertion that every word is Knuth equivalent to the word of some tableau is an easy consequence of Proposition 1. Indeed, if $w = x_1 \ldots x_r$ is any word, Proposition 1 shows that it is Knuth equivalent to the word of the tableau

$$((\ldots ((\boxed{x_1} \leftarrow x_2) \leftarrow x_3) \leftarrow \ldots) \leftarrow x_{r-1}) \leftarrow x_r.$$

We call this the **canonical** procedure for constructing a tableau whose word is Knuth equivalent to a given word, and we denote the resulting tableau by $P(w)$. The uniqueness assertion in the theorem, however, is far from obvious at this point, and will require a new idea; we will prove this in the next chapter. To conclude this chapter let us assume the theorem, and draw a few consequences, including the proofs of the three claims stated in Chapter 1. First, from the proposition and theorem we have

**Corollary 1**  *The rectification of a skew tableau  S  is the unique tableau whose word is Knuth equivalent to the word of  S. If  S  and  S′  are skew tableaux, then*  Rect($S$) = Rect($S′$)  *if and only if*  $w(S) \equiv w(S′)$.

The theorem provides a third way to define the product  $T \cdot U$  of two tableaux. Define  $T \cdot U$  to be the unique tableau whose word is Knuth equivalent to the word  $w(T) \cdot w(U)$,  where the product of two words is defined simply by juxtaposition, writing one after the other.

**Corollary 2**  *The three constructions of the product of two tableaux agree.*

**Proof**  It suffices to show that each of the first two constructions produces a product  $T \cdot U$  with the property that  $w(T \cdot U) = w(T) \cdot w(U)$.  For the first construction, that row-inserts the entries of  $U$  into  $T$,  this is the corollary to Proposition 1. For the second construction, performing the jeu de taquin on  $T * U$,  this follows from Proposition 2.    □

In particular, the three claims made in the last chapter have now been proved – once we have proved the uniqueness in the theorem. There is a nice way to formalize the content of the main theorem, following (up to notation) Knuth, Lascoux, and Schützenberger. Let  $M = M_m$  be the set of Knuth equivalence classes of words on our alphabet  $[m] = \{1, \dots, m\}$.  The juxtaposition of words determines a product on this set, since if  $w \equiv w′$  and  $v \equiv v′$,  then by definition  $w \cdot v \equiv w′ \cdot v \equiv w′ \cdot v′$.  This makes  $M$  into an associative monoid, with unit represented by the empty word  $\emptyset$.  More formally, the words form a free monoid  $F$;  the product is the juxtaposition we have been using; and the unit is the empty word  $\emptyset$.  The map from  $F$  to  $M$  that takes a word to its equivalence class is a homomorphism of monoids;  $M = F/R$,  where  $R$  is the equivalence relation generated by the Knuth relations  $(K′)$  and  $(K″)$.  Lascoux and Schützenberger call  $M$  the ***plactic monoid***. What we have done amounts to saying that the monoid of tableaux is isomorphic to the plactic monoid  $M = F/R$.  The associativity of the product is particularly obvious with this description. Here, we will usually regard the monoid as the set of tableaux with the product defined above.

As with any monoid, one has an associated "group ring." For the monoid of tableaux with entries in  $[m]$,  we denote the corresponding ring by  $R_{[m]}$, and call it the **tableau ring**. This is the free $\mathbb{Z}$-module with basis the tableaux with entries in the alphabet $[m]$, with multiplication determined by the multiplication of tableaux; it is an associative, but not commutative ring. There is a canonical homomorphism from $R_{[m]}$ onto the ring $\mathbb{Z}[x_1, \ldots, x_m]$ of polynomials that takes a tableau $T$ to its monomial $x^T$, where $x^T$ is the product of the variables $x_i$, each occurring as many times in $x^T$ as $i$ occurs in $T$.

## 2.2 Schur polynomials

Define $S_\lambda = S_\lambda[m]$ in the tableau ring $R_{[m]}$ to be the sum of all tableaux $T$ of shape $\lambda$ with entries in $[m]$. The image of $S_\lambda$ in the polynomial ring is the **Schur polynomial** $s_\lambda(x_1, \ldots, x_m)$. In Chapter 5 we will give a general formula for multiplying two arbitrary Schur polynomials. Two important special cases are easy consequences of the Row Bumping Lemma; they are often called "Pieri formulas," since they are the same as formulas found by Pieri for multiplying Schubert varieties in the intersection (cohomology) ring of a Grassmannian. With $(p)$ and $(1^p)$ the Young diagrams with one row and one column of length $p$, these formulas are:

$$(4) \qquad S_\lambda \cdot S_{(p)} = \sum_\mu S_\mu,$$

the sum over all $\mu$'s that are obtained from $\lambda$ by adding $p$ boxes, with no two in the same column; and

$$(5) \qquad S_\lambda \cdot S_{(1^p)} = \sum_\mu S_\mu,$$

the sum over all $\mu$'s that are obtained from $\lambda$ by adding $p$ boxes, with no two in the same row. These facts are translations of the proposition in §1.1, which says that the product of a tableau $T$ times a tableau $V$ whose shape is a row (resp. column) has the shape $\mu$ specified in (4) (resp. (5)), and that any tableau $U$ of this shape factors uniquely into such a product $U = T \cdot V$.

**Exercise 1** If $\lambda = (\lambda_1 \geq \ldots \geq \lambda_k \geq 0)$, verify that the algebraic conditions on $\mu$ in (4) are

$$\mu_1 \geq \lambda_1 \geq \mu_2 \geq \lambda_2 \geq \ldots \geq \mu_k \geq \lambda_k \geq \mu_{k+1} \geq \mu_{k+2} = 0$$

and $\sum \mu_i = \sum \lambda_i + p$. Find similar expressions for (5).

Applying the homomorphism $T \mapsto x^T$ from $R_{[m]}$ to the polynomial ring we deduce from (4) and (5) the formulas

(6)     $$s_\lambda(x_1, \ldots, x_m) \cdot h_p(x_1, \ldots, x_m) = \sum_\mu s_\mu(x_1, \ldots, x_m),$$

where $h_p(x_1, \ldots, x_m)$ is the complete symmetric polynomial of degree $p$, and the sum is over all $\mu$'s that are obtained from $\lambda$ by adding $p$ boxes, with no two in the same column;

(7)     $$s_\lambda(x_1, \ldots, x_m) \cdot e_p(x_1, \ldots, x_m) = \sum_\mu s_\mu(x_1, \ldots, x_m),$$

where $e_p(x_1, \ldots, x_m)$ is the elementary polynomial of degree $p$, and the sum is over all $\mu$'s that are obtained from $\lambda$ by adding $p$ boxes, with no two in the same row.

A tableau $T$ has **content** (or **type** or **weight**) $\mu = (\mu_1 \ldots, \mu_\ell)$ if its entries consist of $\mu_1$ 1's, $\mu_2$ 2's, and so on up to $\mu_\ell$ $\ell$'s. For any partition $\lambda$, and any sequence $\mu = (\mu_1, \ldots, \mu_\ell)$ of nonnegative integers, let $K_{\lambda\mu}$ be the number of tableaux of shape $\lambda$ with content $\mu$. Equivalently, $K_{\lambda\mu}$ is the number of sequences of partitions $\lambda^{(1)} \subset \lambda^{(2)} \subset \ldots \subset \lambda^{(\ell)} = \lambda$, where the skew diagram $\lambda^{(i)}/\lambda^{(i-1)}$ has $\mu_i$ boxes, with no two in the same column:

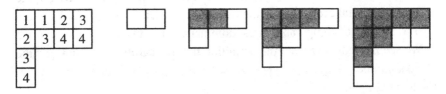

The number $K_{\lambda\mu}$ is called a **Kostka number**, at least when $\mu$ is a partition. As a consequence of (6) we have the formula

(8)     $$h_{\mu_1} \cdot h_{\mu_2} \cdot \ldots \cdot h_{\mu_\ell} = \sum_\lambda K_{\lambda\mu} s_\lambda,$$

the sum over all partitions $\lambda$, where $h_p$ denotes the $p^{\text{th}}$ complete symmetric polynomial in the given variables $x_1, \ldots, x_m$. In fact, the corresponding formula

$$S_{(\mu_1)} \cdot S_{(\mu_2)} \cdot \ldots \cdot S_{(\mu_\ell)} = \sum_\lambda K_{\lambda\mu} S_\lambda$$

is valid in $R_{[m]}$, where it says that, for any sequence $\lambda^{(1)} \subset \ldots \subset \lambda^{(\ell)} = \lambda$ as above, any tableau of shape $\lambda$ can be written uniquely as a product $U_1 \cdot \ldots \cdot U_\ell$, where $U_i$ is a tableau whose shape is a row of length $\mu_i$. This follows by induction on $\ell$ from the proposition in §1.1. Similarly, we have

$$(9) \qquad e_{\mu_1} \cdot e_{\mu_2} \cdot \ldots \cdot e_{\mu_\ell} = \sum_\lambda K_{\tilde{\lambda}\mu} s_\lambda = \sum_\lambda K_{\lambda\mu} s_{\tilde{\lambda}},$$

for the product of the elementary symmetric polynomials, where $\tilde{\lambda}$ denotes the conjugate to a partition $\lambda$. Note for this that a sequence $\lambda^{(1)} \subset \ldots \subset \lambda^{(\ell)} = \lambda$ such that $\lambda^{(i)}/\lambda^{(i-1)}$ has $\mu_i$ boxes, with no two in the same row, corresponds by transposing to a similar sequence for $\tilde{\lambda}$, but with successive differences having no two boxes in the same column. By the proposition in §1.1 any tableau of shape $\lambda$ can be written uniquely as a product $U_1 \cdot \ldots \cdot U_\ell$, where $U_i$ is a tableau whose shape is a column of length $\mu_i$.

There are two important partial orderings on partitions, besides that of inclusion $\mu \subset \lambda$. First is the *lexicographic* ordering, denoted $\mu \leq \lambda$, which means that the first $i$ for which $\mu_i \neq \lambda_i$, if any, has $\mu_i < \lambda_i$. The other is the *dominance* ordering, denoted $\mu \trianglelefteq \lambda$, which means that

$$\mu_1 + \ldots + \mu_i \leq \lambda_1 + \ldots + \lambda_i \quad \text{for all } i,$$

and we say that $\lambda$ *dominates* $\mu$. Note that $\mu \subset \lambda \Rightarrow \mu \trianglelefteq \lambda \Rightarrow \mu \leq \lambda$, but neither implication can be reversed. For example, $(2,2) \trianglelefteq (3,1)$ but $(2,2) \not\subset (3,1)$; and $(3,3) \leq (4,1)$, but $(3,3) \not\trianglelefteq (4,1)$. The lexicographic ordering is a total ordering, but the dominance ordering is not: $(2,2,2)$ and $(3,1,1,1)$ are not comparable in the dominance ordering.

We see immediately from the definition that for partitions $\lambda$ and $\mu$, the Kostka number $K_{\lambda\mu}$ is 1 when $\mu = \lambda$, and $K_{\lambda\mu} = 0$ unless $\mu \leq \lambda$ in the lexicographic ordering.

**Exercise 2** For partitions $\lambda$ and $\mu$ of the same integer, show in fact that $K_{\lambda\mu} \neq 0$ if and only if $\mu \trianglelefteq \lambda$.

If the partitions are ordered lexicographically, the matrix $K_{\lambda\mu}$ is a lower triangular matrix with 1's on the diagonal. This implies that equations (8) and (9) can be solved to write the Schur polynomials in terms of the complete or elementary symmetric functions. (We will give explicit solutions to these equations in Chapter 6.) This implies that the *Schur polynomials are symmetric polynomials*.

It follows from (8), and the fact that the Schur polynomials are linearly independent (see §6.1), that the *number* $K_{\lambda\mu}$ *of tableaux on* $\lambda$ *with* $\mu_1$ 1's, $\mu_2$ 2's, ... *is independent of the order of the numbers* $\mu_1, \ldots, \mu_\ell$, a fact we will prove directly in Chapter 4.

## 2.3 Column words

Although the row words suffice for our study of tableaux, it is sometimes useful
to use a "dual" way to write down a word from a tableau or skew tableau $T$: list
the entries from bottom to top in each column, starting in the left column and
moving to the right. Call this word the ***column word***, and denote it $w_{col}(T)$.
We claim that

(10)        $w_{col}(T) \equiv w_{row}(T) = w(T)$.

This follows by induction on the number of columns from a general fact:

**Lemma 1** *Suppose a skew tableau  $T$  is divided into two skew tableaux  $T'$
and  $T''$  by a horizontal or vertical line. Then*

$$w_{row}(T) \equiv w_{row}(T')\cdot w_{row}(T'').$$

**Proof** The result is obvious for horizontal cuts by the definition of row words.
For a vertical cut, consider the skew tableau  $T' * T''$  obtained from  $T$  by
shifting  $T'$  down until its top row lies below the bottom row of  $T''$.

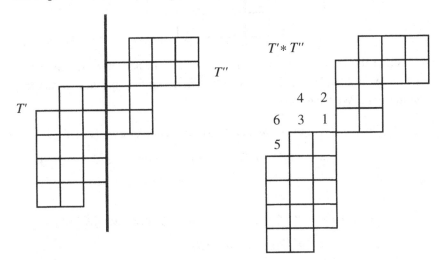

It is clear that  $w_{row}(T'*T'') = w_{row}(T')\cdot w_{row}(T'')$.  One can get from  $T'*T''$
to  $T$  by a sequence of slides, by choosing inside corners directly above  $T'$,
in order from right to left, as numbered in the figure. Proposition 2 of §2.1
then gives the equation  $w_{row}(T) \equiv w_{row}(T'*T'')$.     □

In fact, a stronger result is true, although we will need it only in Appendix A.
Let us call ***K'-equivalence,*** denoted $\equiv'$, the equivalence relation obtained

by using only the transformations $(K')$, i.e., the equivalence relation generated by

$$u{\cdot}y{\cdot}x{\cdot}z{\cdot}v \;\equiv'\; u{\cdot}y{\cdot}z{\cdot}x{\cdot}v \quad \text{if} \quad x < y \le z.$$

**Lemma 2** *If $T$ is any skew tableau, then $w_{\mathrm{col}}(T)$ is $K'$-equivalent to $w_{\mathrm{row}}(T)$.*

**Proof** Let $y$ be the entry in the lower left corner of $T$. Let $X$ be the column in $T$ above $y$, let $Z$ be the row in $T$ to the right of $y$ (either or both of which can be empty), and let $S$ be the skew tableau obtained from $T$ by removing its left column and bottom row.

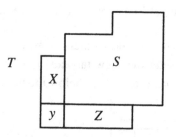

We assume the result for smaller skew tableaux, in particular for the skew tableaux $X \cup S$ and $Z \cup S$ obtained by removing the bottom row and left column of $T$. We have

$$w_{\mathrm{col}}(T) = y{\cdot}w(X){\cdot}w_{\mathrm{col}}(Z \cup S) \;\equiv'\; y{\cdot}w(X){\cdot}w_{\mathrm{row}}(Z \cup S)$$
$$= y{\cdot}w(X){\cdot}w(Z){\cdot}w_{\mathrm{row}}(S) \;\equiv'\; y{\cdot}w(X){\cdot}w(Z){\cdot}w_{\mathrm{col}}(S),$$

and

$$w_{\mathrm{row}}(T) = y{\cdot}w(Z){\cdot}w_{\mathrm{row}}(X \cup S) \;\equiv'\; y{\cdot}w(Z){\cdot}w_{\mathrm{col}}(X \cup S)$$
$$= y{\cdot}w(Z){\cdot}w(X){\cdot}w_{\mathrm{col}}(S).$$

It therefore suffices to show that $y{\cdot}w(X){\cdot}w(Z) \equiv' y{\cdot}w(Z){\cdot}w(X)$. If $w(X) = x_1 \ldots x_p$, and $w(Z) = z_1 \ldots z_q$, we have

$$x_p < \;\cdots\; < x_1 < y \le z_1 \le \;\cdots\; \le z_q,$$

and we must show that, under these conditions,

$$y\,x_1 \ldots x_p\,z_1 \ldots z_q \;\equiv'\; y\,z_1 \ldots z_q\,x_1 \ldots x_p.$$

Since $x_p < x_{p-1} \le z_1$, the $x_p$ and the $z_1$ in the word on the left can be interchanged. Then, since $x_p < z_1 \le z_2$, the $x_p$ and the $z_2$ can be interchanged, and so on, until $x_p$ is moved past all the $z_i$'s, i.e.,

$$y\, x_1 \ldots x_p\, z_1 \ldots z_q \equiv' y\, x_1 \ldots x_{p-1}\, z_1 \ldots z_q\, x_p.$$

By the same argument, $x_{p-1}$ can be moved past all the $z_i$'s, and so on, until each of the $x_j$'s is moved past all the $z_i$'s, and that proves the required equation.   □

**Exercise 3** Using the equivalence relation $\equiv''$ corresponding to $(K'')$, show that $w_{\mathrm{col}}(T) \equiv'' w_{\mathrm{row}}(T)$ for any skew tableau $T$.

There are many other ways to construct a word that is Knuth equivalent to that of a given skew tableau $T$. It follows from Lemma 1 that if $T$ is successively divided into skew tableaux by cutting with vertical or horizontal lines, then $w(T)$ is Knuth equivalent to the product of the words of the pieces, taken in the proper order. For example, if $T$ is the tableau

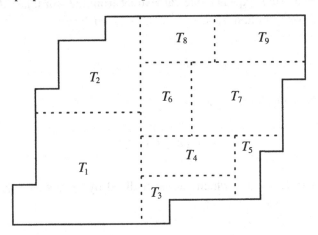

then, with parentheses inserted corresponding to the cuts, and $w_i = w(T_i)$,

$$w(T) \equiv (w_1 w_2)\,(((w_3 w_4) w_5)(w_6 w_7)(w_8 w_9)).$$

As a special case, one may decompose $T$ into a sequence $T_1, \ldots, T_s$ of rows or columns, where $T_1$ is either the bottom row or column of $T$, $T_2$ the bottom row or column of what is left, and so on, until $T$ is exhausted; then $w(T) \equiv w(T_1)\cdot\ldots\cdot w(T_s)$. Similarly one can peel off top rows and columns from the end of the word. As before, these equivalences are actually valid for $K'$-equivalence and $K''$-equivalence.

# 3

# Increasing sequences; proofs of the claims

## 3.1 Increasing sequences in a word

Schensted developed his algorithm to study the lengths of increasing sequences that can be extracted from a word. If $w = x_1 x_2 \ldots x_r$ is a word (as usual, on the alphabet $[m] = \{1, \ldots, m\}$), let $L(w, 1)$ be the length of the longest (weakly) increasing sequence one can extract from the word, i.e., the largest $\ell$ for which one can find $i_1 < i_2 < \cdots < i_\ell$ such that

$$x_{i_1} \leq x_{i_2} \leq \cdots \leq x_{i_\ell}.$$

For example, consider the word

$$w = 1\,3\,4\,2\,3\,4\,1\,2\,2\,3\,3\,2.$$

We have $L(w, 1) = 6$, which can be realized by extracting either of two increasing sequences from $w$:

$$①\,3\,4\,②\,3\,4\,1\,②\,②\,③\,③\,2$$
$$①\,3\,4\,2\,3\,4\,①\,②\,②\,③\,③\,2$$

For any positive integer $k$ let $L(w, k)$ be the largest number that can be realized as the sum of the lengths of $k$ disjoint increasing sequences extracted from $w$. Here are the numbers $L(w, k)$ for the above word $w$, $k \geq 2$, together with a few examples of how to achieve them:

30

$L(w, 2) = 9$:

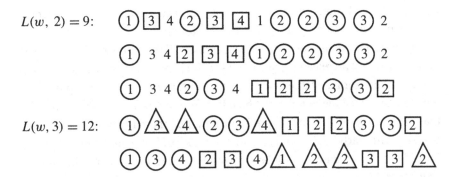

$L(w, 3) = 12$:

And $L(w,k) = 12$ for all $k \geq 3$ (allowing empty sequences). Not only can many collections of $k$ sequences achieve the maximum number, but even the number of elements in the individual sequences can vary. In addition, it may not be possible to add to any of the sequences found at one stage to make sequences for the next stage. In this example, there is no disjoint set of three increasing sequences with 6, 3, and 3 letters.

Let us consider another example, with the same letters occurring with the same multiplicities but in different order:

$$w' = 3\,4\,4\,2\,3\,3\,1\,1\,2\,2\,2\,3.$$

This word has the same numbers $L(w',k)$, but this time maximal sequences can be read off from the sequence, taking blocks starting from the right:

$L(w', 1) = 6$:  3 4 4 2 3 3 ① ① ② ② ② ③

$L(w', 2) = 9$:  3 4 4 ② ③ ③ ① ① ② ② ② ③

$L(w', 3) = 12$:  △3 △4 △4 ② ③ ③ ① ① ② ② ② ③

Note that $w'$ is the (row) word of the tableau

| 1 | 1 | 2 | 2 | 2 | 3 |
|---|---|---|---|---|---|
| 2 | 3 | 3 |   |   |   |
| 3 | 4 | 4 |   |   |   |

The reader may check that this tableau is the one obtained from the first example $w$ by the canonical procedure described after the theorem in the preceding chapter. We will soon see that this is a general fact: if $w$ and $w'$ are Knuth equivalent words, then $L(w,k) = L(w',k)$ for all $k$.

For any word $w$ of a tableau it is easy to read off the numbers $L(w,k)$. The point is that any increasing sequence taken from $w$ must consist of numbers that are taken from the tableau in order from left to right, never going down in the tableau. Since the letters must be taken from different columns, it follows that $L(w,1)$ is just the number of columns, i.e., the number of boxes in the first row. Similarly, $L(w,k)$ is the total number of boxes in the first $k$ rows. Indeed, this sum can certainly be realized by taking the first $k$ rows for the sequences. Conversely, any disjoint union of $k$ sets of boxes in a Young diagram, each of which contains at most one box in a column, can have no more boxes than there are in the first $k$ rows; for example, given $k$ such sets, one can find $k$ sets with the same number of boxes, all taken from the first $k$ rows, by replacing lower boxes by higher boxes in these $k$ rows. This proves

**Lemma 1** *Let $w$ be the word of a tableau $T$ of shape $\lambda$, with $\lambda = (\lambda_1 \geq \lambda_2 \ldots \geq \lambda_\ell \geq \lambda_{\ell+1} = \ldots = 0)$. Then, for all $k \geq 1$,*

$$L(w,k) = \lambda_1 + \lambda_2 + \ldots + \lambda_k.$$

We now verify that the numbers $L(w,k)$ are the same for Knuth equivalent words.

**Lemma 2** *If $w$ and $w'$ are two Knuth equivalent words, then*

$$L(w,k) = L(w',k)$$

*for all $k$.*

**Proof** We simply have to look closely at what happens when $w$ and $w'$ are the left and right sides of an elementary transformation:

$$\text{(i)} \quad u{\cdot}y\,x\,z{\cdot}v \equiv u{\cdot}y\,z\,x{\cdot}v \quad (x < y \leq z)$$
$$\text{(ii)} \quad u{\cdot}x\,z\,y{\cdot}v \equiv u{\cdot}z\,x\,y{\cdot}v \quad (x \leq y < z)$$

where $u$ and $v$ are arbitrary words, and $x, y$, and $z$ are numbers. The inequality $L(w,k) \geq L(w',k)$ is clear, since any collection of $k$ disjoint increasing sequences taken from $w'$ determines the same collection of sequences for $w$. The opposite inequality is not so obvious. To prove it, suppose we have $k$ disjoint increasing sequences taken from the word $w$. It suffices to produce $k$ disjoint increasing sequences taken from $w'$, with the same total number of entries. In most cases, a given collection of sequences for $w$ determines the same collection of sequences for $w'$. In fact, this is the case unless one of the sequences for $w$ includes both of the indicated entries $x$ and $z$, in which case the same sequence will not be increasing for $w'$. So suppose one of the increasing sequences for $w$ has the form $u_1 \cdot x \, z \cdot v_1$, where $u_1$ and $v_1$ are sequences (possibly empty) extracted from $u$ and $v$. If there is no other sequence for $w$ that uses the entry $y$, then we may simply use the sequence $u_1 \cdot y \, z \cdot v_1$ for $w'$ in case (i), and $u_1 \cdot x \, y \cdot v_1$ in case (ii). So the critical case is when we also have a sequence $u_2 \cdot y \cdot v_2$ chosen for $w$. In this case, we can simply replace these two sequences by the sequences $u_2 \cdot y \, z \cdot v_1$ and $u_1 \cdot x \cdot v_2$ in case (i), and by $u_1 \cdot x \, y \cdot v_2$ and $u_2 \cdot z \cdot v_1$ in case (ii). In each case both of these sequences are increasing, and the entries used by both together are the same. Leaving the remaining sequences for $w$ unchanged, one therefore has a collection of sequences for $w'$ with the same total length, as required. □

### 3.2 Proof of the main theorem

From Lemma 1 and Lemma 2 we know the shape of any tableau whose word is Knuth equivalent to a given word. We need a little more if we are to recover the entire tableau by analyzing increasing sequences from its word. Suppose we want to find the box where the largest element should go. In the above example, to tell where the second 4 should go, we can remove that 4 from the word, and apply the above analysis to what is left. This word gives a diagram with one fewer box, and this is the diagram of the tableau obtained by removing that 4 from the given tableau, so the 4 must go in the remaining box. Next, removing both of the 4's from the word, one gets a diagram with two fewer boxes, which tells where the other 4 should go. Continuing in this way, one recovers the entire tableau.

To apply this procedure in general, we need to know that removing the largest letters from Knuth equivalent words leaves Knuth equivalent words. As usual, when letters are equal, we regard those to the right as larger, so the right-most of equals are removed first. For example, to remove the three largest letters from 1 3 4 2 3 1 3, one removes the 4 and the two right-most 3's, leaving the word 1 3 2 1. For later use we state the general case:

**Lemma 3** *If $w$ and $w'$ are Knuth equivalent words, and $w_\circ$ and $w'_\circ$ are the results of removing the $p$ largest and the $q$ smallest letters from each, for any $p$ and $q$, then $w_\circ$ and $w'_\circ$ are Knuth equivalent words.*

**Proof** It suffices by induction to prove that the results of removing the largest or the smallest letters from $w$ and $w'$ are Knuth equivalent. We consider the

case of the largest, the other being symmetric. It suffices to prove this when $w$
and $w'$ are the left and right sides of an elementary transformation (i) or (ii)
as in the proof of Lemma 2. If the element removed from the words is not one
of the letters $x$, $y$, or $z$ in the positions indicated, the Knuth equivalence
of the resulting words is obvious. Otherwise the letter removed must be the
letter $z$ in the place shown, and in this case the resulting words are the same.

□

Now we can complete the proof of the theorem in Chapter 2 – which completes the proofs of all the claims in Chapters 1 and 2 – by showing that if
a word $w$ is Knuth equivalent to the word $w(T)$ of a tableau $T$, then $T$
is uniquely determined by $w$. We proceed by induction on the length of the
word, i.e., the number of boxes in $T$, the result being obvious for words of
length 1. By Lemmas 1 and 2, the shape $\lambda$ of $T$ is determined by $w$:

$$\lambda_k \;=\; L(w,k) \;-\; L(w,k-1).$$

Let $x$ be the largest letter occurring in $w$, and let $w_\circ$ be the word left by
removing the right-most occurrence of $x$ from $w$. Let $T_\circ$ be the tableau
obtained by removing $x$ from $T$, from the position farthest to the right in $T$
if more than one $x$ occurs. Note that $w(T_\circ) = w(T)_\circ$. By Lemma 3, $w_\circ$
is Knuth equivalent to $w(T_\circ)$. By induction on the length of the word, $T_\circ$ is
the unique tableau whose word is Knuth equivalent to $w_\circ$. Since we know
the shape of $T$ and the shape of $T_\circ$, the only possibility for $T$ is that $T$ is
obtained from $T_\circ$ by putting $x$ in the remaining box.     □

Although this procedure gives an algorithm to determine the tableau whose
word is congruent to a given word, the algorithm that is useful in practice is
the opposite: one constructs the tableau by row bumping, and from its shape
one reads off the numbers $L(w,k)$.

**Exercise 1** If $w$ is Knuth equivalent to the word of a tableau $T$, show that
the number of rows of $T$ is the longest strictly decreasing sequence that can be
extracted from $w$. Show that the total number of boxes in the first $k$ columns
of $T$ is the maximum sum of the lengths of $k$ disjoint strictly decreasing
sequences that can be extracted from $w$.

**Exercise 2** Deduce from this a result of Erdös and Szekeres: Any word of
length greater than $n^2$ must contain either an increasing or a strictly decreasing
sequence of length greater than $n$. Show that this is sharp. More generally,

a word of length greater than $m \cdot n$ must contain an increasing sequence of length greater than $n$ or a decreasing sequence of length greater than $m$.

**Exercise 3** Find a word $w$ of length 6 with $L(w,1) = 4$, $L(w,2) = 6$, but which does not have two disjoint increasing sequences of lengths 4 and 2.

The results of this chapter are special cases of much more general results of C. Greene (1976). For any finite partially ordered set $W$ one can define numbers $L(W,k)$ as before to be the largest number of elements in a disjoint union of $k$ increasing subsequences, and take successive differences to get a sequence of nonnegative integers $\lambda_1, \lambda_2, \ldots$. One can do the same for "antichain" sequences, using disjoint unions of $k$ sets such that no two elements in a set are comparable, getting a sequence $\mu_1, \mu_2, \ldots$. Greene's theorem is that these numbers are *always* the lengths of rows and columns of some Young diagram, i.e., that they are both partitions, and they are conjugate to each other. We won't need these generalizations here.

# 4

# The Robinson–Schensted–Knuth correspondence

The row bumping algorithm can be used to give a remarkable one-to-one correspondence between matrices with nonnegative integer entries and pairs of tableaux of the same shape, known as the Robinson–Schensted–Knuth correspondence. In the second section we give an alternative construction, from which the symmetry theorem – that the transpose of a matrix corresponds to the same pair in the opposite order – is evident. See Appendix A.4 for variations of this idea.

## 4.1 The correspondence

For each word $w$ we denote by $P(w)$ the (unique) tableau whose word is Knuth equivalent to $w$. Different (but Knuth equivalent) words determine the same tableau. If $w = x_1 x_2 \ldots x_r$, $P(w)$ can be constructed by the canonical procedure:

$$P(w) = ((\ldots((\boxed{x_1} \leftarrow x_2) \leftarrow x_3) \leftarrow \ldots) \leftarrow x_{r-1}) \leftarrow x_r.$$

We have seen that the Schensted algorithm of row inserting a letter into a tableau is reversible, provided one knows which box has been added to the diagram. This means that we can recover the word $w$ from the tableau $P(w)$ *together with* the numbering of the boxes that arise in the canonical procedure. This can be formalized as follows. At the same time as we construct $P(w)$ we construct another tableau with the same shape, denoted $Q(w)$, and called the **recording tableau** (or **insertion tableau**), whose entries are the integers $1, 2, \ldots, r$. The integer $k$ is simply placed in the box that is added at the $k^{\text{th}}$ step in the construction of $P(w)$. So a 1 is placed in the upper left box,

36

and if $P_k$ is the tableau

$$P_k = ((\ldots((\boxed{x_1} \leftarrow x_2) \leftarrow x_3) \leftarrow \ldots) \leftarrow x_{k-1}) \leftarrow x_k,$$

then a $k$ is placed in the box that is in the diagram of $P_k$ but not in $P_{k-1}$. Since the new box is at an outside corner, the new entry is larger than entries above or to its left. Let $Q_k$ be the tableau that is constructed this way after $k$ steps. For example, if $w$ is the word $w = 5\ 4\ 8\ 2\ 3\ 4\ 1\ 7\ 5\ 3\ 1$, the successive pairs $(P_k, Q_k)$ are:

By reversing the steps in the Schensted algorithm, we can recover our word from the pair of tableaux $(P, Q)$. To go from $(P_k, Q_k)$ to $(P_{k-1}, Q_{k-1})$, take the largest numbered box in $Q_k$, and apply the reverse row-insertion algorithm to $P_k$ with that box. The resulting tableau is $P_{k-1}$, and the element that is bumped out of the top row of $P_k$ is the $k^{\text{th}}$ element of the word $w$. Remove the largest element (which is $k$) from $Q_k$ to form $Q_{k-1}$.

In fact any pair $(P,Q)$ of tableaux on the same shape, with $Q$ standard, arises in this way. One can always perform the procedure of the preceding paragraph. If there are $r$ boxes, one gets a chain of pairs of tableaux

$$(P,Q) \; = \; (P_r,Q_r), \; (P_{r-1},Q_{r-1}), \; \ldots, \; (P_1,Q_1),$$

each pair having the same shape, with each $Q_k$ standard. The letter $x_k$ that occurs once more in $P_k$ than in $P_{k-1}$ is the $k^{\text{th}}$ letter of a word $w$, and the tableaux come from this word, i.e., $P = P(w)$, $Q = Q(w)$. Note that specifying a standard tableau on the given shape is the same as numbering the boxes in such a way that the shape of the boxes with the first $k$ numbers is a Young diagram for all $k \leq r$; this follows from the fact that the largest entry in a standard tableau is in an outside corner.

This sets up a one-to-one correspondence between words $w$ of length $r$ using the letters from $[n]$, and (ordered) pairs $(P,Q)$ of tableaux of the same shape with $r$ boxes, with the entries of $P$ taken from $[n]$ and $Q$ standard. This is the **Robinson–Schensted correspondence**. In case $r = n$, and the letters of $w$ are required to be the numbers $1, \ldots, n$, each occurring once; i.e., $w$ is a **permutation** of $[n]$ (the permutation that takes $i$ to the $i^{\text{th}}$ letter of $w$), $P$ will also be a standard tableau, and conversely. This was the original correspondence of Robinson, rediscovered independently by Schensted, between permutations and pairs of standard tableaux. We will refer to this special case as the **Robinson correspondence**.

Knuth generalized this to describe what corresponds to an arbitrary ordered pair of tableaux $(P,Q)$ of the same shape, say if $P$ has entries from the alphabet $[n]$ and $Q$ from $[m]$. One can still perform the above reverse process, to get a sequence of pairs of tableaux

$$(P,Q) \; = \; (P_r,Q_r), \; (P_{r-1},Q_{r-1}), \; \ldots, \; (P_1,Q_1),$$

with the two tableaux in each pair having the same shape, and each having one fewer box than the preceding. To construct $(P_{k-1},Q_{k-1})$ from $(P_k,Q_k)$, one finds the box in which $Q_k$ has the largest entry; if there are several equal entries, the box that is farthest to the right is selected. Then $P_{k-1}$ is the result of performing the reverse row-insertion to $P_k$ using this box to start, and $Q_{k-1}$ is the result of simply removing the entry of this box from $Q_k$. Let $u_k$ be the entry removed from $Q_k$, and let $v_k$ be the entry that is bumped from the top row of $P_k$. One gets from this a two-rowed array $\begin{pmatrix} u_1 \; u_2 \; \ldots \; u_r \\ v_1 \; v_2 \; \ldots \; v_r \end{pmatrix}$.

Note that if $Q$ is a standard tableau, this array is just $\begin{pmatrix} 1 & 2 & \dots & r \\ v_1 & v_2 & \dots & v_r \end{pmatrix}$, which is the same information as the word $w = v_1 v_2 \dots v_r$. We will say that a two-rowed array is a *word* if its top row consists of the numbers $1, \dots, r$ in order. And when the $v_i$ are also the distinct elements from $[r]$, this array is a common way to write the permutation: it takes the integer $i$ to the integer $v_i$ below it; we call such arrays *permutations*.

Which two-rowed arrays arise this way? First, by construction, the $u_i$'s are in weakly increasing order (taken from $[m]$):

(1)     $u_1 \leq u_2 \leq \dots \leq u_r.$

We claim that for the $v_i$'s (taken from $[n]$),

(2)     $v_{k-1} \leq v_k \quad \text{if} \quad u_{k-1} = u_k.$

This follows from the Row Bumping Lemma in Chapter 1, which describes the bumping routes of successive row-insertions. For if $u_{k-1} = u_k$, then by construction the box $B'$ that is removed from $P_k$ lies strictly to the right of the box $B$ that is removed from $P_{k-1}$ in the next step. This means we are in case (1) of the Row Bumping Lemma, so the entry $v_k$ removed first is at least as large as the entry $v_{k-1}$ removed second, which is what we needed to prove.

We say that a two-rowed array $\omega = \begin{pmatrix} u_1 & u_2 & \dots & u_r \\ v_1 & v_2 & \dots & v_r \end{pmatrix}$ is in *lexicographic order* if (1) and (2) hold. This is the ordering on pairs $\binom{u}{v}$, with the top entry taking precedence: $\binom{u}{v} \leq \binom{u'}{v'}$ if $u < u'$ or if $u = u'$ and $v \leq v'$.

Given any two-rowed array $\omega = \begin{pmatrix} u_1 & u_2 & \dots & u_r \\ v_1 & v_2 & \dots & v_r \end{pmatrix}$ that is in lexicographic order, we can construct a pair of tableaux $(P, Q)$ with the same shape, by essentially the same procedure as before. This $(P, Q)$ is the last in a sequence of pairs $(P_k, Q_k)$, $1 \leq k \leq r$. Start with $P_1 = \boxed{v_1}$ and $Q_1 = \boxed{u_1}$. To construct $(P_k, Q_k)$ from $(P_{k-1}, Q_{k-1})$, row-insert $v_k$ in $P_{k-1}$, getting $P_k$; add a box to $Q_{k-1}$ in the position of the new box in $P_k$, and place $u_k$ in this box, to get $Q_k$. We have seen that each $P_k$ is a tableau. To see inductively that each $Q_k$ is a tableau, we must show that if $u_k$ is placed under an entry $u_i$ in $Q_{k-1}$, then $u_k$ is strictly larger than $u_i$. If not, they must be equal, and by (2) we must have $v_i \leq v_{i+1} \leq \dots \leq v_k$. But then by the Row Bumping Lemma, the added boxes in going from $P_i$ to $P_k$ must be in different columns; that is a contradiction.

**Exercise 1** Show that the pair corresponding to $\begin{pmatrix} 1\,1\,1\,2\,2\,3\,3\,3\,3 \\ 1\,2\,2\,1\,2\,1\,1\,1\,2 \end{pmatrix}$ is $(P,Q)$, with

$$P = \begin{array}{|c|c|c|c|c|c|} \hline 1 & 1 & 1 & 1 & 1 & 2 \\ \hline 2 & 2 & 2 \\ \cline{1-3} \end{array} \qquad Q = \begin{array}{|c|c|c|c|c|c|} \hline 1 & 1 & 1 & 2 & 3 & 3 \\ \hline 2 & 3 & 3 \\ \cline{1-3} \end{array}$$

It is clear from the constructions that the two processes just described are inverse to each other. We denote by $(P(\omega), Q(\omega))$ the pair of tableaux constructed from an array $\omega$. Summarizing, we have proved:

**R–S–K Theorem** *The above operations set up a one-to-one correspondence between two-rowed lexicographic arrays $\omega$ and (ordered) pairs of tableaux $(P,Q)$ with the same shape. Under this correspondence:*

(i) *$\omega$ has $r$ entries in each row $\Longleftrightarrow$ $P$ and $Q$ each has $r$ boxes. The entries of $P$ are the elements of the bottom row of $\omega$ and the entries of $Q$ are the elements of the top row of $\omega$.*

(ii) *$\omega$ is a word $\Longleftrightarrow$ $Q$ is a standard tableau.*

(iii) *$\omega$ is a permutation $\Longleftrightarrow$ $P$ and $Q$ are standard tableaux.*

An arbitrary two-rowed array determines a unique lexicographic array by putting its vertical pairs in lexicographic order; two arrays are identified if they consist of the same pairs, each occurring with the same multiplicity, which is the same as saying their corresponding lexicographic arrays are the same. By means of this we may associate an ordered pair of tableaux to an arbitrary two-rowed array.

Although the construction of $P$ and $Q$ from an array treats the two rows of an array very differently – using bumping for the bottom row and merely placing the entries in the top row – this is something of an illusion. In fact we have the following result of Schützenberger, generalized by Knuth:

**Symmetry Theorem** *If an array $\begin{pmatrix} u_1 & u_2 & \ldots & u_r \\ v_1 & v_2 & \ldots & v_r \end{pmatrix}$ corresponds to the pair of tableaux $(P,Q)$, then the array $\begin{pmatrix} v_1 & v_2 & \ldots & v_r \\ u_1 & u_2 & \ldots & u_r \end{pmatrix}$ corresponds to the pair $(Q,P)$.*

**Exercise 2** With $P$ and $Q$ as found in the preceding exercise, verify that the array $\begin{pmatrix} 1\,2\,2\,1\,2\,1\,1\,1\,2 \\ 1\,1\,1\,2\,2\,3\,3\,3\,3 \end{pmatrix} = \begin{pmatrix} 1\,1\,1\,1\,2\,2\,2\,2 \\ 1\,2\,3\,3\,3\,1\,1\,2\,3 \end{pmatrix}$ corresponds to the pair $(Q, P)$.

**Corollary** *If $\omega$ is a permutation, then $P(\omega^{-1}) = Q(\omega)$ and $Q(\omega^{-1}) = P(\omega)$.*

Before turning to the proof of the symmetry theorem, let us observe that this all has a simple translation into the language of matrices. An equivalence class of two-rowed arrays can be identified with a collection of pairs of elements $(i, j)$, $i \in [m]$, $j \in [n]$, each occurring with some nonnegative multiplicity. This data can be described simply by an $m \times n$ matrix $A$ whose $(i, j)$ entry is the number of times $\binom{i}{j}$ occurs in the array. For the array $\begin{pmatrix} 1\,1\,1\,2\,2\,3\,3\,3\,3 \\ 1\,2\,2\,1\,2\,1\,1\,1\,2 \end{pmatrix}$, this matrix is

$$\begin{bmatrix} 1 & 2 \\ 1 & 1 \\ 3 & 1 \end{bmatrix}.$$

The R–S–K correspondence is then a correspondence between matrices $A$ with nonnegative integer entries and ordered pairs $(P, Q)$ of tableaux of the same shape. If $A$ is an $m \times n$ matrix, then $P$ has entries in $[n]$ and $Q$ has entries in $[m]$. The $i^{\text{th}}$ row sum of the matrix is the number of times $i$ occurs in the top row of the array, so is the number of times $i$ appears in $Q$. Similarly, the $j^{\text{th}}$ column sum is the number of times $j$ occurs in $P$. A matrix is the matrix of a word when it has just one $1$ in each row, with the other entries $0$, and it is the matrix of a permutation when it is a permutation matrix; this follows the convention that the matrix associated to a permutation $w$ has a $1$ in the $w(i)^{\text{th}}$ column of the $i^{\text{th}}$ row of the matrix.

In terms of matrices, turning an array upside down corresponds to taking the *transpose* of the matrix. The symmetry theorem then says that if the matrix $A$ corresponds to the tableau pair $(P, Q)$, then the transpose $A^{\tau}$ corresponds to $(Q, P)$. In particular, symmetric matrices correspond to the pairs of the form $(P, P)$. This implies that involutions in the symmetric group $S_n$ correspond to pairs $(P, P)$ with $P$ a standard tableau with $n$ boxes; so there is a one-to-one correspondence between involutions and standard tableaux. It should be interesting to investigate how other natural operations on matrices correspond to operations on pairs of tableaux, and vice versa.

## 4.2 Matrix-ball construction

We will now give a more "geometric" prescription for going directly from a matrix or an array to the ordered pair of tableaux. For most arrays, this gives a much faster way to construct the tableaux, since there is no successive rewriting of tableaux, as is necessary with the row bumping method. In addition, with this construction, the symmetry theorem will be evident. We give a recipe for assigning to a matrix $A$ a pair $(P, Q) = (P(A), Q(A))$ of tableaux, which we call the "matrix-ball" construction.

We use the following notation to describe the relative position of two boxes in a diagram, or to compare positions of entries in a matrix. Let us say that a box $B'$ is *West* of $B$ if the column $B'$ is strictly to the left of the column of $B$, and we say that $B'$ is *west* of $B$ if the column of $B'$ is left of or equal to the column of $B$. We use the corresponding notations for other compass directions, and we combine them, using capital and small letters to denote strict or weak inequalities. For example, we say that $B'$ is *northWest* of $B$ if the row of $B'$ is above or equal to the row of $B$, and the column of $B'$ is strictly left of the column of $B$.

Given an $m$ by $n$ matrix $A = (a(i, j))$ with nonnegative integer entries, put $a(i, j)$ balls in the $(i, j)$ position of the matrix, and order the balls arbitrarily in each position, by arranging them diagonally from NorthWest to SouthEast. Let us say that one ball is *northwest* of another if it is in the same position and NorthWest in this arrangement, or it is in a position that is north and west of the other: i.e., the row and column numbers of the first ball are less than or equal to those of the second, with at least one inequality strict. Now number all the balls in the matrix, working from upper left to lower right, giving a ball the smallest number that is larger than all numbers that number balls to the northwest. Each ball is numbered with a positive integer, and the balls in a given position are numbered with consecutive integers. A ball is numbered "1" if there are no balls northwest of it. A ball is numbered "$k$" if $k-1$ is the number of the preceding ball in the same position, or if the ball is the first one in a given position, and $k-1$ is the largest number occurring in a ball northwest of the given position. This numbering can be carried out quickly by first numbering the balls in the first row and column of the matrix, then the remaining balls in the second row and column, etc. We call this configuration of numbered balls in a matrix $A^{(1)}$. For example,

$$A = \begin{bmatrix} 1 & 2 \\ 1 & 1 \\ 3 & 1 \end{bmatrix} \qquad A^{(1)} = $$

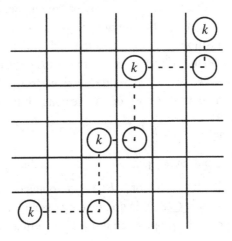

The first row of $P = P(A)$ is read off from this by simply listing the left-most columns where each number occurs in this figure, and the first row of $Q = Q(A)$ lists the top-most row where each number occurs. That is, the $i^{\text{th}}$ entry of the first row of $P$ is the number of the left-most column where a ball numbered $i$ occurs, and the $i^{\text{th}}$ entry of the first row of $Q$ is the number of the upper-most row where a ball numbered $i$ occurs. In the example, this gives the first row of $P$ as $(1,1,1,1,1,2)$ and the first row of $Q$ as $(1,1,1,2,3,3)$.

To continue, form a new matrix of balls, as follows. Whenever there are $\ell > 1$ balls with the same number $k$ in the given matrix, put $\ell-1$ balls in the new matrix, by the following rule: the $\ell$ balls in $A^{(1)}$ lie in a string from SouthWest to NorthEast. Put a ball to the right of each ball in the string but the last, directly under the next ball:

This puts $\ell-1$ balls in a new matrix. Do this for each $k$, which gives a matrix of balls. Define the "derived" matrix $A^\flat$ of $A$ to have for its $(i,j)$ entry the number of balls in the $(i,j)$ position of this new matrix of balls. Then number the balls of the result by the same rule as before, getting a matrix $A^{(2)}$ of numbered balls. For the example, this gives

$$A^\flat = \begin{bmatrix} 0 & 0 \\ 0 & 1 \\ 0 & 2 \end{bmatrix} \qquad A^{(2)} = $$

From $A^{(2)}$ one reads the second rows of $P$ and $Q$ by the same procedure as before. Here one gets $(2,2,2)$ for $P$ and $(2,3,3)$ for $Q$.

This process is repeated, constructing $A^{(3)}$ from $A^{(2)}$, and so on, stopping when no two balls in $A^{(p)}$ have the same label. For the preceding example, we see that $P(A)$ and $Q(A)$ agree with the pair $(P,Q)$ found in Exercise 1. We must prove in general that this "matrix-ball" construction agrees with that described earlier by using the two-rowed array.

**Proposition 1** *If the matrix $A$ corresponds to a two-rowed array $\omega$, then* $(P(A),Q(A)) = (P(\omega),Q(\omega))$.

We will prove this by induction on the total of the entries in the matrix $A$, which is the number of pairs in the array or the number of balls in $A^{(1)}$. The assertion is obvious when this number is $0$ or $1$. Let us call the *last* position of $A$ the position $(x,y)$, where the $x^{\text{th}}$ row of $A$ is the lowest row that is nonzero, and the $y^{\text{th}}$ entry of this row is the right-most entry of this row that is nonzero. Let $A_\circ$ be the matrix obtained from $A$ by subtracting $1$ from corresponding entry $a(x,y)$ of $A$, leaving all the other entries the same. If $\omega = \begin{pmatrix} u_1 & u_2 & \dots & u_r \\ v_1 & v_2 & \dots & v_r \end{pmatrix}$ is the lexicographic array corresponding to $A$, then $\binom{x}{y} = \binom{u_r}{v_r}$ and $\omega_\circ = \begin{pmatrix} u_1 & \dots & u_{r-1} \\ v_1 & \dots & v_{r-1} \end{pmatrix}$ is the lexicographic array corresponding to $A_\circ$. By induction, $P(A_\circ)$ is obtained by row bumping $v_1 \leftarrow \dots \leftarrow v_{r-1}$, and $Q(A_\circ)$ is obtained by placing $u_1, \dots, u_{r-1}$ in the new boxes. To prove the proposition it therefore suffices to prove the

**Claim**   $P(A) = P(A_\circ) \leftarrow y$, *and* $Q(A)$ *is obtained from* $Q(A_\circ)$ *by placing* $x$ *in the box that is in* $P(A)$ *but not in* $P(A_\circ)$.

Now $A^{(1)}$ has one ball that is not in $A_\circ^{(1)}$, and which we suppose has the number $k$; that is, $k$ is the largest number of a ball in the $(x,y)$ position of $A^{(1)}$. Suppose first that there are no other balls in $A^{(1)}$ numbered $k$. In this case $A^\flat = (A_\circ)^\flat$, so all rows of $P(A)$ and $P(A_\circ)$ (and of $Q(A)$ and $Q(A_\circ)$) after the top row are the same. There are no balls in $A^{(1)}$ numbered larger than $k$, since any such balls would have to be located southeast of the position $(x,y)$, and there are no such nonzero entries in $A$. It follows that the first row of $P(A)$ is obtained from that of $P(A_\circ)$ by adding a $y$ to the end, and $Q(A)$ is obtained from $Q(A_\circ)$ by adding an $x$ to the end of the first row. Since the other entries in the first row of $P(A)$ label the left-most columns of balls numbered from 1 to $k-1$, these entries are all no larger than $y$; this implies that $P(A) = P(A_\circ) \leftarrow y$, and the claim is evident.

For the rest of the proof we may therefore suppose there are other balls in $A^{(1)}$ numbered with a $k$. They are all NorthEast of the given ball, since there are no nonzero entries in succeeding rows. Let the next one to the NorthEast be in the $x'$ row and the $y'$ column (so $x' < x$ is maximal and $y' > y$ is minimal for all such balls):

The fact that the first row of $P(A)$ is the first row of $P(A_\circ) \leftarrow y$ is a consequence of

**Subclaim 1**   *When* $y$ *is row-inserted into the first row of* $P(A_\circ)$, *the element* $y'$ *is bumped from the* $k^{\text{th}}$ *box.*

**Proof** Indeed, the entries of the first row of $P(A_\circ)$ give the numbers of the left-most columns of $A_\circ^{(1)}$ containing balls numbered $1, 2, \ldots$ . The first $k-1$ entries of this first row are less than or equal to $y$, but the $k^{\text{th}}$ entry is $y'$, which is larger than $y$, so it is the entry that is bumped.     □

The part of $P(A)$ below its first row is by definition $P(A^b)$, and the part of $P(A_\circ)$ below its first row is likewise $P((A_\circ)^b)$. To prove the claim we are reduced to showing that $P(A^b) = P((A_\circ)^b) \leftarrow y'$, and that the new box from this row-insertion is the box in $Q(A^b)$ that is not in $Q((A_\circ)^b)$. This follows from the claim, assumed by induction for the matrix $A_\circ$, provided we know the following:

**Subclaim 2** *The last position of $A^b$ is $(x, y')$, and $(A^b)_\circ = (A_\circ)^b$.*

**Proof** From the construction we see that the $(x, y')$ entry of $A^b$ is positive, and that $A^b$ has no entries below the $x$ row. Any other nonzero entry of $A^b$ in the $x$ row comes from two balls of $A^{(1)}$ labelled with $\ell < k$, the first in the $x$ row, and the second the next ball labelled $\ell$ that lies NorthEast of that. Since the ball labelled $k$ in the $(x', y')$ position must have a ball labelled $\ell$ NorthWest of it, this second ball cannot lie in a column labelled larger than $y'$. The entry of $A^b$ arising from these two balls therefore lies west of the position $(x, y')$. This proves that the last position of $A^b$ is $(x, y')$, and the equation $(A^b)_\circ = (A_\circ)^b$ follows immediately.     □

This finishes the proof of the proposition. The Symmetry Theorem is clear from this proposition, since the matrix-ball construction is symmetric in the rows and columns of the matrix $A$. For the case of permutations, so $A$ is a permutation matrix, our "matrix-ball construction" reduces to the "shadow" construction of Viennot (1977).

**Exercise 3** Use this algorithm to calculate the tableau pair corresponding to the matrix

$$\begin{bmatrix} 1 & 2 & 1 \\ 2 & 2 & 0 \\ 1 & 1 & 1 \end{bmatrix}.$$

The matrix $A$ is symmetric if and only if $P = Q$, so there is a one-to-one correspondence between symmetric matrices and tableaux.

**Exercise 4** If a symmetric matrix $A$ corresponds to a tableau $P$, show that

$$\text{Trace}(A) + \text{Trace}(A^\flat) = \text{length of first row of } P.$$

Deduce that the trace of $A$ is the number of odd columns of the Young diagram of $P$. In particular, if an involution corresponds to a standard Young tableau, the number of fixed points of the involution is the number of odd columns of the tableau.

We conclude this section by showing how to go directly from a pair $(P,Q)$ of Young tableaux of the same shape to its matrix, without going through the long reverse row bumping process. Using this, in fact, one can work out the entire R–S–K correspondence without any row insertions at all. This is not required for the rest of these notes, however, so we will give only a brief explanation. Our basic operation was to assign to a matrix $A$ a pair of rows of the same length – the first rows of the pair $(P,Q)$ – together with a matrix $B = A^\flat$. If these rows are $(v_1, \ldots, v_r)$ and $(u_1, \ldots, u_r)$, the matrix $B$ has the property that, when its entries are replaced by balls, and these are numbered from northwest to southeast as described above, all balls numbered $k$ lie SouthEast of the box in the $u_k{}^{\text{th}}$ row and the $v_k{}^{\text{th}}$ column. We must show how, given a pair of row tableaux $(v_1, \ldots, v_r)$ and $(u_1, \ldots, u_r)$ and a matrix $B$ with this property, to recover the matrix $A$.

For this, we number the balls in $B$ differently, with numbers that may be larger than in the preceding numbering, proceeding from southeast to northwest. A ball in $B$ with no ball lying south or east of it is numbered with the *largest* integer $k$ such that the box in the $u_k{}^{\text{th}}$ row and the $v_k{}^{\text{th}}$ column is strictly NorthWest of it. Once all the balls southeast of a given ball have been numbered, it is numbered with the largest $k$ such that all balls south or east of it have numbers larger than $k$, and such that the box in the $u_k{}^{\text{th}}$ row and $v_k{}^{\text{th}}$ column is strictly NorthWest of it. For an example, suppose the row tableaux are $(1,1,1,1,2,3,3,4)$ and $(1,1,2,2,2,2,3,3)$, and $B$ is the matrix

$$\begin{bmatrix} 0 & 0 & 0 & 0 \\ 0 & 0 & 1 & 0 \\ 0 & 1 & 1 & 3 \\ 0 & 2 & 1 & 2 \end{bmatrix}.$$

We replace the entries of $B$ by balls, and we record the number $k$ in the $(u_k, v_k)$ box, which assists the numbering of the balls:

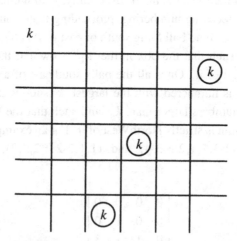

In the result, we have several configurations of the form

with $\ell-1$ balls labelled $k$. The ball matrix of $A$ is recovered by replacing each such configuration by $\ell$ balls as indicated:

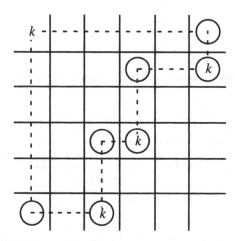

If $\ell = 1$, one just puts a ball in the position of the $k$. Label each of these balls with a $k$. The result is a matrix of balls, numbered with its usual numbering from northwest to southeast. It is clear that this matrix $A$ produces the given pair of rows, and that $A^\flat = B$. In the example, we get

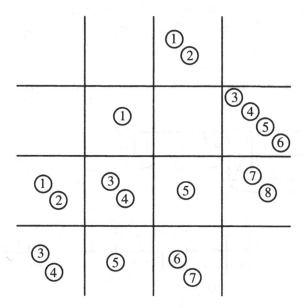

which gives the matrix $A$ to be

$$\begin{bmatrix} 0 & 0 & 2 & 0 \\ 0 & 1 & 0 & 4 \\ 2 & 2 & 1 & 2 \\ 2 & 1 & 2 & 0 \end{bmatrix}.$$

**Exercise 5** Use this algorithm to find the matrix corresponding to the tableau pair

| 1 | 1 | 1 | 1 | 2 | 3 | 3 | 4 |
|---|---|---|---|---|---|---|---|
| 2 | 2 | 2 | 3 | 4 | 4 | 4 |
| 3 | 3 | 4 | 4 |

| 1 | 1 | 2 | 2 | 2 | 2 | 3 | 3 |
|---|---|---|---|---|---|---|---|
| 2 | 3 | 3 | 3 | 3 | 4 | 4 |
| 3 | 4 | 4 | 4 |

## 4.3 Applications of the R–S–K correspondence

This section contains some applications of the R–S–K correspondence to some counting problems. They are not needed for the continued study of tableaux, but they will be used in Part II. To start, the R–S–K correspondence gives a direct proof of the following basic result, which we have seen in §2.2:

**Proposition 2** *The number of tableaux on a given shape* $\lambda$ *with* $m_1$ *1's,* $m_2$ *2's, ...,* $m_n$ *n's, is the same as the number of tableaux on* $\lambda$ *with* $m_{\sigma(1)}$ *1's,* $m_{\sigma(2)}$ *2's, ...,* $m_{\sigma(n)}$ *n's, for any permutation* $\sigma \in S_n$.

**Proof** The case of tableaux with only two distinct entries can be seen directly by hand:

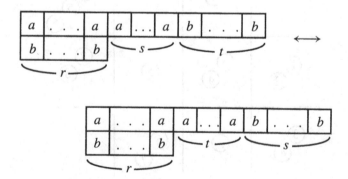

Here $\lambda = (r+s+t, r)$, $m_1 = r+s$, $m_2 = r+t$, with $r$, $s$, and $t$ nonnegative integers.

Given a general $\lambda$, fix any tableau $P$ of shape $\lambda$. Using the R–S–K correspondence between pairs $(P, Q)$ and matrices $A$, the proposition is equivalent to the assertion that the two sets

$$\{A : P(A) = P \quad \text{and} \quad A \text{ has row sums } m_1, \ldots, m_n\}$$

and

$$\{A : P(A) = P \quad \text{and} \quad A \text{ has row sums } m_{\sigma(1)}, \ldots, m_{\sigma(n)}\}$$

have the same cardinality. It suffices to prove this when $\sigma$ is the transposition of $k$ and $k+1$, for $1 \le k < n$, since these transpositions generate $S_n$. Given $A$ in the first set, write

$$A = \begin{pmatrix} B \\ C \\ D \end{pmatrix},$$

where $B$ consists of the first $k-1$ rows of $A$, $C$ the next two rows, and $D$ the rest.

It follows immediately from the construction of $P(A)$, by row bumping from the corresponding two-rowed array, that

$$P(A) = P(B) \cdot P(C) \cdot P(D).$$

By the case considered first – translated to the language of matrices – there is a one-to-one correspondence between matrices $C$ with row sums $m_k$ and $m_{k+1}$ and matrices $C'$ with row sums $m_{k+1}$ and $m_k$, and with $P(C) = P(C')$. Then

$$A' = \begin{pmatrix} B \\ C' \\ D \end{pmatrix},$$

is the corresponding matrix in the second set. □

Since $x^T = x_1^{m_1} \cdot \ldots \cdot x_n^{m_n}$, this proves again the basic fact:

**Corollary** *The Schur polynomials $s_\lambda(x_1, \ldots, x_n)$ are symmetric.*

The R–S–K correspondence leads to a direct proof of a formula of Cauchy and Littlewood (cf. Knuth [1970]):

$$(3) \qquad \prod_{i=1}^{n}\prod_{j=1}^{m}\frac{1}{1-x_iy_j} = \sum_{\lambda} s_\lambda(x_1,\ldots,x_n)s_\lambda(y_1,\ldots,y_m),$$

where the sum is over all partitions $\lambda$. Indeed, on the right one has the sum of all $x^P y^Q$ over all pairs $(P,Q)$ of the same shape, with $P$ having its entries in $[n]$ and $Q$ in $[m]$. On the left one has the sum of all products of $(x_j y_i)^{a(i,j)}$ over all $m$ by $n$ matrices $A = (a(i,j))$ with nonnegative integer entries. And if $A$ corresponds to $(P,Q)$, this product is $x^P y^Q$.

The R–S–K correspondence also leads to a number of important combinatorial identities. Let $f^\lambda$ denote the number of standard tableaux on a given shape $\lambda$, and let $d_\lambda(m)$ denote the number of tableaux on the shape $\lambda$ whose entries are taken from $[m]$. Since the number of permutations of $n$ elements is $n!$, the Robinson correspondence gives

$$(4) \qquad n! = \sum_{\lambda \vdash n} (f^\lambda)^2.$$

Since the number of words with $n$ letters taken from the alphabet $[m]$ is $m^n$, the Robinson–Schensted correspondence gives

$$(5) \qquad m^n = \sum_{\lambda \vdash n} d_\lambda(m) \cdot f^\lambda.$$

From the symmetry theorem, the pairs $(P,P)$ of standard tableaux with $n$ boxes correspond to permutations in $S_n$ that are equal to their inverses.

**Exercise 6** Show that the number of involutions in $S_n$ is

$$\sum_{k=0}^{[n/2]} \frac{n!}{(n-2k)! \cdot 2^k k!}.$$

From this correspondence we therefore have the formula

$$(6) \qquad \sum_{\lambda \vdash n} f^\lambda = \sum_{k=0}^{[n/2]} \frac{n!}{(n-2k)! \cdot 2^k k!}.$$

Similarly, the R–S–K correspondence implies that the number of $m$ by $n$ matrices with nonnegative integer entries that sum to $r$ is the sum of $d_\lambda(m) \cdot d_\lambda(n)$ over partitions $\lambda$ of $r$.

**Exercise 7** Show that the number of $k$-tuples $(a_1, \ldots, a_k)$ of nonnegative integers that sum to $r$ is $\binom{r+k-1}{k-1} = \binom{r+k-1}{r}$.

This gives the identity

(7) $$\binom{r+mn-1}{r} = \sum_{\lambda \vdash r} d_\lambda(m) \cdot d_\lambda(n).$$

**Exercise 8** If $r = (r_1, \ldots, r_m)$ and $c = (c_1, \ldots, c_n)$, show that the number of $m$ by $n$ matrices with nonnegative integer entries and row sums $r_1, \ldots, r_m$ and column sums $c_1, \ldots, c_n$ is $\sum K_{\lambda r} K_{\lambda c}$, the sum taken over all partitions $\lambda$ of $\sum r_i = \sum c_j$, where $K_{\lambda r}$ and $K_{\lambda c}$ are the Kostka numbers. Show that the number of symmetric $n$ by $n$ matrices with nonnegative integer entries and row sums $r_1, \ldots, r_n$ is $\sum K_{\lambda r}$, the sum over all partitions $\lambda$ of $\sum r_i$.

There are some remarkable closed formulas for the numbers $f^\lambda$ and $d_\lambda(m)$, in terms of "hook lengths" of the corresponding diagram. Although we will refer to them occasionally, and they are certainly useful for calculations, they are not essential for the rest of these notes. For a Young diagram $\lambda$, each box determines a **hook**, which consists of that box and all boxes in its row to the right of the box or in its column below the box:

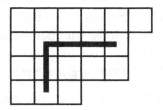

The **hook length** of a box is the number of boxes in its hook; denote by $h(i, j)$ the hook length of the box in the $i^{\text{th}}$ row and $j^{\text{th}}$ column. Labelling each box with its hook length, in this example we have

| 9 | 8 | 7 | 5 | 4 | 1 |
|---|---|---|---|---|---|
| 7 | 6 | 5 | 3 | 2 |   |
| 6 | 5 | 4 | 2 | 1 |   |
| 3 | 2 | 1 |   |   |   |

**Hook length formula** (Frame, Robinson, and Thrall) *If $\lambda$ is a Young diagram with $n$ boxes, then the number $f^\lambda$ of standard tableaux with shape $\lambda$ is $n!$ divided by the product of the hook lengths of the boxes.*

For small diagrams or simple shapes, this is easy to verify by hand. For example, $f^{(n)} = 1$ for all $n$. And $f^{(2,1)} = 2$, as seen by the two tableaux

$$\begin{array}{|c|c|}\hline 1 & 2 \\\hline 3 \\\cline{1-1}\end{array} \quad \text{and} \quad \begin{array}{|c|c|}\hline 1 & 3 \\\hline 2 \\\cline{1-1}\end{array}$$

For the above example, the hook length formula gives

$$f^{(6,5,5,3)} = (19)!/9{\cdot}8{\cdot}7{\cdot}5{\cdot}4{\cdot}1{\cdot}7{\cdot}6{\cdot}5{\cdot}3{\cdot}2{\cdot}6{\cdot}5{\cdot}4{\cdot}2{\cdot}1{\cdot}3{\cdot}2{\cdot}1$$

$$= 19{\cdot}17{\cdot}16{\cdot}13{\cdot}11{\cdot}9 = 6,651,216.$$

There is a quick way to "see" (and remember) this remarkable formula. Consider all $n!$ ways to number the $n$ boxes of the diagram with the integers from 1 up to $n$. A numbering will be a tableau exactly when, in each hook, the corner box of the hook is the smallest among the entries in the hook. The probability of this happening is $1/h$, where $h$ is the hook length. If these probabilities were independent (which they certainly are not!), then the proportion of tableaux among all numberings would be 1 over the product of the hook lengths, and this is the assertion of the proposition. Greene, Nijenhuis, and Wilf (1979) have given a short probabilistic proof of the hook length formula that comes close to justifying this heuristic argument; this proof is also given in Sagan (1991).

An equivalent form of the formula is given in the following exercise:

**Exercise 9** Suppose $\lambda = (\lambda_1 \geq \ldots \geq \lambda_k \geq 0)$, and set $\ell_i = \lambda_i + k - i$, so that

$$\ell_1 = \lambda_1 + k - 1 > \ell_2 = \lambda_2 + k - 2 > \ldots > \ell_k = \lambda_k \geq 0.$$

Show that the hook length formula is equivalent to the formula

$$f^\lambda = \frac{n! \cdot \prod_{i<j}(\ell_i - \ell_j)}{\ell_1! \cdot \ell_2! \cdot \ldots \cdot \ell_k!}.$$

We will see that the formula in the preceding exercise (and hence the hook length formula) is a consequence of the Frobenius character formula, which is proved in Chapter 7. There is also a fairly straightforward inductive proof, as follows. The numbers $f^\lambda$ satisfy an obvious inductive formula, since giving a standard tableau with $n$ boxes is the same as giving one with $n-1$ boxes and saying where to put the $n^{\text{th}}$ box. In other words,

$$(8) \qquad f^{(\lambda_1,\ldots,\lambda_k)} = \sum_{i=1}^{k} f^{(\lambda_1,\ldots,\lambda_i-1,\ldots,\lambda_k)}$$

where $f^{(\lambda_1,\ldots,\lambda_i-1,\ldots,\lambda_k)}$ is defined to be zero if the sequence is not weakly decreasing, i.e., if $\lambda_i = \lambda_{i+1}$.

**Exercise 10** Give an inductive proof of the hook length formula by showing that if $F(\ell_1,\ldots,\ell_k)$ is the expression on the right side of the formula in Exercise 9, then

$$F(\ell_1,\ldots,\ell_k) = \sum_{i=1}^{k} F(\ell_1,\ldots,\ell_i-1,\ldots,\ell_k).$$

Show that this is equivalent to the formula

$$n{\cdot}\Delta(\ell_1,\ldots,\ell_k) = \sum_{i=1}^{k} \ell_i{\cdot}\Delta(\ell_1,\ldots,\ell_i-1,\ldots,\ell_k),$$

where we write $\Delta(\ell_1,\ldots,\ell_k)$ for $\Pi_{i<j}(\ell_i-\ell_j)$. Deduce this formula from the identity

$$\sum_{i=1}^{k} x_i\,\Delta(x_1,\ldots,x_i+t,\ldots,x_k)$$

$$= \left(x_1 + \ldots + x_k + \tbinom{k}{2}t\right)\cdot\Delta(x_1,\ldots,x_k).$$

Prove this identity.

There is also a hook length formula for the numbers $d_\lambda(m)$, due to Stanley (1971), but also a special case of the Weyl character formula:

$$(9) \qquad d_\lambda(m) = \prod_{(i,j)\in\lambda} \frac{m+j-i}{h(i,j)} = \frac{f^\lambda}{n!} \prod_{(i,j)\in\lambda} (m+j-i),$$

where in each case the product is over the boxes in the $i^{\text{th}}$ row and $j^{\text{th}}$ column of the Young diagram of $\lambda$. Note that the numbers in the numerator are obtained by putting the numbers $m$ down the diagonal of the Young diagram, putting $m \pm p$ in boxes that are $p$ steps above or below the diagonal. For the diagram considered above, if $m = 5$, we have

| 5 | 6 | 7 | 8 | 9 | 10 |
|---|---|---|---|---|----|
| 4 | 5 | 6 | 7 | 8 |    |
| 3 | 4 | 5 | 6 | 7 |    |
| 2 | 3 | 4 |   |   |    |

so that

$$d_\lambda(5) = 10 \cdot 9 \cdot 8^2 \cdot 7^3 \cdot 6^3 \cdot 5^3 \cdot 4^3 \cdot 3^2 \cdot 2 \, / \, 9 \cdot 8 \cdot 7 \cdot 5 \cdot 4 \cdot 1 \cdot 7 \cdot 6 \cdot 5 \cdot 3 \cdot 2 \cdot 6 \cdot 5 \cdot 4 \cdot 2 \cdot 1 \cdot 3 \cdot 2 \cdot 1$$

$$= 10 \cdot 8 \cdot 7 \cdot 6 \cdot 4 \cdot 3 \, / \, 2 \cdot 3 \cdot 2 \; = \; 3,360.$$

We will prove formula (9) in Chapter 6.

The final exercises have a few more applications of these ideas.

**Exercise 11** Show that for any $k \geq 2$,

$$\sum \frac{\prod_{i<j}(\ell_i - \ell_j)^2}{\ell_1!^2 \cdot \ell_2!^2 \cdot \ldots \cdot \ell_k!^2} \; = \; 1,$$

the sum over all $k$-tuples $\ell_1, \ldots, \ell_k$ of nonnegative integers whose sum is $(k+1)k/2$.

**Exercise 12** (a) Show that the number of permutations in $S_n$ whose longest increasing sequence has length $\ell$ and whose longest decreasing sequence has length $k$ is $\sum (f^\lambda)^2$, the sum over all partitions $\lambda$ of $n$ that have exactly $k$ rows and $\ell$ columns. (b) Find the number of permutations of $1, \ldots, 21$ whose longest increasing sequence has length $15$ and whose longest decreasing sequence is $4$.

**Exercise 13** Let $m$ and $n$ be positive integers with $m \leq n \leq 2m$. Show that the number of sequences of 1's and 2's of length $n$, such that the longest nondecreasing subsequence has length $m$, is

$$\frac{n! \cdot (2m - n + 1)^2}{(m + 1)! \cdot (n - m)!}.$$

**Exercise 14** Prove an identity of Schur:

$$\prod_{i=1}^{m}(1 - x_i)^{-1} \cdot \prod_{1 \leq i < j \leq m}(1 - x_i x_j)^{-1} \; = \; \sum_\lambda s_\lambda(x_1, \ldots, x_m).$$

**Exercise 15** Show that the number of tableaux whose entries are taken from a given set $S$ of positive integers and whose entries sum to a given integer $k$ is the coefficient of $t^k$ in the power series

$$\prod_{i \in S}(1 - t^i)^{-1} \cdot \prod_{\substack{i,j \in S \\ i < j}}(1 - t^{i+j})^{-1}.$$

**Exercise 16** Let $T$ be a tableau of shape $\lambda$. Show that there are exactly $f^\lambda$ words that are Knuth equivalent to $w(T)$.

**Exercise 17** To any permutation $\omega = v_1 \ldots v_n$ one can assign an ***up–down sequence***, which is a sequence of $n-1$ plus or minus signs, the $i^{\text{th}}$ being $+$ if $v_i < v_{i+1}$ and $-$ if $v_i > v_{i+1}$. Show that $Q(\omega)$ determines the up–down sequence of $\omega$.

# 5

# The Littlewood–Richardson rule

This chapter constructs correspondences between tableaux, that will translate to give results known as Littlewood–Richardson rules for representations and symmetric polynomials. In the language of tableaux, the main problem is to give formulas for the number of ways a given tableau can be written as a product of two tableaux of given shapes, and for the number of skew tableaux of a given shape with a given rectification. This is studied in the first section, with the standard formulas for these rules given in the second, and a few variations in the third. (For more variations on this theme, see Appendix A.3.)

## 5.1 Correspondences between skew tableaux

The key to the Littlewood–Richardson rule will be the following fact:

**Proposition 1** *Suppose* $\begin{pmatrix} u_1 & \ldots & u_m \\ v_1 & \ldots & v_m \end{pmatrix}$ *is an array in lexicographic order, corresponding by the R–S–K correspondence to the pair* $(P, Q)$ *of tableaux. Let* $T$ *be any tableau, and perform the row-insertions*

$$(\ldots ((T \leftarrow v_1) \leftarrow v_2) \leftarrow \ldots) \leftarrow v_m,$$

*and place* $u_1, \ldots, u_m$ *successively in the new boxes. Then the entries* $u_1, \ldots, u_m$ *form a skew tableau* $S$ *whose rectification is* $Q$.

For example, if the array and the tableau are $\begin{pmatrix} 1 & 1 & 2 & 2 & 3 \\ 2 & 2 & 1 & 1 & 1 \end{pmatrix}$ and $\begin{array}{|c|c|} \hline 2 & 3 \\ \hline 3 \\ \cline{1-1} \end{array}$, bumping the bottom row into $T$ and placing the top row into the new boxes, we find

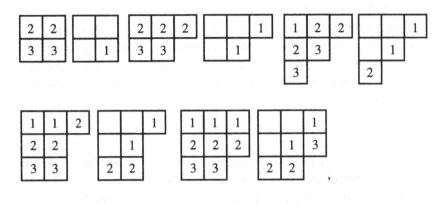

so $S =$ 

|   |   |   |
|---|---|---|
|   | 1 |   |
|   | 1 | 3 |
| 2 | 2 |   |

. The array $\begin{pmatrix} 1\,1\,2\,2\,3 \\ 2\,2\,1\,1\,1 \end{pmatrix}$ corresponds to the pair

$(P,Q)$ with $P =$

| 1 | 1 | 1 |
|---|---|---|
| 2 | 2 |   |

and $Q =$

| 1 | 1 | 3 |
|---|---|---|
| 2 | 2 |   |

. The theorem

asserts that $S$ rectifies to $Q$, which is easily verified in this example.


**Proof** Take any tableau $T_\circ$ with the same shape as $T$, using an alphabet whose letters are all smaller than the letters $u_i$ in $S$ (e.g., using negative integers). The pair $(T,T_\circ)$ corresponds to some lexicographic array $\begin{pmatrix} s_1 \, \ldots \, s_n \\ t_1 \, \ldots \, t_n \end{pmatrix}$.

The lexicographic array $\begin{pmatrix} s_1 \, \ldots \, s_n \; u_1 \, \ldots \, u_m \\ t_1 \, \ldots \, t_n \; v_1 \, \ldots \, v_m \end{pmatrix}$ corresponds to a pair $(T \cdot P, V)$, where $T \cdot P$ is the result of successively row-inserting $v_1, \ldots, v_m$ into $T$, and $V$ is therefore the tableau whose entries $s_1, \ldots, s_n$ make up $T_\circ$, and whose entries $u_1, \ldots, u_m$ make up $S$.

Now invert this array and put the result in lexicographic order. By the definition of lexicographic order, the terms $\binom{v_i}{u_i}$ will occur in lexicographic order in this array, with terms $\binom{t_j}{s_j}$ interspersed (and also in lexicographic order). By the Symmetry Theorem, this array corresponds to $(V, T \cdot P)$, and the array with the $\binom{t_j}{s_j}$'s removed corresponds to $(Q, P)$. The word on the bottom of this array is therefore Knuth equivalent to $w_{\text{row}}(V)$, and when the $s_j$'s are removed from this word, we get a word Knuth equivalent to $w_{\text{row}}(Q)$. However, removing the $n$ smallest letters from $w_{\text{row}}(V)$ obviously leaves the word $w_{\text{row}}(S)$. We saw in Lemma 3 of §3.2 that removing the $n$ smallest letters of Knuth equivalent words gives Knuth equivalent words. Therefore $w_{\text{row}}(S)$ is Knuth equivalent to $w_{\text{row}}(Q)$, which means that the rectification of $S$ is $Q$.   □

Fix three partitions (Young diagrams) $\lambda$, $\mu$, and $\nu$, and let $n$, $m$, and $r$ be the number of boxes in $\lambda$, $\mu$, and $\nu$. We want to know how many ways a given tableau $V$ of shape $\nu$ can be written as a product $T \cdot U$ of a tableau $T$ of shape $\lambda$ and a tableau $U$ of shape $\mu$. (This number will clearly be zero unless $r = n + m$ and $\nu$ contains $\lambda$.) The special cases when $\mu$ consists of one row or one column are essentially translations of the row bumping algorithm, and were given in Chapter 2. The general case will be more complicated, but one feature that is evident in these examples will generalize: the number of ways to factor a given tableau will depend only on the shapes, and not on the given tableau. By one of our constructions of the product of two tableaux, we see that the number of ways to factor $V$ is the same as the number of skew tableaux on the shape

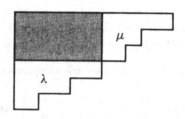

whose rectification is $V$. We denote this skew shape by $\lambda * \mu$. Remarkably, as we shall see, this number is also the number of skew tableaux on the shape $\nu/\lambda$ whose rectification is a given tableau of shape $\mu$.

For any tableau $U_\circ$ with shape $\mu$, set

$$S(\nu/\lambda, U_\circ) = \{\text{skew tableaux } S \text{ on } \nu/\lambda : \text{Rect}(S) = U_\circ\}.$$

For any tableau $V_\circ$ with shape $\nu$, set[1]

$$\mathcal{T}(\lambda,\mu,V_\circ) = \{[T,U]: T \text{ is a tableau on } \lambda, U \text{ is a}$$
$$\text{tableau on } \mu, \text{ and } T \cdot U = V_\circ\}.$$

**Proposition 2** *For any tableaux $U_\circ$ on $\mu$ and $V_\circ$ on $\nu$, there is a canonical one-to-one correspondence*

$$\mathcal{T}(\lambda,\mu,V_\circ) \longleftrightarrow \mathcal{S}(\nu/\lambda,U_\circ).$$

**Proof** Given $[T,U]$ in $\mathcal{T}(\lambda,\mu,V_\circ)$ consider the lexicographic array corresponding to the pair $(U,U_\circ)$:

$$(U,U_\circ) \longleftrightarrow \begin{pmatrix} u_1 \ \ldots \ u_m \\ v_1 \ \ldots \ v_m \end{pmatrix}.$$

Successively row-insert $v_1, \ldots, v_m$ into $T$, and let $S$ be the skew tableau obtained by successively placing $u_1, \ldots, u_m$ into the new boxes. Since $T \cdot U = T \leftarrow v_1 \leftarrow \ldots \leftarrow v_m = V_\circ$ has shape $\nu$, Proposition 1 states precisely that $S$ is in $\mathcal{S}(\nu/\lambda,U_\circ)$.

Conversely, starting with $S$ in $\mathcal{S}(\nu/\lambda,U_\circ)$, choose an arbitrary tableau $T_\circ$ on $\lambda$ such that all the letters in the alphabet of $T_\circ$ come before all letters in the alphabet of $S$. Let $(T_\circ)_S$ be the tableau on $\nu$ that is simply $T_\circ$ on $\lambda$ and $S$ on $\nu/\lambda$. By the R–S–K correspondence, the pair $(V_\circ,(T_\circ)_S)$ of shape $\nu$ corresponds to a unique lexicographic array

$$(1) \qquad (V_\circ,(T_\circ)_S) \longleftrightarrow \begin{pmatrix} t_1 \ \ldots \ t_n \ u_1 \ \ldots \ u_m \\ x_1 \ \ldots \ x_n \ v_1 \ \ldots \ v_m \end{pmatrix}.$$

Consider the tableau pairs corresponding to the two pieces. We claim that

$$(2) \qquad \begin{pmatrix} t_1 \ \ldots \ t_n \\ x_1 \ \ldots \ x_n \end{pmatrix} \longleftrightarrow (T,T_\circ),$$

and

$$(3) \qquad \begin{pmatrix} u_1 \ \ldots \ u_m \\ v_1 \ \ldots \ v_m \end{pmatrix} \longleftrightarrow (U,U_\circ),$$

---

[1] Since the notation $(P,Q)$ has been used in the R–S–K correspondence for pairs of tableaux of the same shape, we use a different notation $[T,U]$ for the pairs of this correspondence.

for some tableaux $T$ and $U$ of shapes $\lambda$ and $\mu$ with $T \cdot U = V_o$. Indeed, the fact that $T \cdot U = V_o$ follows from the construction of the product by row-insertion, which also shows that the second tableau in (2) is $T_o$. The fact that $U_o$ is the second tableau in (3) is the content of Proposition 1. This gives us the pair $[T, U]$ in $\mathfrak{T}(\lambda, \mu, V_o)$, and the two constructions are clearly inverse to each other.   □

Since neither set in the correspondence of the proposition knows about the tableau used to define the other, we have

**Corollary 1** *The cardinalities of the sets* $\mathcal{S}(\nu/\lambda, U_o)$ *and* $\mathfrak{T}(\lambda, \mu, V_o)$ *are independent of choice of* $U_o$ *or* $V_o$, *and depend only on the shapes* $\lambda$, $\mu$, *and* $\nu$.

The number in this corollary will be denoted $c_{\lambda\mu}^{\nu}$, and called a **Littlewood–Richardson** number.

**Corollary 2** *The following sets also have cardinality* $c_{\lambda\mu}^{\nu}$:

  (i)  $\mathcal{S}(\nu/\mu, T_o)$ *for any tableau* $T_o$ *on* $\lambda$;
 (ii)  $\mathfrak{T}(\mu, \lambda, V_o)$ *for any tableau* $V_o$ *on* $\nu$;
(iii)  $\mathcal{S}(\tilde{\nu}/\tilde{\lambda}, \tilde{U}_o)$ *for any tableau* $\tilde{U}_o$ *on the conjugate diagram* $\tilde{\mu}$;
 (iv)  $\mathfrak{T}(\tilde{\lambda}, \tilde{\mu}, \tilde{V}_o)$ *for any tableau* $\tilde{V}_o$ *on the conjugate diagram* $\tilde{\nu}$;
  (v)  $\mathcal{S}(\lambda * \mu, V_o)$ *for any tableau* $V_o$ *on* $\nu$.

**Proof** We know that $\mathcal{S}(\lambda * \mu, V_o)$ corresponds to $\mathfrak{T}(\lambda, \mu, V_o)$, by the discussion before the proposition, which takes care of (v). Taking $U_o$ to be a standard tableau on $\mu$, there is an obvious bijection between $\mathcal{S}(\nu/\lambda, U_o)$ and $\mathcal{S}(\tilde{\nu}/\tilde{\lambda}, U_o^\tau)$ by taking transposes, so (iii) has cardinality $c_{\lambda\mu}^{\nu}$. The sets in (iii) and (iv) have the same cardinality by the proposition. The conjugate diagram for $\lambda * \mu$ is the diagram $\tilde{\mu} * \tilde{\lambda}$, so $\mathcal{S}(\tilde{\mu} * \tilde{\lambda}, \tilde{V}_o)$ has cardinality $c_{\lambda\mu}^{\nu}$ for any $\tilde{V}_o$ on $\tilde{\nu}$. Applying what we have just proved to $\tilde{\mu}$ and $\tilde{\lambda}$, this cardinality is the same as that of $\mathfrak{T}(\tilde{\mu}, \tilde{\lambda}, \tilde{V}_o)$, or of $\mathfrak{T}(\mu, \lambda, V_o)$, which proves that (ii) has cardinality $c_{\lambda\mu}^{\nu}$. Finally, (i) and (ii) have the same cardinality by the proposition.   □

In particular, $c_{\mu\lambda}^{\nu} = c_{\tilde{\lambda}\tilde{\mu}}^{\tilde{\nu}} = c_{\lambda\mu}^{\nu}$. Note that $c_{\lambda\mu}^{\nu} = 0$ unless $|\lambda| + |\mu| = |\nu|$ and $\nu$ contains $\lambda$ and $\mu$. Corollary 2 can also be used to see quickly that some of these numbers are zero. For example, if $\lambda = (2, 2)$, $\mu = (3, 2)$, and $\nu = (3,3,3)$, the shape $\nu$ juts through $\lambda * \mu$, so $c_{\lambda\mu}^{\nu} = 0$ by (v) of the corollary.

From this we deduce a formula for multiplying the elements $S_\lambda = S_\lambda[m]$, which are the sums of all tableaux of shape $\lambda$, in the tableau ring $R_{[m]}$:

**Corollary 3** *With the integers* $c_{\lambda\mu}^{\nu}$ *defined as above, the identity*

$$S_\lambda \cdot S_\mu = \sum_\nu c_{\lambda\mu}^{\nu} S_\nu$$

*holds in the tableau ring* $R_{[m]}$.

This is precisely the assertion that each $V$ of shape $\nu$ can be written exactly $c_{\lambda\mu}^{\nu}$ ways as a product of a tableau of shape $\lambda$ times a tableau of shape $\mu$. The identity $c_{\mu\lambda}^{\nu} = c_{\lambda\mu}^{\nu}$ implies that the subring of $R_{[m]}$ generated by these elements $S_\lambda$, as $\lambda$ varies over all partitions, is *commutative*. Similarly, define $S_{\nu/\lambda} = S_{\nu/\lambda}[m]$ in $R_{[m]}$ to be the sum of all $\mathrm{Rect}(T)$, as $T$ varies over all skew tableaux on the alphabet $[m]$ with shape $\nu/\lambda$. The fact that each tableau of shape $\mu$ occurs precisely $c_{\lambda\mu}^{\nu}$ times as the rectification of a skew tableau of shape $\nu/\lambda$ translates to the following corollary. (We will discuss the corresponding identities among the Schur polynomials in the next section.)

**Corollary 4** *With the integers* $c_{\lambda\mu}^{\nu}$ *defined as above, the identity*

$$S_{\nu/\lambda} = \sum_\mu c_{\lambda\mu}^{\nu} S_\mu$$

*holds in the tableau ring* $R_{[m]}$.

## 5.2 Reverse lattice words

A word $w = x_1 \ldots x_r$ is called a **reverse lattice word**, or sometimes a **Yamanouchi word,** if, when it is read backwards from the end to any letter, the sequence $x_r, x_{r-1}, \ldots, x_s$ contains at least as many 1's as it does 2's, at least as many 2's as 3's, and so on for all positive integers. For example, 2 1 3 2 1 2 1 is a reverse lattice word; but 1 2 3 2 1 2 1 is not, since the last six letters of this word contain more 2's than 1's. Let us call a skew tableau $T$ a **Littlewood–Richardson skew tableau** if its word $w_{\mathrm{row}}(T)$ is a reverse lattice word. For example, it is straightforward to verify that the following are all of the Littlewood–Richardson skew tableaux on the skew shape

(5,4,3,2)/(3,3,1):

A skew tableau is said to have **content** $\mu = (\mu_1, \ldots, \mu_\ell)$ if its entries consist of $\mu_1$ 1's, $\mu_2$ 2's, and so on up to $\mu_\ell$ $\ell$'s; $\mu$ is also called the **type** or **weight**. Usually here $\mu$ will be a partition.

**Proposition 3** *The number $c_{\lambda\mu}^{\nu}$ is the number of Littlewood–Richardson skew tableaux on the shape $\nu/\lambda$ of content $\mu$.*

In the above example, with $\nu = (5,4,3,2)$ and $\lambda = (3,3,1)$ fixed, this gives the numbers $c_{\lambda\mu}^{\nu}$, as $\mu$ varies, as

    1   for $\mu =$ (5,2), (5,1,1), (4,1,1,1), and (2,2,2,1);

    2   for $\mu =$ (4,3), (3,2,1,1), (3,3,1), and (3,2,2);

3  for $\mu = (4,2,1)$;

0  for all other $\mu$.

This gives the decomposition

$$S_{(5,4,3,2)/(3,3,1)} = S_{(5,2)} + S_{(5,1,1)} + S_{(4,1,1,1)} + S_{(2,2,2,1)} + 2S_{(4,3)}$$
$$+ 2S_{(3,2,1,1)} + 2S_{(3,3,1)} + 2S_{(3,2,2)} + 3S_{(4,2,1)}.$$

The reason for the proposition is easily discovered when one computes the rectifications of these Littlewood–Richardson skew tableaux: in each case, it is the Young tableau of shape $\mu$ whose $i^{\text{th}}$ row consists entirely of the letter $i$. For any partition $\mu$, let $U(\mu)$ denote this tableau on $\mu$ whose $i^{\text{th}}$ row contains only the letter $i$ for all $i$.

| 1 | 1 | 1 | 1 |
|---|---|---|---|
| 2 | 2 | 2 | 2 |
| 3 | 3 | 3 |   |
| 4 | 4 |   |   |

$U(\mu), \quad \mu = (4, 4, 3, 2).$

To prove Proposition 3, in light of Proposition 2, it suffices to prove

**Lemma 1** *A skew tableau $S$ is a Littlewood–Richardson skew tableau of content $\mu$ if and only if its rectification is the tableau $U(\mu)$.*

**Proof** Consider first the simple case where the skew tableau is a tableau. In this case the lemma says that the only tableau on a given Young diagram whose word is a reverse lattice word is the one whose first row consists of 1's, its second row of 2's, and so on. This is straightforward, for if the word is to be a reverse lattice word, the last entry in the first row must be a 1, and to be a tableau all the entries in the first row must therefore be 1's. The last entry in the second row must be a 2, since it must be larger than 1 to be a tableau, and it must be a 2 for the word to be a reverse lattice word. The second row must be all 2's to be a tableau, and so on row by row.

To conclude the proof, it suffices to show that a skew tableau is a Littlewood–Richardson skew tableau if and only if its rectification is a Littlewood–Richardson tableau. Since the rectification process preserves the Knuth equivalence class of the word, this is an immediate consequence of the following lemma.

□

**Lemma 2** *If  w  and  w′  are Knuth equivalent words, then  w  is a reverse lattice word if and only if  w′  is a reverse lattice word.*

**Proof** This too is perfectly straightforward. Consider an elementary Knuth transformation:

$$w \;=\; u\,x\,z\,y\,v \;\mapsto\; u\,z\,x\,y\,v \;=\; w' \quad \text{with } x \le y < z.$$

We need to consider possible changes in the numbers of consecutive integers $k$ and $k+1$, reading from right to left. If  $x < y < z$  there is no change, and the only case to check is when  $x = y = k$  and  $z = k+1$. For either to be a reverse lattice word, the number of  $k$'s  appearing in  $v$  must be at least as large as the number of  $(k+1)$'s  appearing in  $v$. In this case both words  $x\,z\,y\,v$  and  $z\,x\,y\,v$  are reverse lattice words, thus taking care of this elementary transformation. If

$$w \;=\; u\,y\,x\,z\,v \;\mapsto\; u\,y\,z\,x\,v \;=\; w' \quad \text{with } x < y \le z,$$

again the only nontrivial case is when  $x = k$  and  $y = z = k+1$. This time neither will be a reverse lattice word unless the number of  $k$'s in  $v$  is strictly larger than the number of  $k+1$'s,  and if this is the case, both words  $y\,x\,z\,v$  and  $y\,z\,x\,v$  will have at least as many  $k$'s  as  $(k+1)$'s.  □

It follows from Lemma 2 and §2.3 that column words could be used instead of row words in the definition of Littlewood–Richardson skew tableaux.

**Exercise 1** Show that  $c_{\lambda\mu}^{\nu}$  is the number of reverse lattice words consisting of  $\nu_1$  1's,  $\nu_2$  2's,  $\dots$ , of the form  $t \cdot u$, where  $t$  and  $u$  are the words of tableaux of shapes  $\lambda$  and  $\mu$.

**Exercise 2** Let  $\lambda$  and  $\mu$  be partitions, and let  $\nu_i = \lambda_i + \mu_i$  for all  $i$. Show that  $c_{\lambda\mu}^{\nu} = 1$.

**Exercise 3** Show that  $c_{\lambda\mu}^{\nu}$  is the number of tableaux  $T$  on  $\lambda$  such that  $T \cdot U(\mu) = U(\nu)$.

Applying the homomorphism from the tableau ring  $R_{[k]}$  to the polynomials  $\mathbb{Z}[x_1, \dots, x_k]$,  the identity

$$(4) \qquad s_\lambda(x_1, \dots, x_k) \cdot s_\mu(x_1, \dots, x_k) \;=\; \sum_\nu c_{\lambda\mu}^{\nu} s_\nu(x_1, \dots, x_k)$$

follows from Corollary 3. For example, using the Littlewood–Richardson rule, one finds that

$$s_{(2,1)} \cdot s_{(2,1)} = s_{(4,2)} + s_{(4,1,1)} + s_{(3,3)} + 2s_{(3,2,1)}$$

$$+ s_{(3,1,1,1)} + s_{(2,2,2)} + s_{(2,2,1,1)}.$$

Defining $s_{v/\lambda}(x_1, \ldots, x_k)$ to be the image of $S_{v/\lambda}[k]$ by the same homomorphism, Corollary 4 implies the formula

$$s_{v/\lambda}(x_1, \ldots, x_k) = \sum_{\mu} c_{\lambda\mu}^{v} s_{\mu}(x_1, \ldots, x_k).$$

**Exercise 4** Show that for variables $x_1, \ldots, x_k, y_1, \ldots, y_\ell$ and any partition $v$,

$$s_v(x_1, \ldots, x_k, y_1, \ldots, y_\ell) = \sum_{\lambda \subset v} s_\lambda(x_1, \ldots, x_k) s_{v/\lambda}(y_1, \ldots, y_\ell)$$

$$= \sum_{\lambda,\mu} c_{\lambda\mu}^{v} s_\lambda(x_1, \ldots, x_k) s_\mu(y_1, \ldots, y_\ell).$$

The existence of an identity $S_\lambda \cdot S_\mu = \sum_v c_{\lambda\mu}^{v} S_v$ in the tableau ring implies that the linear span of the elements $S_\lambda$ forms a subring of the tableau ring, and that this subring maps isomorphically onto the ring of symmetric polynomials by the map $T \mapsto x^T$. (Here we are using the fact that the Schur polynomials form a basis for the symmetric polynomials, which is verified in §6.1.)

**Exercise 5** Show that the number of skew tableaux of content $(1, \ldots, 1)$ on $v/\lambda$ is $\sum_{\mu} c_{\lambda\mu}^{v} f^{\mu}$.

**Exercise 6** Given a sequence $s$ of $n-1$ plus and minus signs, construct a skew diagram $v(s)/\lambda(s)$, consisting of a connected string of $n$ boxes, starting in the first column and ending in the first row, moving right for each $+$ sign, and up for each $-$ sign. For example, the sequence $+ + - + - + +$ corresponds to the skew diagram

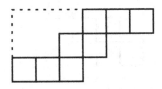

(a) Show that the number of permutations with given up–down sequence $s$ is the number of skew tableaux of content $(1, \ldots, 1)$ on $v(s)/\lambda(s)$, i.e., $\sum_{\mu} c_{\lambda(s)\mu}^{v(s)} f^{\mu}$.   (b) Show that the number of permutations with given up–down sequence $s$ whose inverse has up–down sequence $t$ is $\sum_{\mu} c_{\lambda(s)\mu}^{v(s)} c_{\lambda(t)\mu}^{v(t)}$.

(c)   As an example, use this to show that there are 917 permutations with up–down sequence $- + + - + - +$, among which 16 have inverses with up–down sequence $+ - + - - + -$.

## 5.3 Other formulas for Littlewood–Richardson numbers

There are several other descriptions of the numbers $c_{\lambda\mu}^{\nu}$ that, although not needed in the rest of these notes, are included because they can be deduced rather easily from the preceding discussion. Note first that there is a canonical one-to-one correspondence between reverse lattice words and standard tableaux. To construct this, given a reverse lattice word $w = x_r \ldots x_1$, put the number $p$ in the $x_p^{\text{th}}$ row of the standard tableau, for $p = 1, \ldots, r$. Denote this standard tableau by $U(w)$. For example, the reverse lattice word $1\ 1\ 2\ 3\ 1\ 2\ 1$ corresponds to the standard tableau

$$
\begin{array}{|c|c|c|c|}
\hline
1 & 3 & 6 & 7 \\
\hline
2 & 5 \\
\cline{1-2}
4 \\
\cline{1-1}
\end{array}
\qquad U(w), \quad w = 1\ 1\ 2\ 3\ 1\ 2\ 1.
$$

The reverse lattice property of the word translates into the fact that the boxes numbered by the last $s$ integers form a Young diagram for each $s$. The shape of this tableau is $\mu$, where $\mu_k$ is the number of $k$'s in the word.

Given a skew shape $\nu/\lambda$, number the boxes from right to left in each row, working from the top to the bottom; call it the ***reverse numbering*** of the shape. For example, the reverse numbering of $(5,4,3,2) / (3,3,1)$ is

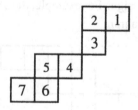

Remmel and Whitney (1984) give the following prescription for the Littlewood–Richardson numbers:

**Proposition 4** *The number $c_{\lambda\mu}^{\nu}$ is the number of standard tableaux $U$ on the shape $\mu$ that satisfy the following two properties:*

(i) *If* $k-1$ *and* $k$ *appear in the same row of the reverse numbering of* $\nu/\lambda$, *then* $k$ *occurs weakly above and strictly right of* $k-1$ *in* $U$.

(ii) *If* $k$ *appears in the box directly below* $j$ *in the reverse numbering of* $\nu/\lambda$, *then* $k$ *occurs strictly below and weakly left of* $j$ *in* $U$.

In reverse numbering          In tableau $U$
of skew diagram

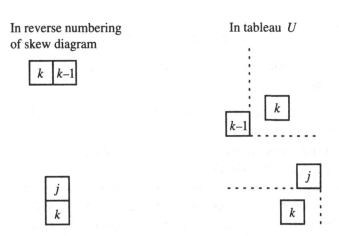

**Proof** To each Littlewood–Richardson skew tableau $S$ on $\nu/\lambda$, we have a reverse lattice word $w(S) = x_r \ldots x_1$, so a standard tableau $U(w(S))$. The fact that $S$ is a skew tableau translates precisely to the conditions (i) and (ii) for $U(w(S))$. In fact, if $k-1$ and $k$ are in the same row of the reverse numbering, for $S$ to be a skew tableau we must have $x_k \leq x_{k-1}$, and this translates to the condition that $k$ is entered in a row of $U$ at least as high as $k-1$; since $k$ enters $U$ after $k-1$, if it goes in weakly above $k-1$ it automatically goes in strictly right of $k-1$. Similarly, if $j$ is directly above $k$ in the reverse numbering, the tableau condition is that $x_j < x_k$, which translates to the fact that $k$ goes in a lower row (and therefore weakly left) of $j$ in $U$.  □

For a given skew diagram, one can construct all the tableaux that satisfy the two properties (i) and (ii) inductively by starting with a 1 in the upper left corner and using the properties to see how to enlarge it. In the above example with $\nu/\lambda = (5,4,3,2)/(3,3,1)$, the 2 must go to the right of the 1, and the 3 below the 2, so all must have 

| 1 | 2 |
|---|---|
| 3 |   |

for the first three entries. The

4 can go in any of the three possible places:

| 1 | 2 |
|---|---|
| 3 |   |
| 4 |   |

| 1 | 2 |
|---|---|
| 3 | 4 |

| 1 | 2 | 4 |
|---|---|---|
| 3 |   |   |

then the 5 must go to the right of the 4, and so on.

**Exercise 7** Continue this to compute the fifteen possible tableaux satisfying the conditions of Proposition 4.

Combined with Corollary 2(v) from §5.1, this gives another prescription (Chen, Garsia, and Remmel [1984]) for the Littlewood–Richardson numbers: $c_{\lambda\mu}^{\nu}$ is the number of standard tableaux on $\nu$ whose entries satisfy the conditions of Proposition 4 for the skew diagram $\lambda * \mu$.

Zelevinsky (1981) defined a **picture** between two skew diagrams to be a bijection between their boxes such that if a box $A$ is weakly above and weakly left of a box $B$ in either diagram, the corresponding boxes $A'$ and $B'$ of the other diagram are in order in its reverse row numbering.

**Corollary 5** _The Littlewood–Richardson number_ $c_{\lambda\mu}^{\nu}$ _is the number of pictures between_ $\mu$ _and_ $\nu/\lambda$.

**Proof** A bijection is given by numbering the boxes of $\mu$ so the corresponding boxes of $\nu/\lambda$ have the reverse row numbering. The condition that the map from $\mu$ to $\nu/\lambda$ takes the upper left to lower right ordering into the reverse row numbering says that this numbering of $\mu$ is a tableau. The same condition on the inverse map from $\nu/\lambda$ to $\mu$ is precisely the requirement that this tableau satisfies the conditions of Proposition 4.     □

**Exercise 8** Show that the number of sequences $v_1, \dots, v_{2n}$ such that the word $v_1 \dots v_{2n}$ is a reverse lattice word with $n$ 1's and $n$ 2's is the number $f^{(n,n)}$, which is $(2n)!/(n+1)! \cdot n!$. (This is the number of binary trees with $n$ nodes.)

**Exercise 9** If $T$ is a standard tableau on shape $(n,n)$, let $P$ be the subtableau of $T$ consisting of its smallest $n$ entries, and let $S$ be the remaining skew tableau. Let $Q$ be the standard tableau obtained by rotating $S$ by $180°$ and replacing $n+i$ by $n+1-i$. Show that any such pair $(P,Q)$ arises uniquely in this way. Deduce that the number of permutations $\sigma$ in $S_n$ containing no decreasing sequence of length three $(\sigma(i) > \sigma(j) > \sigma(k)$ for $i < j < k)$ is $(2n)!/(n+1)! \cdot n!$.

**Exercise 10** (a) Given $\lambda \leq \nu$, and $r = (r_1, \ldots, r_p)$, show that the number of skew tableaux on $\nu/\lambda$ with content $r$ is $\sum_{\mu} K_{\mu r} c^{\nu}_{\lambda \mu}$. In particular, this number is independent of order of $r_1, \ldots, r_p$. (b) For $r = (r_1, \ldots, r_p)$ and $s = (s_1, \ldots, s_q)$, show that $K_{\nu(r,s)} = \sum_{\lambda, \mu} K_{\lambda r} K_{\mu s} c^{\nu}_{\lambda \mu}$, where $(r, s) = (r_1, \ldots, r_p, s_1, \ldots, s_q)$.

**Exercise 11** Let $\mathcal{A}$ and $\mathcal{B}$ be alphabets, with all letters in $\mathcal{A}$ being smaller than all letters in $\mathcal{B}$. A word $w$ on the alphabet $\mathcal{A} \cup \mathcal{B}$ is a *shuffle* of a word $u$ on $\mathcal{A}$ and a word $v$ on $\mathcal{B}$ if $u$ (resp. $v$) is the word obtained from $w$ by removing the letters in $\mathcal{B}$ (resp. $\mathcal{A}$). Let $u_o$ and $v_o$ be fixed words on $\mathcal{A}$ and $\mathcal{B}$. Show that the number of words whose tableaux have shape $\nu$ and that are shuffles of words $u$ and $v$ with $u \equiv u_o$ and $v \equiv v_o$ is $c^{\nu}_{\lambda \mu} f^{\nu}$, where $\lambda$ and $\mu$ are the shapes of $P(u_o)$ and $P(v_o)$.

# 6

# Symmetric polynomials

The first section contains the facts about symmetric polynomials that will be used to study the representations of the symmetric groups. These include formulas expressing the Schur polynomials in terms of other natural bases of symmetric polynomials. A proof of the Jacobi–Trudi formula for Schur polynomials is also sketched. In the second section the number of variables is allowed to grow arbitrarily, and the formulas become identities in the ring of "symmetric functions." For a thorough survey on symmetric functions see Macdonald (1979).

## 6.1 More about symmetric polynomials

To start, we fix a positive integer $m$ and consider polynomials $f(x) = f(x_1, \ldots, x_m)$ in $m$ variables. For each partition $\lambda = (\lambda_1 \geq \ldots \geq \lambda_k)$, we have the Schur polynomials $s_\lambda(x) = s_\lambda(x_1, \ldots, x_m)$, and polynomials

(1a)     $h_\lambda(x) = h_{\lambda_1}(x) \cdot \ldots \cdot h_{\lambda_k}(x)$

(1b)     $e_\lambda(x) = e_{\lambda_1}(x) \cdot \ldots \cdot e_{\lambda_k}(x),$

where $h_p(x)$ and $e_p(x)$ are the $p^{\text{th}}$ **complete** and **elementary** symmetric polynomials in variables $x_1, \ldots, x_m$. We will also need the **monomial** symmetric polynomial $m_\lambda(x)$, which is the sum of all distinct monomials obtained from $x_1^{\lambda_1} \cdot \ldots \cdot x_m^{\lambda_m}$ by permuting all the variables; this is defined provided $\lambda_i = 0$ for $i > m$.

We have seen that the Schur polynomials $s_\lambda(x_1, \ldots, x_m)$ are symmetric. The next thing to observe is the fact that, as $\lambda$ varies over all partitions of $n$ into at most $m$ parts, these Schur polynomials form a basis (over $\mathbb{Z}$) for the symmetric polynomials of degree $n$ in variables $x_1, \ldots, x_m$.

72

**Proposition 1** *The following are bases over $\mathbb{Z}$ of the homogeneous symmetric polynomials of degree $n$ in $m$ variables:*

(i) $\{m_\lambda(x) : \lambda$ *a partition of $n$ with at most $m$ rows*$\}$;

(ii) $\{s_\lambda(x) : \lambda$ *a partition of $n$ with at most $m$ rows*$\}$;

(iii) $\{e_\lambda(x) : \lambda$ *a partition of $n$ with at most $m$ columns*$\}$;

(iv) $\{h_\lambda(x) : \lambda$ *a partition of $n$ with at most $m$ columns*$\}$;

(v) $\{h_\lambda(x) : \lambda$ *a partition of $n$ with at most $m$ rows*$\}$.

**Proof** The proof for (i) is the standard one for bases of symmetric polynomials: given a symmetric polynomial, suppose $x^\lambda = x_1^{\lambda_1} \cdot \ldots \cdot x_m^{\lambda_m}$ occurs with a nonzero coefficient $a$, with $\lambda = (\lambda_1, \ldots, \lambda_m)$ maximal in the lexicographic ordering of $m$-tuples. By symmetry this $m$-tuple $\lambda$ will be a partition, and subtracting $a \cdot m_\lambda(x)$ from the polynomial gives a symmetric polynomial that is smaller with respect to this ordering. The same idea shows that the $m_\lambda(x)$'s form a basis, for if $\sum a_\lambda m_\lambda(x) = 0$, and $\lambda$ is maximal with $a_\lambda \neq 0$, then the coefficient of $x^\lambda$ in $\sum a_\lambda m_\lambda$ is $a_\lambda$, a contradiction.

Since all the sets in (i)–(v) have the same cardinality, it suffices to show that each set spans the homogeneous polynomials of degree $n$ in the $m$ variables. For the $s_\lambda(x)$ in (ii), the proof is the same as for the $m_\lambda(x)$, since $x^\lambda$ is the leading monomial in $s_\lambda(x)$. Similarly, (iii) follows from the fact that the leading monomial in $e_\lambda(x)$ is $x^\mu$, where $\mu$ is the conjugate partition to $\lambda$.

That the sets in (iv) and (iii) have the same span amounts to showing that $\mathbb{Z}[h_1(x), \ldots, h_m(x)] = \mathbb{Z}[e_1(x), \ldots, e_m(x)]$. This follows from the identities

$$h_k(x) - e_1(x)\, h_{k-1}(x) + e_2(x)\, h_{k-2}(x) - \ldots + (-1)^k e_k(x) = 0,$$

which in turn follow from the identity

$$\left(\sum h_p(x)\, t^p\right) \cdot \left(\sum (-1)^q e_q(x)\, t^q\right) = \prod_{i=1}^{m} \frac{1}{1 - x_i t} \cdot \prod_{i=1}^{m} (1 - x_j t) = 1.$$

That the $h_\lambda(x)$'s in (v) have the same span as the $s_\lambda(x)$'s in (ii) follows from equation (5) below. $\square$

The same is true, with the same proof, when $\mathbb{Z}$ is replaced by any commutative ground ring. If rational coefficients are allowed, there is another basis that is particularly useful for calculations, called the Newton *power sums*

(2) $\qquad p_\lambda(x) = p_{\lambda_1}(x) \cdot \ldots \cdot p_{\lambda_k}(x), \qquad p_r(x) = x_1^r + \ldots + x_m^r.$

We will need the following exercise:

**Exercise 1** Prove the following identities:

$$ne_n(x) - p_1(x)e_{n-1}(x) + p_2(x)e_{n-2}(x) - \ldots + (-1)^n p_n(x) = 0;$$

$$nh_n(x) - p_1(x)h_{n-1}(x) - p_2(x)h_{n-2}(x) - \ldots - p_n(x) = 0.$$

For any partition $\lambda$, define the integer $z(\lambda)$ by the formula

(3)      $$z(\lambda) = \prod_r r^{m_r} \cdot m_r!,$$

where $m_r$ is the number of times $r$ occurs in $\lambda$.

**Lemma 1** *For any positive integers* $m$ *and* $n$,

$$h_n(x_1, \ldots, x_m) = \sum_{\lambda \vdash n} \frac{1}{z(\lambda)} p_\lambda(x_1, \ldots, x_m).$$

**Proof** This follows from the following identities of formal power series:

$$\sum_{n=0}^{\infty} h_n(x)t^n = \prod_{i=1}^{m} \frac{1}{1 - x_i t} = \prod_{i=1}^{m} \exp(-\log(1 - x_i t))$$

$$= \prod_{i=1}^{m} \exp\left(\sum_{r=1}^{\infty} \frac{(x_i t)^r}{r}\right) = \exp\left(\sum_{r=1}^{\infty} \sum_{i=1}^{m} \frac{(x_i t)^r}{r}\right)$$

$$= \exp\left(\sum_{r=1}^{\infty} \frac{p_r(x)t^r}{r}\right) = \prod_{r=1}^{\infty} \exp\left(\frac{p_r(x)t^r}{r}\right)$$

$$= \prod_{r=1}^{\infty} \sum_{m_r=0}^{\infty} \frac{(p_r(x)t^r)^{m_r}}{m_r! \cdot r^{m_r}} = \sum_{\lambda} \frac{1}{z(\lambda)} p_\lambda(x) t^{|\lambda|}. \qquad \square$$

**Proposition 2** *The power series* $\prod_{i=1}^{m} \prod_{j=1}^{\ell} \dfrac{1}{1 - x_i y_j}$ *is equal to each of the following sums, adding over all partitions* $\lambda$:

(i)   $$\sum_\lambda h_\lambda(x_1, \ldots, x_m) m_\lambda(y_1, \ldots, y_\ell);$$

(ii)  $$\sum_\lambda \frac{1}{z(\lambda)} p_\lambda(x_1, \ldots, x_m) p_\lambda(y_1, \ldots, y_\ell);$$

(iii) $$\sum_\lambda s_\lambda(x_1, \ldots, x_m) s_\lambda(y_1, \ldots, y_\ell).$$

**Proof** The power series is equal to $\prod_j \left(\sum_n h_n(x)y_j^n\right)$, from which (i) follows. Part (ii) follows from the lemma, when applied to the $m\ell$ variables $x_i y_j$. Part (iii) is the Cauchy formula (3) from §4.3. □

We saw in §2.2 that the complete symmetric and elementary polynomials can be expressed in terms of the Schur polynomials by the formulas

$$(4) \qquad h_\mu(x) = \sum_\lambda K_{\lambda\mu} s_\lambda(x), \quad e_\mu(x) = \sum_\lambda K_{\tilde\lambda\mu} s_\lambda(x).$$

The coefficients are the Kostka numbers, which give an upper triangular change of coordinates between these bases, so they can be solved for the Schur polynomials in terms of the others. In fact, these formulas can be expressed compactly as determinantal formulas: if $\lambda = (\lambda_1 \geq \ldots \geq \lambda_k \geq 0)$,

$$(5) \qquad s_\lambda(x) = \det(h_{\lambda_i+j-i}(x))_{1 \leq i,j \leq k}.$$

This is the determinant of the matrix whose diagonal elements are $h_{\lambda_1}(x)$, $h_{\lambda_2}(x), \ldots, h_{\lambda_k}(x)$, with the subscripts to the right or left of the diagonal increased or decreased by the horizontal distance from the diagonal (with $h_p(x) = 0$ if $p < 0$). The dual formula is

$$(6) \qquad s_\lambda(x) = \det(e_{\mu_i+j-i}(x))_{1 \leq i,j \leq \ell},$$

where $\mu = (\mu_1 \geq \ldots \geq \mu_\ell \geq 0)$ is the conjugate partition to $\lambda$.

The following *Jacobi–Trudi* formula was the original definition of the Schur polynomials:

$$(7) \qquad s_\lambda(x_1, \ldots, x_m) = \frac{\det((x_j)^{\lambda_i+m-i})_{1 \leq i,j \leq m}}{\det((x_j)^{m-i})_{1 \leq i,j \leq m}}.$$

The denominator is the Vandermonde determinant $\prod_{1 \leq i < j \leq m}(x_i - x_j)$. Jacobi proved that the right side of (7) is equal to the right side of (5).

Tableau-theoretic proofs can be given for these three formulas, cf. Sagan (1991), Proctor (1989), or the references. The following exercises sketch short algebraic proofs, cf. Macdonald (1979). These formulas are useful for calculations, but are not essential for reading the rest of these notes.

**Exercise 2** Let $t_\lambda$ be the right side of (7). Deduce (7) from the fact that these functions satisfy the "Pieri formula"

$$(8) \qquad h_p(x) \cdot t_\lambda = \sum_\nu t_\nu,$$

the sum over all $\nu$'s that are obtained from $\lambda$ by adding $p$ boxes, with no two in a column. For any $\ell_1 > \ell_2 > \ldots > \ell_m \geq 0$, let $a(\ell_1, \ldots, \ell_m) = |(x_j)^{\ell_i}|$. Show that (8) is equivalent to

$$(9) \qquad a(\ell_1, \ldots, \ell_m) \cdot \prod_{i=1}^{m} (1 - x_i)^{-1} = \sum a(n_1, \ldots, n_m),$$

the sum over all $n_1 \geq \ell_1 > n_2 \geq \ell_2 > \ldots > n_m \geq \ell_m$. Prove (9) by induction on $m$, expanding the determinant $a(\ell_1, \ldots, \ell_m)$ along the top row.

The following gives a quick proof of (5), following Macdonald (1979, §I.3), starting with the identities $\sum h_n(x) t^n = \prod (1 - x_i t)^{-1}$ and $\sum e_n(x) t^n = \prod (1 + x_i t)$.

**Exercise 3** For any $p$ between 1 and $m$ let $e_r^{(p)}$ denote the $r^{\text{th}}$ symmetric polynomial in the variables $x_1, \ldots, x_{p-1}, x_{p+1}, \ldots, x_m$. Show that

$$\left( \sum h_i(x) t^i \right) \cdot \left( \sum e_r^{(p)} (-t)^r \right) = (1 - x_p t)^{-1}.$$

Deduce the formula $\sum_{j=1}^{m} h_{q+j-m}(x)(-1)^{m-j} e_{m-j}^{(p)} = (x_p)^q$, and hence the matrix identity

$$\left( h_{\lambda_i + j - i}(x) \right)_{1 \leq i,j \leq m} \cdot \left( (-1)^{m-j} e_{m-j}^{(p)} \right)_{1 \leq j, p \leq m}$$

$$= \left( (x_p)^{\lambda_i + m - i} \right)_{1 \leq i, p \leq m}.$$

Deduce (5) from this identity.

**Exercise 4** Show that (5) is equivalent to the formula

$$\sum_{\sigma \in S_m} \text{sgn}(\sigma) K_{\nu\,(\lambda_1 + \sigma(1) - 1, \ldots, \lambda_m + \sigma(m) - m)} = \begin{cases} 1 & \text{if} \quad \nu = \lambda \\ 0 & \text{otherwise}, \end{cases}$$

and deduce (6) from this.

**Exercise 5** Show that

$$s_\lambda(1, x, x^2, \ldots, x^{m-1}) = x^r \prod_{i < j} \frac{x^{\lambda_i - \lambda_j + j - i} - 1}{x^{j-i} - 1},$$

where $r = \lambda_2 + 2\lambda_3 + \ldots = \sum (i-1)\lambda_i$, the product over all $1 \leq i < j \leq m$.

**Exercise 6** Show that

$$s_\lambda(1, \ldots, 1) = \prod_{i < j} \frac{\lambda_i - \lambda_j + j - i}{j - i}.$$

Deduce formula (9) of §4.3.

**Exercise 7** Prove the following generalizations to skew shapes $\lambda/\mu$:

$$\text{(i)} \quad s_{\lambda/\mu}(x) = \det(h_{\lambda_i - \mu_j + j - i}(x))_{1 \le i, j \le k}$$

$$\text{(ii)} \quad s_{\lambda/\mu}(x) = \det(e_{\tilde{\lambda}_i - \tilde{\mu}_j + j - i}(x))_{1 \le i, j \le \ell}$$

where $\tilde{\lambda}$ and $\tilde{\mu}$ denote the conjugate partitions, and $k$ and $\ell$ are the numbers of rows and columns of $\lambda$.

**Exercise 8** For any variables $x_1, \ldots, x_m, y_1, \ldots, y_n$, and any partition $\lambda$, define a "super Schur polynomial" by the formula

$$s_\lambda(x_1, \ldots, x_m; y_1, \ldots, y_n) = \det(c_{\lambda_i + j - i})_{1 \le i, j \le \text{length}(\lambda)},$$

where $c_k$ is the coefficient of $t^k$ in $\prod_{i=1}^{m}(1 - x_i t)^{-1} \prod_{j=1}^{n}(1 + y_j t)$. Show that

$$s_\lambda(x_1, \ldots, x_m; y_1, \ldots, y_n) = s_{\tilde{\lambda}}(y_1, \ldots, y_n; x_1, \ldots, x_m).$$

## 6.2 The ring of symmetric functions

For most purposes it does not matter how many variables are used. The basic reason for this is that all of them specialize: the Schur polynomials and the other polynomials just discussed all satisfy the property that, if $\ell < m$, then

$$p(x_1, \ldots, x_\ell, 0, \ldots, 0) = p(x_1, \ldots x_\ell).$$

It is sometimes important, however, that the number of variables is sufficiently large; for example, the Schur polynomials vanish if the number of variables is smaller than the number of parts of the partition. For this it is convenient to define a *symmetric function of degree* $n$ to be a collection of symmetric polynomials $p(x_1, \ldots, x_m)$ of degree $n$, one for each $m$, that satisfy the displayed identity for all $\ell < m$. We let $\Lambda_n$ be the $\mathbb{Z}$-module of all such functions with integer coefficients. For each partition $\lambda$ of $n$ let $s_\lambda$, $h_\lambda$, $e_\lambda$, $m_\lambda$, and $p_\lambda$ be the corresponding symmetric functions. The first four of these sets form $\mathbb{Z}$-bases for $\Lambda_n$, as $\lambda$ varies over all partitions of $n$, while the power sums form a $\mathbb{Q}$-basis of the corresponding polynomials $\Lambda_n \otimes \mathbb{Q}$ with rational coefficients. We set

$$\Lambda = \bigoplus_{n=0}^{\infty} \Lambda_n,$$

the graded ring of symmetric functions. The ring $\Lambda$ can be identified with the ring of polynomials in the variables $h_1, h_2, \ldots$, or with the ring of polynomials in the variables $e_1, e_2, \ldots$. The identities proved at the finite

level, all being compatible with setting some variables equal to zero, extend
to identities in $\Lambda$. For example, (4) gives

(10)        $$h_\mu = \sum_\lambda K_{\lambda\mu} s_\lambda, \quad e_\mu = \sum_\lambda K_{\tilde{\lambda}\mu} s_\lambda = \sum_\lambda K_{\lambda\mu} s_{\tilde{\lambda}},$$

and the Littlewood–Richardson formula becomes $s_\lambda \cdot s_\mu = \sum_\nu c_{\lambda\mu}^\nu s_\nu$.

One can define a symmetric inner product $\langle \, , \rangle$ on $\Lambda_n$ by requiring that the
Schur functions $s_\lambda$ form an orthonormal basis, i.e., requiring that $\langle s_\lambda, s_\lambda \rangle = 1$
and $\langle s_\lambda, s_\mu \rangle = 0$ if $\mu \neq \lambda$.

**Proposition 3** (1) $\langle h_\lambda, m_\lambda \rangle = 1$ *and* $\langle h_\lambda, m_\mu \rangle = 0$ *if* $\mu \neq \lambda$;
(2) $\langle p_\lambda, p_\lambda \rangle = z(\lambda)$ *and* $\langle p_\lambda, p_\mu \rangle = 0$ *if* $\mu \neq \lambda$.

**Proof** Write $h_\lambda = \sum a_{\lambda\nu} s_\nu$ and $m_\lambda = \sum b_{\lambda\nu} s_\nu$. The equality of (i) and
(iii) in Proposition 2 implies that $(a_{\lambda\nu})$ and $(b_{\lambda\nu})$ are inverse matrices, from
which (1) follows. The proof of (2) is similar, comparing (ii) and (iii) of
Proposition 2.     □

For partitions $\lambda$ and $\mu$ of the same integer, define integers $\chi_\mu^\lambda$ and $\xi_\mu^\lambda$
by the formulas

(11)        $$p_\mu = \sum_\lambda \chi_\mu^\lambda s_\lambda \quad \text{and} \quad p_\mu = \sum_\lambda \xi_\mu^\lambda m_\lambda.$$

From Proposition 3, we have the equivalent formulas

(12)        $$s_\lambda = \sum_\mu \frac{1}{z(\mu)} \chi_\mu^\lambda p_\mu \quad \text{and} \quad h_\lambda = \sum_\mu \frac{1}{z(\mu)} \xi_\mu^\lambda p_\mu.$$

Define an involution $\omega \colon \Lambda \to \Lambda$ to be the additive homomorphism that
takes $s_\lambda$ to $s_{\tilde{\lambda}}$, where $\tilde{\lambda}$ is the conjugate to $\lambda$. In particular, taking $\lambda = (p)$, $\omega(h_p) = e_p$.

**Corollary 1** (1) *The involution* $\omega$ *is a ring homomorphism and an isometry;*
(2) $\omega(h_\lambda) = e_\lambda$, *and* $\omega(p_\mu) = (-1)^{\Sigma(\mu_i - 1)} p_\mu$.

**Proof** The map is an isometry since it takes an orthonormal basis to an orthonormal basis. The fact that $\omega(h_\lambda) = e_\lambda$ follows from (10), and this shows
that $\omega$ preserves products, so is a ring homomorphism. For the last, it therefore suffices to show that $\omega$ takes $p_r$ to $(-1)^{r-1} p_r$, and this follows easily
from Exercise 1.     □

# Part II

## Representation theory

In this part we describe some uses of tableaux in studying representations of the symmetric group $S_n$ and the general linear group $GL_m(\mathbb{C})$. We will see that to each partition $\lambda$ of $n$ one can construct an irreducible representation $S^\lambda$ of the symmetric group $S_n$ (called a *Specht module*) and an irreducible representation $E^\lambda$ of $GL(E)$ for $E$ a finite dimensional complex vector space (called a *Schur* or *Weyl module*). The space $S^\lambda$ will have a basis with one element $v_T$ for each standard tableau $T$ of shape $\lambda$. If $e_1, \ldots, e_m$ is a basis for $E$, then $E^\lambda$ will have a basis with one element $e_T$ for each (semistandard) tableau $T$ on $\lambda$ with entries from $[m]$. These basis vectors $e_T$ will be eigenvectors for the diagonal matrix with entries $x_1, \ldots, x_m$, with eigenvalue $x^T$; the character of the representation will be the Schur polynomial $s_\lambda(x_1, \ldots, x_m)$.

Two extreme cases of these constructions should be familiar in some version, corresponding to the two extreme partitions $\lambda = (n)$ and $\lambda = (1^n)$. We describe these here in order to fix some notation, as well as to motivate the general story.

For $\lambda = (n)$, the representation $S^\lambda$ of $S_n$ is the one-dimensional *trivial representation* $\mathbb{I}_n$, i.e., the vector space $\mathbb{C}$ with the action $\sigma \cdot z = z$ for all $\sigma$ in $S_n$ and all $z$ in $\mathbb{C}$. The Schur module $E^{(n)}$ is the $n^{\text{th}}$ **symmetric power** $\operatorname{Sym}^n E$, which is defined to be the quotient space of the tensor product $E^{\otimes n}$ of $E$ with itself $n$ times, dividing by the subspace generated by all differences $v_1 \otimes \ldots \otimes v_n - v_{\sigma(1)} \otimes \ldots \otimes v_{\sigma(n)}$ for $v_i$ in $E$ and $\sigma$ in $S_n$. The image of $v_1 \otimes \ldots \otimes v_n$ in $\operatorname{Sym}^n E$ is denoted $v_1 \cdot \ldots \cdot v_n$. The map $E^{\times n} \to \operatorname{Sym}^n E$ is multilinear and symmetric. The vector space $\operatorname{Sym}^n E$ is determined by a universal property: for any vector space $F$, and any map $\varphi: E^{\times n} \to F$ that is multilinear and symmetric, there is a unique linear map $\tilde{\varphi}: \operatorname{Sym}^n E \to F$ such that $\varphi(v_1 \times \ldots \times v_n) = \tilde{\varphi}(v_1 \cdot \ldots \cdot v_n)$. Note that

$GL(E)$ acts on $\mathrm{Sym}^n E$ by $g\cdot(v_1\cdot\ldots\cdot v_n) = (g\cdot v_1)\cdot\ldots\cdot(g\cdot v_n)$, so the symmetric powers are representations of $GL(E)$. If $e_1, \ldots, e_m$ is a basis for $E$, the products $e_{i_1}\cdot\ldots\cdot e_{i_n}$, where the indices vary over weakly increasing sequences $1 \le i_1 \le i_2 \le \ldots \le i_n \le m$, form a basis for $\mathrm{Sym}^n E$.

For the other extreme $\lambda = (1^n)$, the representation $S^\lambda$ of $S_n$ is the one-dimensional **alternating representation** $\mathbb{U}_n$, which is the vector space $\mathbb{C}$ with the action $\sigma\cdot z = \mathrm{sgn}(\sigma)z$ for all $\sigma$ in $S_n$ and all $z$ in $\mathbb{C}$; here $\mathrm{sgn}(\sigma)$ is $+1$ if $\sigma$ is an even permutation, and $-1$ if $\sigma$ is odd. In this case the representation $E^\lambda$ is the $n^{\mathrm{th}}$ **exterior power** $\bigwedge^n E$, which is the quotient space of $E^{\otimes n}$ by the subspace generated by all $v_1\otimes\ldots\otimes v_n$ with $v_i = v_{i+1}$ for some $i$. (In characteristic zero, where we usually will be, this is the subspace generated by all differences $v_1\otimes\ldots\otimes v_n - \mathrm{sgn}(\sigma)v_{\sigma(1)}\otimes\ldots\otimes v_{\sigma(n)}$ for $v_i$ in $E$ and $\sigma$ in $S_n$.) The image of $v_1\otimes\ldots\otimes v_n$ in $\bigwedge^n E$ is denoted $v_1 \wedge \ldots \wedge v_n$. The map $E^{\times n} \to \bigwedge^n E$ is multilinear and alternating, and $\bigwedge^n E$ is determined by the corresponding universal property. If $e_1, \ldots, e_m$ is a basis for $E$, then $\bigwedge^n E$ has a basis of the form $e_{i_1} \wedge \ldots \wedge e_{i_n}$, for all strictly increasing sequences $1 \le i_1 < i_2 < \ldots < i_n \le m$.

Although the general case will be worked out in the text, it may be a useful exercise now to work out by hand the first case of this story that does not fall under these two extremes, namely, when $\lambda = (2,1)$. In this case the representation $S^{(2,1)}$ of $S_3$ is the "standard" two-dimensional representation of $S_3$ on the hyperplane $z_1 + z_2 + z_3 = 0$ in the space $\mathbb{C}^3$ with the action

$$\sigma\cdot(z_1,z_2,z_3) = (z_{\sigma^{-1}(1)}, z_{\sigma^{-1}(2)}, z_{\sigma^{-1}(3)}).$$

The space $E^{(2,1)}$ can be realized as the quotient space of $\bigwedge^2 E\otimes E$ by the subspace $W$ generated by all vectors of the form[1]

$$(u \wedge v) \otimes w \; - \; (w \wedge v) \otimes u \; - \; (u \wedge w) \otimes v.$$

**Exercise** (a) Show that if $e_1, \ldots, e_m$ is a basis for $E$, then the images of the vectors $(e_i \wedge e_j) \otimes e_k$, for all $i < j$ and $i \le k$, form a basis for $E^{(2,1)}$. Note that these $(i,j,k)$ correspond to tableaux on the shape $(2,1)$.

---

[1] The symmetric form $(u \wedge v) \otimes w + (v \wedge w) \otimes u + (w \wedge u) \otimes v = 0$ of these relations identifies $E^{(2,1)}$ with the third graded piece of the free Lie algebra on $E$. The displayed form, however, is the one we will generalize to construct the general representations $E^\lambda$.

(b) Construct isomorphisms

$$E^{\otimes 2} \;\cong\; \mathrm{Sym}^2 E \oplus \wedge^2 E; \quad \wedge^2 E \otimes E \;\cong\; \wedge^3 E \oplus E^{(2,1)};$$

$$\mathrm{Sym}^2 E \otimes E \;\cong\; \mathrm{Sym}^3 E \oplus E^{(2,1)};$$

$$E^{\otimes 3} \;\cong\; \wedge^3 E \oplus \mathrm{Sym}^3 E \oplus E^{(2,1)} \oplus E^{(2,1)}.$$

Although we work primarily over the complex numbers, we emphasize methods that make sense over the integers or in positive characteristics; however, we do not consider special features of positive characteristics. In addition, we emphasize constructions that are intrinsic, i.e., that do not depend on a choice of a fixed standard tableau of given shape.

**Notation** In this part we will need the notion of an ***exchange***. This depends on a choice of two columns of a Young diagram $\lambda$, and a choice of a set of the same number of boxes in each column. For any filling $T$ of $\lambda$ (with entries in any set), the corresponding exchange is the filling $S$ obtained from $T$ by interchanging the entries in the two chosen sets of boxes, maintaining the vertical order in each; the entries outside these chosen boxes are unchanged. For example, if $\lambda = (4,3,3,2)$, and the chosen boxes are the top two in the third column, and the second and fourth box in the second column, the exchange takes

$$
T = 
\begin{array}{|c|c|c|c|}
\hline
1 & 5 & 2 & 1 \\
\hline
1 & 3 & 4 \\
\cline{1-3}
2 & 4 & 5 \\
\cline{1-3}
3 & 5 \\
\cline{1-2}
\end{array}
\quad \text{to} \quad
S = 
\begin{array}{|c|c|c|c|}
\hline
1 & 5 & 3 & 1 \\
\hline
1 & 2 & 5 \\
\cline{1-3}
2 & 4 & 5 \\
\cline{1-3}
3 & 4 \\
\cline{1-2}
\end{array}
\;.
$$

# 7

# Representations of the symmetric group

The first section describes the action of the symmetric group $S_n$ on the numberings of a Young diagram with the integers $1, 2, \ldots, n$ with no repeats. A basic combinatorial lemma is proved that will be used in the rest of the chapter. In the second section the Specht modules are defined. They are seen to give all the irreducible representations of $S_n$, and they have bases corresponding to standard tableaux with $n$ boxes. The third section uses symmetric functions to prove some of the main theorems about these representations, including the character formula of Frobenius, Young's rule, and the branching formula. The last section contains a presentation of the Specht module as a quotient of a simpler representation; this will be useful in the next two chapters.

For brevity we assume a few of the basic facts about complex representations (always assumed to be finite-dimensional) of a finite group: that the number of irreducible representations, up to isomorphism, is the number of conjugacy classes; that the sum of the squares of the dimensions of these representations is the order of the group; that every representation decomposes into a sum of irreducible representations, each occurring with a certain multiplicity; that representations are determined by their characters. The orthogonality of characters, which is used to prove some of these facts, and the notion of induced representations, will also be assumed.

## 7.1 The action of $S_n$ on tableaux

The symmetric group $S_n$ is the automorphism group of the set $[n]$, acting on the left, so $(\sigma \cdot \tau)(i) = \sigma(\tau(i))$. In this chapter $T$ and $T'$ will denote numberings of a Young diagram with $n$ boxes with the numbers from 1 to $n$, *with no repeats allowed*. The symmetric group $S_n$ acts on the set of such numberings, with $\sigma \cdot T$ being the numbering that puts $\sigma(i)$ in the box in which $T$ puts $i$. For a numbering $T$ we have a subgroup $R(T)$ of $S_n$, the **row group** of $T$, which consists of those permutations that permute the entries of each row among themselves. If $\lambda = (\lambda_1 \geq \ldots \geq \lambda_k > 0)$ is the shape of $T$, then $R(T)$ is a product of symmetric groups $S_{\lambda_1} \times S_{\lambda_2} \times \ldots \times S_{\lambda_k}$. Such a subgroup of $S_n$ is usually called a **Young subgroup**. Similarly we have the **column group** $C(T)$ of permutations preserving the columns. These subgroups are compatible with the action:

$$(1) \qquad R(\sigma \cdot T) = \sigma \cdot R(T) \cdot \sigma^{-1} \quad \text{and} \quad C(\sigma \cdot T) = \sigma \cdot C(T) \cdot \sigma^{-1}.$$

The following lemma is the basic tool needed for representation theory.

**Lemma 1** *Let $T$ and $T'$ be numberings of the shapes $\lambda$ and $\lambda'$. Assume that $\lambda$ does not strictly dominate $\lambda'$. Then exactly one of the following occurs:*

(i) *There are two distinct integers that occur in the same row of $T'$ and the same column of $T$;*

(ii) *$\lambda' = \lambda$ and there is some $p'$ in $R(T')$ and some $q$ in $C(T)$ such that $p' \cdot T' = q \cdot T$.*

**Proof** Suppose (i) is false. The entries of the first row of $T'$ must occur in different columns of $T$, so there is a $q_1 \in C(T)$ so that these entries occur in the first row of $q_1 \cdot T$. The entries of the second row of $T'$ occur in different columns of $T$, so also of $q_1 \cdot T$, so there is a $q_2$ in $C(q_1 \cdot T) = C(T)$, not moving the entries equal to those in the first row of $T'$, so that these entries all occur in the first two rows of $q_2 \cdot q_1 \cdot T$. Continuing in this way, we get $q_1, \ldots, q_k$ in $C(T)$ such that the entries in the first $k$ rows of $T'$ occur in the first $k$ rows of $q_k \cdot q_{k-1} \cdot \ldots \cdot q_1 \cdot T$. In particular, since $T$ and $q_k \cdot \ldots \cdot q_1 \cdot T$ have the same shape, it follows that $\lambda_1' + \ldots \lambda_k' \leq \lambda_1 + \ldots + \lambda_k$. Since this is true for all $k$, this means that $\lambda' \trianglelefteq \lambda$.

Since we have assumed that $\lambda$ does not strictly dominate $\lambda'$, we must have $\lambda = \lambda'$. Taking $k$ to be the number of rows in $\lambda$, and $q = q_k \cdot \ldots \cdot q_1$, we see that $q \cdot T$ and $T'$ have the same entries in each row. This means that there is a $p'$ in $R(T')$ such that $p' \cdot T' = q \cdot T$. $\quad \square$

Define a linear ordering on the set of all numberings with $n$ boxes, by saying that $T' > T$ if either 1) the shape of $T'$ is larger than the shape of $T$ in the lexicographic ordering, or 2) $T'$ and $T$ have the same shape, and the largest entry that is in a different box in the two numberings occurs earlier in the column word of $T'$ than in the column word of $T$. For example, this ordering puts the standard tableaux of shape $(3,2)$ in the following order:

$$
\begin{array}{|c|c|c|}\hline 1 & 2 & 3 \\\hline 4 & 5 \\\hline\end{array} >
\begin{array}{|c|c|c|}\hline 1 & 2 & 4 \\\hline 3 & 5 \\\hline\end{array} >
\begin{array}{|c|c|c|}\hline 1 & 3 & 4 \\\hline 2 & 5 \\\hline\end{array} >
\begin{array}{|c|c|c|}\hline 1 & 2 & 5 \\\hline 3 & 4 \\\hline\end{array} >
\begin{array}{|c|c|c|}\hline 1 & 3 & 5 \\\hline 2 & 4 \\\hline\end{array}
$$

An important property of this ordering, which follows from the definition, is that if $T$ is a standard tableau, then for any $p \in R(T)$ and $q \in C(T)$,

(2)     $p{\cdot}T \geq T$  and  $q{\cdot}T \leq T$.

Indeed, the largest entry of $T$ moved by $p$ is moved to the left, and the largest entry moved by $q$ is moved up.

**Corollary** *If $T$ and $T'$ are standard tableaux with $T' > T$, then there is a pair of integers in the same row of $T'$ and the same column of $T$.*

**Proof** Since $T' > T$, the shape of $T$ cannot dominate the shape of $T'$. If there is no such pair, we are in case (ii) of the lemma: $p'{\cdot}T' = q{\cdot}T$. Since $T$ and $T'$ are tableaux, by (2) we have $q{\cdot}T \leq T$, and $p'{\cdot}T' \geq T'$; this contradicts the assumption $T' > T$.     □

## 7.2 Specht modules

A *tabloid* is an equivalence class of numberings of a Young diagram (with distinct numbers $1, \ldots, n$), two being equivalent if corresponding rows contain the same entries. The tabloid determined by a numbering $T$ is denoted $\{T\}$. So $\{T'\} = \{T\}$ exactly when $T' = p{\cdot}T$ for some $p$ in $R(T)$. Tabloids are sometimes displayed by omitting the vertical lines between boxes, emphasizing that only the content of each row matters. For example,

$$
\begin{array}{|ccc|}\hline 1 & 4 & 7 \\\hline 3 & 6 \\\hline 2 & 5 \\\hline\end{array}
\;=\;
\begin{array}{|ccc|}\hline 4 & 7 & 1 \\\hline 6 & 3 \\\hline 2 & 5 \\\hline\end{array}
$$

The symmetric group $S_n$ acts on the set of tabloids by the formula

$$
\sigma{\cdot}\{T\} \;=\; \{\sigma{\cdot}T\}.
$$

As a set with $S_n$-action, the orbit of $\{T\}$ is isomorphic to the left coset $S_n/R(T)$.

Let $A = \mathbb{C}[S_n]$ denote the group ring of $S_n$, which consists of all complex linear combinations $\sum x_\sigma \sigma$, with multiplication determined by composition in $S_n$; a representation of $S_n$ is the same as a left $A$-module. Given a numbering $T$ of a diagram with $n$ boxes (with integers from 1 to $n$ each occurring once), define $a_T$ and $b_T$ in $A$ by the formulas

$$(3)\qquad a_T = \sum_{p \in R(T)} p, \quad b_T = \sum_{q \in C(T)} \mathrm{sgn}(q)q.$$

These elements, and the product

$$c_T = b_T \cdot a_T,$$

are called *Young symmetrizers*.

The four exercises of this section will be used later.

**Exercise 1** (a) For $p$ in $R(T)$ and $q$ in $C(T)$, show that

$$p \cdot a_T = a_T \cdot p = a_T \quad \text{and} \quad q \cdot b_T = b_T \cdot q = \mathrm{sgn}(q)b_T.$$

(b) Show that $a_T \cdot a_T = \#R(T) \cdot a_T$ and $b_T \cdot b_T = \#C(T) \cdot b_T$.

We want to produce one irreducible representation for each conjugacy class in $S_n$. The conjugacy classes correspond to partitions $\lambda$ of $n$, the conjugacy class $C(\lambda)$ consisting of those permutations that, when decomposed into cycles, have cycles of lengths $\lambda_1, \lambda_2, \ldots, \lambda_k$, where $\lambda = (\lambda_1 \geq \ldots \geq \lambda_k > 0)$.

**Exercise 2** Show that the number of elements in $C(\lambda)$ is $n!/z(\lambda)$, where $z(\lambda)$ is the integer defined in (3) of §6.1.

Define $M^\lambda$ to be the complex vector space with basis the tabloids $\{T\}$ of shape $\lambda$, with $\lambda$ a partition of $n$. Since $S_n$ acts on the set of tabloids, it acts on $M^\lambda$, making $M^\lambda$ into a left $A$-module. For each numbering $T$ of $\lambda$ there is an element $v_T$ in $M^\lambda$ defined by the formula

$$(4)\qquad v_T = b_T \cdot \{T\} = \sum_{q \in C(T)} \mathrm{sgn}(q)\{q \cdot T\}.$$

**Exercise 3** Show that $\sigma \cdot v_T = v_{\sigma \cdot T}$ for all $T$ and all $\sigma \in S_n$.

**Lemma 2** *Let $T$ and $T'$ be numberings of shapes $\lambda$ and $\lambda'$, and assume that $\lambda$ does not strictly dominate $\lambda'$. If there is a pair of integers in the same row of $T'$ and the same column of $T$, then $b_T \cdot \{T'\} = 0$. If there is no such pair, then $b_T \cdot \{T'\} = \pm v_T$.*

**Proof** If there is such a pair of integers, let $t$ be the transposition that permutes them. Then $b_T \cdot t = -b_T$, since $t$ is in the column group of $T$, but $t \cdot \{T'\} = \{T'\}$ since $t$ is in the row group of $T'$. This leads to

$$b_T \cdot \{T'\} \;=\; b_T \cdot (t \cdot \{T'\}) \;=\; (b_T \cdot t) \cdot \{T'\} \;=\; -b_T \cdot \{T'\},$$

so $b_T \cdot \{T'\} = 0$. If there is no such pair, let $p'$ and $q$ be as in (ii) of Lemma 1 in §7.1. Then

$$b_T \cdot \{T'\} = b_T \cdot \{p' \cdot T'\} \;=\; b_T \cdot \{q \cdot T\}$$

$$= b_T \cdot q \cdot \{T\} \;=\; \mathrm{sgn}(q) b_T \cdot \{T\} \;=\; \mathrm{sgn}(q) \cdot v_T. \qquad \square$$

From the Corollary in §7.1 we deduce

**Corollary** *If $T$ and $T'$ are standard tableaux with $T' > T$, then $b_T \cdot \{T'\} = 0$.*

Define the **Specht module** $S^\lambda$ to be the subspace of $M^\lambda$ spanned by the elements $v_T$, as $T$ varies over all numberings of $\lambda$. It follows from Exercise 3 that $S^\lambda$ is preserved by $S_n$; i.e., it is an $A$-submodule of $M^\lambda$. It follows in fact that $S^\lambda = A \cdot v_T$ for any such numbering $T$.

**Exercise 4** For $\lambda = (n)$, show that $S^{(n)}$ is the trivial representation $\mathbb{1}_n$ of $S_n$, and for $\lambda = (1^n)$, $S^{(1^n)}$ is the alternating representation $\mathbb{U}_n$.

No $v_T$ is zero, so the modules $S^\lambda$ are all nonzero. No two of them are isomorphic. In fact, Lemma 2 (with Exercise 1) implies that for any numbering $T$ of $\lambda$ we have

(5a) $\qquad b_T \cdot M^\lambda \;=\; b_T \cdot S^\lambda \;=\; \mathbb{C} \cdot v_T \;\neq\; 0;$

(5b) $\qquad b_T \cdot M^{\lambda'} \;=\; b_T \cdot S^{\lambda'} \;=\; 0 \quad \text{if} \quad \lambda' > \lambda.$

These same equations imply that each $S^\lambda$ is irreducible. Indeed, irreducibility in characteristic zero is the same as indecomposability, and if $S^\lambda = V \oplus W$, then $\mathbb{C} \cdot v_T = b_T \cdot S^\lambda = b_T \cdot V \oplus b_T \cdot W$, so one of $V$ or $W$ must contain $v_T$. If $V$ contains $v_T$, then $S^\lambda = A \cdot v_T = V$.

We have therefore produced an irreducible representation for each partition of $n$, and since there are the same number of partitions of $n$ as there are conjugacy classes in $S_n$, and the number of conjugacy classes is always the number of irreducible complex representations of a finite group, these are all of them. This proves the following proposition:

**Proposition 1** *For each partition* $\lambda$ *of* $n$, $S^\lambda$ *is an irreducible representation of* $S_n$. *Every irreducible representation of* $S_n$ *is isomorphic to exactly one* $S^\lambda$.

It follows from the construction of $M^\lambda$ and $S^\lambda$ that these complex representations arise from corresponding representations defined over $\mathbb{Q}$. In particular, it follows that the characters of all representations of $S_n$ take on only rational values.

**Lemma 3** *Let* $\vartheta: M^\lambda \to M^{\lambda'}$ *be a homomorphism of representations of* $S_n$. *If* $S^\lambda$ *is not contained in the kernel of* $\vartheta$, *then* $\lambda' \trianglelefteq \lambda$.

**Proof** Let $T$ be a numbering of $\lambda$. Since $v_T$ is not in the kernel of $\vartheta$, $b_T \cdot \vartheta(\{T\}) = \vartheta(v_T) \neq 0$. Therefore $b_T \cdot \{T'\} \neq 0$ for some numbering $T'$ of $\lambda'$. If $\lambda \neq \lambda'$ and $\lambda$ does not dominate $\lambda'$, then we are in case (i) of Lemma 1 of §7.1, and this contradicts Lemma 2.     $\square$

**Corollary** *There are nonnegative integers* $k_{\nu\lambda}$, *for* $\nu \triangleright \lambda$, *such that*

$$M^\lambda \cong S^\lambda \oplus \bigoplus_{\nu \triangleright \lambda} (S^\nu)^{\oplus k_{\nu\lambda}}.$$

**Proof** For each $\nu$, let $k_{\nu\lambda}$ be the number of times the irreducible representation $S^\nu$ occurs in the decomposition of $M^\lambda$. To see that $k_{\lambda\lambda} = 1$, take any numbering $T$ of $\lambda$ and use equation (5a). Since every $S^\nu$ occurs in $M^\nu$, there is a projection from $M^\nu$ to $S^\nu$. Suppose $S^\nu$ also occurs in the decomposition of $M^\lambda$. Then the projection from $M^\nu$ to $S^\nu$ followed by an imbedding of $S^\nu$ in $M^\lambda$ is a homomorphism $\vartheta$ from $M^\nu$ to $M^\lambda$ that does not contain $S^\nu$ in its kernel. Lemma 3 implies that $\lambda \trianglelefteq \nu$, concluding the proof.     $\square$

**Proposition 2** *The elements* $v_T$, *as* $T$ *varies over the standard tableaux on* $\lambda$, *form a basis for* $S^\lambda$.

**Proof** The element $v_T$ is a linear combination of $\{T\}$, with coefficient 1, and elements $\{q \cdot T\}$, for $q \in C(T)$, with coefficients $\pm 1$. Note that when $T$ is a tableau, $q \cdot T < T$, in the ordering we defined in §7.1, for each such nontrivial element $q$. It follows easily that the elements $v_T$ are linearly independent, by looking at the maximal $T$ occurring with nonzero coefficient in a relation $\sum x_T v_T = 0$. This shows in particular that the dimension of $S^\lambda$ is at least the number $f^\lambda$ of standard tableaux of shape $\lambda$.

There are effective ways to show that these elements span $S^\lambda$, one of which we will give in §7.4, but the fact itself can be deduced easily from the fact that the sum of the squares of the dimensions of the representations is the order of the group:

$$n! = \sum_\lambda (\dim(S^\lambda))^2 \geq \sum_\lambda (f^\lambda)^2 = n!,$$

the last by equation (4) of §4.3. It follows that $\dim(S^\lambda) = f^\lambda$ for all $\lambda$, which shows that the elements $v_T$ must also span $S^\lambda$.  □

At this point it is far from obvious how to compute the character of the representation $S^\lambda$, but it is straightforward to compute the character of $M^\lambda$. Since $M^\lambda$ is the representation associated to the action of $S_n$ on the set of tabloids, the trace of an element $\sigma$ is the number of tabloids that are fixed by $\sigma$. Writing $\sigma$ as a product of cycles, a tabloid will be fixed exactly when all elements of each cycle occur in the same row. The number of such tabloids can be expressed as follows. Let $\sigma$ be in the conjugacy class $C(\mu)$, and let $m_q$ be the multiplicity with which the integer $q$ occurs in $\mu$. Let $r(p,q)$ be the number of cycles of length $q$ whose elements lie in the $p^{\text{th}}$ row of a tabloid. The number of tabloids fixed by $\sigma$ is therefore

$$(*) \qquad \sum \prod_{q=1}^{n} \frac{m_q!}{r(1,q)! \cdot \ldots \cdot r(n,q)!},$$

the sum over all collections $(r(p,q))_{1 \leq p,q \leq n}$ of nonnegative integers satisfying

$$r(p,1) + 2r(p,2) + 3r(p,3) + \ldots + nr(p,n) = \lambda_p$$

$$r(1,q) + r(2,q) + \ldots + r(n,q) = m_q.$$

Now for any $q$ we have by the binomial expansion

$$(x_1^q + \ldots + x_n^q)^{m_q} = \sum \frac{m_q!}{r(1,q)! \cdot \ldots \cdot r(n,q)!} x_1^{q\,r(1,q)} \cdot \ldots \cdot x_n^{q\,r(n,q)},$$

the sum over all $(r(p,q))_{1 \leq p \leq n}$ such that $\sum_p r(p,q) = m_q$. The number $(*)$ is therefore the coefficient of $x^\lambda = x_1^{\lambda_1} \cdot \ldots \cdot x_n^{\lambda_n}$ in the polynomial $p_\mu(x_1, \ldots, x_n) = \prod_{q=1}^n (x_1^q + \ldots + x_n^q)^{m_q}$. This number in turn is the integer $\xi_\mu^\lambda$ defined in equation (11) of §6.2. This proves

**Lemma 4** *The value of the character of $M^\lambda$ on the conjugacy class $C(\mu)$ is the coefficient $\xi_\mu^\lambda$ of $x^\lambda$ in $p_\mu$.*

For any numbering $T$ of $\lambda$, we have a Young subgroup $R(T)$ of $S_n$. The fact that $M^\lambda$ has a basis of elements of the form $\sigma \cdot \{T\}$, as $\sigma$ ranges over representatives of $S_n/R(T)$, means that $M^\lambda$ is isomorphic to the induced representation of the trivial representation $\mathbb{I}$ from $R(T)$ to $S_n$:

$$(6) \qquad M^\lambda \;=\; \mathrm{Ind}^{S_n}_{R(T)}(\mathbb{I}) \;=\; \mathbb{C}[S_n] \otimes_{\mathbb{C}[R(T)]} \mathbb{C}.$$

## 7.3 The ring of representations and symmetric functions

Let $R_n$ be the free abelian group on the isomorphism classes of irreducible representations of $S_n$. A representation $V$ of $S_n$ determines a class $[V]$ in $R_n$ by $[V] = \sum m_\lambda [S^\lambda]$ if $V \cong \oplus (S^\lambda)^{\oplus m_\lambda}$. Equivalently, $R_n$ is the Grothendieck group of representations of $S_n$, i.e., the free abelian group on the set of isomorphism classes $[V]$ of all representations $V$ of $S_n$, modulo the subgroup generated by all $[V \oplus W] - [V] - [W]$. Let $R = \bigoplus_{n=0}^\infty R_n$, where $R_0 = \mathbb{Z}$. Define a product $R_n \times R_m \to R_{n+m}$, denoted $\circ$, by the formula

$$(7) \qquad [V] \circ [W] \;=\; \left[ \mathrm{Ind}^{S_{n+m}}_{S_n \times S_m} V \otimes W \right].$$

Here the tensor product $V \otimes W = V \otimes_{\mathbb{C}} W$ is regarded as a representation of $S_n \times S_m$ in the obvious way: $(\sigma \times \tau) \cdot (v \otimes w) = \sigma \cdot v \otimes \tau \cdot w$; and $S_n \times S_m$ is regarded as a subgroup of $S_{n+m}$ in the usual way (with $S_n$ acting on the first $n$ integers, and $S_m$ on the last $m$ integers). The induced representation can be defined quickly by the formula

$$(8) \qquad \mathrm{Ind}^{S_{n+m}}_{S_n \times S_m} V \otimes W \;=\; \mathbb{C}[S_{n+m}] \otimes_{\mathbb{C}[S_n \times S_m]} (V \otimes W).$$

It is straightforward to verify that this product is well defined, and makes $R$ into a commutative, associative, graded ring with unit. (Note that the usual tensor product makes each $R_n$ into a ring in its own right, which is the representation ring of the fixed $S_n$, but this is *not* what is used here.)

There is a symmetric inner product $\langle \, , \, \rangle$ on $R_n$, determined by requiring that the irreducible representations $[S^\lambda]$ form an orthonormal basis. If $V$ and $W$ are representations of $S_n$, it follows that

$$(9) \qquad \langle [V], [W] \rangle \;=\; \sum m_\lambda n_\lambda, \quad \text{where } V \cong \oplus (S^\lambda)^{\oplus m_\lambda},$$

$$W \cong \oplus (S^\lambda)^{\oplus n_\lambda}.$$

The orthogonality of characters of the finite group $S_n$ says that this inner product can be given by the formula

$$\langle [V], [W] \rangle = \frac{1}{n!} \sum_{\sigma \in S_n} \chi_V(\sigma) \chi_W(\sigma^{-1}),$$

where $\chi_V$ is the character of $V$, i.e., $\chi_V(\sigma) = \mathrm{Trace}(\sigma\colon V \to V)$. Since the inverse of a permutation in $C(\mu)$ is in $C(\mu)$, and the order of $C(\mu)$ is $n!/z(\mu)$ by Exercise 2, this gives the formula

$$(10) \qquad \langle [V], [W] \rangle = \sum_\mu \frac{1}{z(\mu)} \chi_V(C(\mu)) \chi_W(C(\mu)).$$

There is also an additive involution $\omega\colon R_n \to R_n$ that takes $[V]$ to $[V \otimes \mathbb{U}_n]$, where $\mathbb{U}_n$ is the alternating representation of $S_n$.

Since the polynomials $h_\lambda$ form a basis of the ring $\Lambda$ of symmetric functions, we may define an additive homomorphism $\varphi\colon \Lambda \to R$ by the formula

$$(11) \qquad \varphi(h_\lambda) = [M^\lambda].$$

**Theorem** (1) *The homomorphism $\varphi$ is a homomorphism of graded rings, and is an isometric isomorphism of $\Lambda$ with $R$.* (2) $\varphi(s_\lambda) = [S^\lambda]$.

**Proof** The map takes $h_n$ to the class of the trivial representation $M^{(n)} = \mathbb{I}_n$ of $S_n$. Since $\Lambda$ is a polynomial ring in the variables $h_n$, to show that $\varphi$ is a homomorphism it suffices to verify that

$$M^{(\lambda_1)} \circ M^{(\lambda_2)} \circ \dots \circ M^{(\lambda_k)} = M^{(\lambda)},$$

for $\lambda = (\lambda_1 \geq \dots \geq \lambda_k > 0)$. This follows from the description of $M^\lambda$ as an induced representation given at the end of the preceding section, using the numbering $T$ by rows, from left to right and top to bottom. It follows from the corollary to Lemma 3 that the $[M^\lambda]$'s form a basis for $R$, so $\varphi$ is an isomorphism of $\mathbb{Z}$-algebras.

To prove the rest of the theorem, and for later applications, we need a formula for the inverse map $\psi$ from $R$ to $\Lambda$. For computations it is useful to express the result in terms of the power sum polynomials; this means that we want a homomorphism

$$\psi\colon R \to \Lambda \otimes \mathbb{Q}$$

such that the composite $\psi \circ \varphi$ is the inclusion of $\Lambda$ in $\Lambda \otimes \mathbb{Q}$. Since $\varphi(h_\lambda) = [M^\lambda]$, we know from equation (12) of §6.2 that $[M^\lambda]$ should map by $\psi$ to $h_\lambda = \sum_\mu \frac{1}{z(\mu)} \xi_\mu^\lambda p_\mu$. By Lemma 4 the coefficient $\xi_\mu^\lambda$ is the character of $M^\lambda$ on the conjugacy class $C(\mu)$. This tells us what the formula for $\psi$ must be:

$$(12) \qquad \psi([V]) = \sum_\mu \frac{1}{z(\mu)} \chi_V(C(\mu)) p_\mu.$$

From this definition it is clear that $\psi$ is an additive homomorphism, and by the above remarks, the composite $\psi \circ \varphi$ is the inclusion of $\Lambda$ in $\Lambda \otimes \mathbb{Q}$. Since $\varphi$ is an isomorphism of $\Lambda$ onto $R$, it follows in fact that $\psi$ is the inverse isomorphism from $R$ onto $\Lambda$.

We may show that $\varphi$ is an isometry by showing that its inverse $\psi$ is one. By the definition of $\psi$,

$$\langle \psi([V]), \psi([W]) \rangle = \sum_{\lambda,\mu} \frac{1}{z(\lambda)z(\mu)} \chi_V(C(\lambda)) \chi_W(C(\mu)) \langle p_\lambda, p_\mu \rangle.$$

By (2) of Proposition 3 of §6.2, the sum on the right is

$$\sum_\mu \frac{1}{z(\mu)} \chi_V(C(\mu)) \chi_W(C(\mu)) = \langle [V], [W] \rangle.$$

We know from (4) of §6.1 and the corollary to Lemma 3 in §7.2 that $h_\lambda = s_\lambda + \sum K_{\nu\lambda} s_\nu$ and $[M^\lambda] = [S^\lambda] + \sum k_{\nu\lambda}[S^\nu]$, both sums over $\nu \triangleright \lambda$. Since $\varphi(h_\lambda) = [M^\lambda]$, it follows that $\varphi(s_\lambda) = [S^\lambda] + \sum m_{\nu\lambda}[S^\nu]$ for some integers $m_{\nu\lambda}$. But now the fact that $\varphi$ is an isometry implies that

$$1 = \langle s_\lambda, s_\lambda \rangle = \langle \varphi(s_\lambda), \varphi(s_\lambda) \rangle = 1 + \sum (m_{\nu\lambda})^2.$$

The coefficients $m_{\nu\lambda}$ must therefore all vanish, giving the desired equation $\varphi(s_\lambda) = [S^\lambda]$. $\quad\square$

The fact that $\varphi$ is a ring isomorphism, with the Schur functions corresponding to the irreducible representations, means that we can transfer what we know about symmetric functions to deduce corresponding facts about representations. For example, equation (4) of Chapter 6 leads to

**Corollary 1** (Young's rule) $M^\lambda \cong S^\lambda \oplus \bigoplus_{\nu \triangleright \lambda} (S^\nu)^{\oplus K_{\nu,\lambda}}$, *where* $K_{\nu\lambda}$ *is the Kostka number.*

The formula $s_\lambda \cdot s_\mu = \sum_\nu c_{\lambda\mu}^\nu s_\nu$ yields

**Corollary 2** (Littlewood–Richardson rule) $S^\lambda \circ S^\mu \cong \bigoplus_\nu (S^\nu)^{\oplus c_{\lambda\mu}^\nu}$.

Taking $\mu = (1)$, and noting that the inclusion $S_n \times S_1 \subset S_{n+1}$ is the usual inclusion of $S_n$ in $S_{n+1}$, this specializes to

**Corollary 3** (Branching rule) *If $\lambda$ is a partition of $n$, then the representation induced by $S^\lambda$ from $S_n$ to $S_{n+1}$ is the direct sum of one copy of each $S^{\lambda'}$ for which $\lambda'$ is obtained from $\lambda$ by adding one box.*

Frobenius reciprocity says that if $H$ is a subgroup of a finite group $G$, then the number of times an irreducible representation $W$ of $H$ occurs in the restriction of an irreducible representation $V$ of $G$ is the same as the number of times $V$ occurs in $\operatorname{Ind}_H^G W$. (This follows from the isomorphism $\operatorname{Hom}_{\mathbb{C}[G]}(\mathbb{C}[G] \otimes_{\mathbb{C}[H]} W, V) = \operatorname{Hom}_{\mathbb{C}[H]}(W, V)$.) Corollary 3 is therefore equivalent to saying that the restriction of $S^\lambda$ from $S_n$ to $S_{n-1}$ is a sum of those $S^{\lambda'}$ for which $\lambda'$ is obtained from $\lambda$ by removing one box.

**Exercise 5** For the inclusion of $S_n$ in $S_{n+m}$ deduce that the multiplicity of a representation $S^{\lambda'}$ in $\operatorname{Ind}(S^\lambda)$ is the number of skew tableaux on $\lambda'/\lambda$ using the numbers $1, \ldots, m$ without repeats.

**Corollary 4** (Frobenius character formula) *The character of $S^\lambda$ on the conjugacy class $C(\mu)$ is the integer $\chi_\mu^\lambda$ defined in equation (11) of §6.2.*

**Proof** From the definition of $\psi$ in the proof of the theorem we know that $[S^\lambda]$ corresponds to the element $\sum_\mu \frac{1}{z(\mu)} \chi_{S^\lambda}(C(\mu)) p_\mu$ in $\Lambda$, as well as to the element $s_\lambda$. An appeal to equation (12) of §6.2 finishes the proof. $\square$

**Exercise 6** If $\lambda = (\lambda_1 \geq \ldots \geq \lambda_k \geq 0)$, define $\ell_i = \lambda_i + k - i$. Show that the number $\chi_\mu^\lambda$ is the coefficient of $x_1^{\ell_1} \cdot \ldots \cdot x_k^{\ell_k}$ in the polynomial

$$\prod_{1 \leq i < j \leq k} (x_i - x_j) \cdot p_\mu(x_1, \ldots, x_k).$$

**Exercise 7** Use the Frobenius formula with $\mu = (1^n)$ to show that the dimension of $S^\lambda$ is the number given for $f^\lambda$ in Exercise 9 of §4.3. In particular, this gives another proof of the hook length formula.

We have defined involutions $\omega$ for the ring $\Lambda$ and the ring $R$.

**Proposition 3** *The isomorphism of $\Lambda$ with $R$ commutes with the involutions $\omega$.*

**Proof** It suffices to show that $\psi(\omega([M^\lambda])) = \omega(\psi([M^\lambda]))$. Since the character of the alternating representation on $C(\mu)$ is $(-1)^{\Sigma(\mu_i - 1)}$, this follows from

$$\psi(\omega([M^\lambda])) = \psi(([M^\lambda \otimes \mathbb{U}_n]))$$

$$= \sum_\mu \frac{1}{z(\mu)} \chi_{M^\lambda}(C(\mu)) \cdot \chi_{\mathbb{U}_n}(C(\mu)) p_\mu$$

$$= \sum_\mu \frac{1}{z(\mu)} \chi_{M^\lambda}(C(\mu)) \cdot (-1)^{\Sigma(\mu_i - 1)} p_\mu$$

$$= \sum_\mu \frac{1}{z(\mu)} \chi_{M^\lambda}(C(\mu)) \cdot \omega(p_\mu)$$

$$= \omega(\psi([M^\lambda]),$$

where we have used part (2) of the corollary in §6.2 to calculate $\omega(p_\mu)$.   $\square$

In particular, it follows that $\omega \colon R \to R$ is a homomorphism of rings. Since the involution on $\Lambda$ takes $s_\lambda$ to $s_{\tilde\lambda}$, we have the following corollary:

**Corollary** $S^\lambda \otimes \mathbb{U}_n \cong S^{\tilde\lambda}$, *where* $\mathbb{U}_n$ *is the alternating representation of* $S_n$ *and* $\tilde\lambda$ *the conjugate partition to* $\lambda$.

**Exercise 8** Find an explicit isomorphism of $S^{\tilde\lambda}$ with $S^\lambda \otimes \mathbb{U}_n$.

**Exercise 9** For any numbering $T$ of $\lambda$, show that the map $\{\sigma \cdot T\} \mapsto \sigma \cdot a_T$ determines an isomorphism of $M^\lambda$ with the left ideal $A \cdot a_T$. Show that this maps $S^\lambda$ isomorphically onto the ideal $A \cdot c_T$.

The following exercise will be used in the next section.

**Exercise 10** Show that, for any numbering $T$, the map $x \mapsto x \cdot \{T\}$ defines an $A$-linear surjection from $A$ onto $M^\lambda$, with kernel the left ideal generated by all $p - 1$, $p \in R(T)$. This ideal is also generated by those $p - 1$ for which $p$ is a transposition in $R(T)$.

**Exercise 11** (a) Show that $S^{(n-1\,1)}$ is isomorphic to the standard representation $V_n = \{(z_1, \ldots, z_n) \in \mathbb{C}_n : \sum z_i = 0\}$, with $S_n$ acting by $\sigma \cdot (z_1, \ldots, z_n) = (z_{\sigma^{-1}(1)}, \ldots, z_{\sigma^{-1}(n)})$. (b) Use the branching rule to show that $S^{(n-p\,1^p)}$ is isomorphic to $\wedge^p(V_n)$.

## 7.4 A dual construction and a straightening algorithm

There is a dual construction of Specht modules, using column tabloids in place of row tabloids. This will be used to give an algorithm for writing a general element in $S^\lambda$ in terms of the basis $\{v_T\}$ given by standard tableaux of shape $\lambda$. We want column tabloids to be "alternating," however, so that, if two elements in a column are interchanged, the tabloid changes sign. A *column tabloid* will be an equivalence class of numberings of a Young diagram, with two being equivalent if they have the same entries in each column; two equivalent numberings have the same or opposite **orientation** according to whether the permutation taking one to the other has positive or negative sign. These can be pictured as follows:

$$
\begin{array}{|c|c|}
\hline
2 & 1 \\
\hline
3 & 4 \\
\hline
5 \\
\cline{1-1}
\end{array}
\;=\; -\;
\begin{array}{|c|c|}
\hline
2 & 1 \\
\hline
5 & 4 \\
\hline
3 \\
\cline{1-1}
\end{array}
\;=\;
\begin{array}{|c|c|}
\hline
5 & 1 \\
\hline
2 & 4 \\
\hline
3 \\
\cline{1-1}
\end{array}
\;=\; -\;
\begin{array}{|c|c|}
\hline
5 & 4 \\
\hline
2 & 1 \\
\hline
3 \\
\cline{1-1}
\end{array}
$$

We denote by $[T]$ the oriented column tabloid defined by a numbering $T$, and write $-[T]$ for that defined by an odd permutation preserving its columns.

For a partition $\lambda$ of $n$, $\widetilde{M}^\lambda$ denotes the vector space that is a sum of copies of $\mathbb{C}$, one for each column tabloid, but with the corresponding basis element defined only up to sign, depending on orientation. Equivalently, take $\widetilde{M}^\lambda$ to be the vector space with basis $[T]$ for each numbering $T$ of $\lambda$, modulo the subspace generated by all $[T] - \mathrm{sgn}(q)[T]$ for $q \in C(T)$.

The symmetric group $S_n$ acts on $\widetilde{M}^\lambda$ by the rule $\sigma \cdot [T] = [\sigma T]$. Define $\widetilde{S}^\lambda \subset \widetilde{M}^\lambda$ to be the submodule spanned by all elements

$$
\tilde{v}_T \;=\; a_T \cdot [T] \;=\; \sum_{p \in R(T)} [pT].
$$

All the results of §7.2 have analogues in this dual setting, and this provides an alternative construction of the irreducible representations of $S_n$. This is carried out in the following two exercises, which will be used later.

**Exercise 12** (a) Show that $\tilde{v}_{\sigma T} = \sigma \cdot \tilde{v}_T$ for all $\sigma$ in $S_n$.

(b) If there is a pair of integers in the same row of $T'$ and the same column of $T$, show that $a_{T'} \cdot [T] = 0$.

(c) If $T$ and $T'$ have the same shape and there is no such pair, show that $a_{T'} \cdot [T] = \pm \tilde{v}_{T'}$.

(d) If $\vartheta : \widetilde{M}^{\lambda'} \to \widetilde{M}^{\lambda}$ is a homomorphism of $S_n$-modules whose kernel does not contain $\widetilde{S}^{\lambda'}$, show that $\lambda' \trianglelefteq \lambda$.

(e) Show that $\widetilde{S}^{\lambda}$ is an irreducible representation, and that every irreducible representation of $S_n$ is isomorphic to exactly one $\widetilde{S}^{\lambda}$.

(f) Show that the $\tilde{v}_T$'s, as $T$ varies over the standard tableaux on $\lambda$, form a basis for $\widetilde{S}^{\lambda}$.

**Exercise 13** (a) Show that, for a numbering $T$ on $\lambda$, $\widetilde{M}^{\lambda} \cong \mathrm{Ind}^{S_n}_{C(T)}(\mathbb{U})$, where $\mathbb{U}$ is the restriction of the alternating representation from $S_n$ to $C(T)$. Equivalently, $\widetilde{M}^{\lambda} \cong S^{(1^{\mu_1})} \circ \ldots \circ S^{(1^{\mu_\ell})} = \mathbb{U}_{\mu_1} \circ \ldots \circ \mathbb{U}_{\mu_\ell}$, with $\mu = \tilde{\lambda}$.

(b) Show that

$$\widetilde{M}^{\lambda} \cong S^{\lambda} \oplus \bigoplus_{\tilde{\nu} \triangleright \tilde{\lambda}} (S^{\nu})^{\oplus K_{\tilde{\nu}\tilde{\lambda}}}.$$

(c) Show that, for any numbering $T$, the map $x \mapsto x \cdot [T]$ defines an $A$-linear surjection from $A$ onto $\widetilde{M}^{\lambda}$, with kernel the left ideal generated by all $q - \mathrm{sgn}(q) \cdot 1$, $q \in C(T)$. This ideal is also generated by those elements $q + 1$ where $q$ is a transposition in $C(T)$.

(d) For $T$ of shape $\lambda$, construct an isomorphism of $\widetilde{M}^{\lambda}$ with $A \cdot b_T$ taking $\widetilde{S}^{\lambda}$ to $A \cdot a_T \cdot b_T$.

The dual constructions are particularly useful for realizing the Specht modules as quotient modules of the tabloid modules. There are canonical surjections

$$\alpha : \widetilde{M}^{\lambda} \to S^{\lambda}, \quad [T] \mapsto v_T$$

$$\beta : M^{\lambda} \to \widetilde{S}^{\lambda}, \quad \{T\} \mapsto \tilde{v}_T.$$

These are well-defined homomorphisms of $S_n$-modules. For example, the formulas $\sigma \cdot [T] = [\sigma T]$ and $v_{\sigma T} = \sigma \cdot v_T$ show first that $\alpha$ is well defined (since $\sigma \cdot v_T = \mathrm{sgn}(\sigma) \cdot v_T$ for $\sigma \in C(T)$), and then that $\alpha$ commutes with the action of the symmetric group.

**Lemma 5** *The composites* $S^{\lambda} \hookrightarrow M^{\lambda} \to \widetilde{S}^{\lambda}$ *and* $\widetilde{S}^{\lambda} \hookrightarrow \widetilde{M}^{\lambda} \to S^{\lambda}$ *are isomorphisms.*

**Proof** In fact, we show that the composites of the two maps in the lemma, $S^{\lambda} \to \widetilde{S}^{\lambda} \to S^{\lambda}$ and $\widetilde{S}^{\lambda} \to S^{\lambda} \to \widetilde{S}^{\lambda}$, are multiplication by the same positive integer $n_\lambda$. The map $\beta$ takes $v_T = b_T\{T\}$ to $b_T \cdot \tilde{v}_T$, and $\alpha$ takes $b_T \cdot \tilde{v}_T = b_T \cdot a_T[T]$ to $b_T \cdot a_T \cdot v_T$. So we must find $n_\lambda$ so that $(b_T \cdot a_T) \cdot v_T = n_\lambda \cdot v_T$. Take any numbering $T$ of $\lambda$, and define $n_\lambda$ to be the cardinality of the set

$$\{(p_1, q_1, p_2, q_2): p_i \in R(T), \; q_i \in C(T),$$

$$p_1 q_1 p_2 q_2 = 1, \; \mathrm{sgn}(q_1) = \mathrm{sgn}(q_2)\}.$$

This is independent of choice of $T$, since replacing $T$ by $\sigma T$ replaces the groups $R(T)$ and $C(T)$ by conjugate groups $\sigma R(T)\sigma^{-1}$ and $\sigma C(T)\sigma^{-1}$.

By definition, since $v_T = \sum \mathrm{sgn}(q)\{qT\}$, the sum over $q$ in $C(T)$, we have

$$(b_T \cdot a_T) \cdot v_T \;=\; \sum \mathrm{sgn}(q_1 q_2)\{q_1 p_1 q_2 T\},$$

the sum over $q_1$, $q_2 \in C(T)$ and $p_1 \in R(T)$. To show that this is equal to $n_\lambda \cdot v_T$ is equivalent to showing that, for fixed $q$, there are $n_\lambda$ solutions to the equation $\{q_1 p_1 q_2 T\} = \{qT\}$ with $\mathrm{sgn}(q_1 q_2) = \mathrm{sgn}(q)$. Now $\{qT\} = \{\sigma T\}$ exactly when there is some $p$ in $R(T)$ with $q = \sigma p$, so the conclusion follows from the obvious fact that there are $n_\lambda$ solutions to the equation $q_1 p_1 q_2 p_2 = q$ with $\mathrm{sgn}(q_1 q_2) = \mathrm{sgn}(q)$.

Similarly, that the composite $\tilde{S}^\lambda \to S^\lambda \to \tilde{S}^\lambda$ is multiplication by $n_\lambda$ amounts to the equation $(a_T \cdot b_T) \cdot \tilde{v}_T = n_\lambda \cdot \tilde{v}_T$. By taking inverses, one sees that $n_\lambda$ is also the cardinality of the set of quadruples $(q_1, p_1, q_2, p_2)$ such that $q_1 p_1 q_2 p_2 = 1$ and $\mathrm{sgn}(q_1) = \mathrm{sgn}(q_2)$, and the conclusion follows as before.   $\square$

In particular, it follows from Lemma 5 that $\tilde{S}^\lambda$ is isomorphic to $S^\lambda$, so this dual construction also gives all the irreducible representations of the symmetric group (as we saw also in Exercise 12). Our goal in this section, however, is to describe the kernel of the epimorphism $\alpha$ from $\tilde{M}^\lambda$ to $S^\lambda$, which amounts to finding the relations satisfied by the generating elements $v_T$. Let $\mu = \tilde{\lambda}$ be the conjugate partition of $\lambda$, and let $\ell = \lambda_1$ be the length of $\mu$. For any $1 \le j \le \ell - 1$, and $1 \le k \le \mu_{j+1}$, and any numbering $T$ of $\lambda$, define

$$\pi_{j,k}(T) \;=\; \sum [S] \in \tilde{M}^\lambda,$$

where the sum is over all $S$ that are obtained from $T$ by exchanging the top $k$ elements in the $(j+1)^{st}$ column of $T$ with $k$ elements in the $j^{th}$ column of $T$, preserving the vertical orders of each set of $k$ elements. For example,

$$\pi_{1,2}\left(\begin{array}{|c|c|} \hline 1 & 2 \\ \hline 4 & 3 \\ \hline 5 & 6 \\ \hline \end{array}\right) = \left[\begin{array}{|c|c|} \hline 2 & 1 \\ \hline 3 & 4 \\ \hline 5 & 6 \\ \hline \end{array}\right] + \left[\begin{array}{|c|c|} \hline 2 & 1 \\ \hline 4 & 5 \\ \hline 3 & 6 \\ \hline \end{array}\right] + \left[\begin{array}{|c|c|} \hline 1 & 4 \\ \hline 2 & 5 \\ \hline 3 & 6 \\ \hline \end{array}\right]$$

Define $Q^\lambda \subset \tilde{M}^\lambda$ to be the subspace spanned by all elements of the form

$$[T] - \pi_{j,k}(T),$$

as $T$ varies over all numberings $T$ of $\lambda$, and $j$ and $k$ vary as above. This subspace is an $S_n$-submodule, since $\pi_{j,k}(\sigma T) = \sigma \cdot \pi_{j,k}(T)$.

**Lemma 6** *The classes* $[T]$, *as* $[T]$ *varies over the standard tableaux on* $\lambda$, *span the quotient space* $\tilde{M}^\lambda / Q^\lambda$.

**Proof** For this we need a different ordering of the numberings $T$ of $\lambda$ from that defined in §7.1. For this ordering, say that $T' \succ T$ if, in the *right-most* column which is different in the two numberings, the *lowest* box which has different entries has a larger entry in $T'$ than in $T$. It suffices to show that, given a numbering $T$ that is not standard, one can use the relations in $Q^\lambda$ to write $[T]$ as a linear combination of classes $[S]$, with $S \succ T$. First, we may assume the entries in $T$ are increasing in each column, since, if they are not, and $T'$ is the result of putting the column entries in order, then $[T'] = \pm[T]$ and $T' \succ T$. If the columns of $T$ are increasing but $T$ is not a tableau, suppose the $k^{th}$ entry of the $j^{th}$ column of $T$ is larger than the $k^{th}$ entry in the $(j+1)^{st}$ column. Each of the numberings $S$ appearing in $\pi_{j,k}(T)$ is then strictly larger than $T$ in the ordering, so this completes the proof. ☐

The proof of the lemma gives a simple procedure, or "straightening algorithm," for writing a given element of $\tilde{M}^\lambda / Q^\lambda$ as a linear combination of the classes of standard tableaux. For example, with $T$ as in the preceding example, one takes $j = 1$ and $k = 2$, and one sees that

$$
\left[\begin{array}{|c|c|}\hline 1 & 2 \\\hline 4 & 3 \\\hline 5 & 6 \\\hline\end{array}\right] \equiv \left[\begin{array}{|c|c|}\hline 2 & 1 \\\hline 3 & 4 \\\hline 5 & 6 \\\hline\end{array}\right] - \left[\begin{array}{|c|c|}\hline 2 & 1 \\\hline 3 & 5 \\\hline 4 & 6 \\\hline\end{array}\right] + \left[\begin{array}{|c|c|}\hline 1 & 4 \\\hline 2 & 5 \\\hline 3 & 6 \\\hline\end{array}\right]
$$

For each of the first two numberings in the result, one takes $j = 1$ and $k = 1$, and performing the exchanges, rearranging the columns, and cancelling, one finds

$$
\left[\begin{array}{|c|c|}\hline 1 & 2 \\\hline 4 & 3 \\\hline 5 & 6 \\\hline\end{array}\right] \equiv \left[\begin{array}{|c|c|}\hline 1 & 2 \\\hline 3 & 4 \\\hline 5 & 6 \\\hline\end{array}\right] - \left[\begin{array}{|c|c|}\hline 1 & 3 \\\hline 2 & 4 \\\hline 5 & 6 \\\hline\end{array}\right] - \left[\begin{array}{|c|c|}\hline 1 & 4 \\\hline 2 & 5 \\\hline 3 & 6 \\\hline\end{array}\right] - \left[\begin{array}{|c|c|}\hline 1 & 2 \\\hline 3 & 5 \\\hline 4 & 6 \\\hline\end{array}\right] + \left[\begin{array}{|c|c|}\hline 1 & 3 \\\hline 2 & 5 \\\hline 4 & 6 \\\hline\end{array}\right]
$$

We claim next that these generators and relations give a presentation of the Specht module, i.e., that there is a canonical isomorphism of $\widetilde{M}^\lambda/Q^\lambda$ with $S^\lambda$. This is the content of

**Proposition 4**  $Q^\lambda$ *is the kernel of the map* $\alpha\colon \widetilde{M}^\lambda \to S^\lambda$ *that takes* $[T]$ *to* $v_T$.

**Proof**  It suffices to show that each of the generators of $Q^\lambda$ is in the kernel of $\alpha$. For then $\alpha$ determines a surjection $\widetilde{M}^\lambda/Q^\lambda \to S^\lambda$. Lemma 6 shows that the dimension of $\widetilde{M}^\lambda/Q^\lambda$ is at most the number $f^\lambda$ of standard tableaux on $\lambda$, and since we know $\dim(S^\lambda) = f^\lambda$, this map must be an isomorphism, and the proposition will follow.

We will show first that some other elements are in the kernel of $\alpha$. For each nonempty subset $Y$ of the $(j{+}1)^{\text{st}}$ column of a numbering $T$ of $\lambda$, define $\gamma_Y(T) \in \widetilde{M}^\lambda$ by the formula

$$
\gamma_Y(T) = \sum \varepsilon_{(S,T)}[S],
$$

where the sum is over all numberings $S$ obtained from $T$ by interchanging a subset of $Y$ (possibly empty) with a subset of the $j^{\text{th}}$ column, preserving the descending order of the sets exchanged. Here $\varepsilon_{(S,T)}$ is 1 if an even number of entries is exchanged, and $-1$ if an odd number is exchanged; equivalently, $\varepsilon_{(S,T)}$ is the sign of the permutation $\sigma$ such that $S = \sigma T$. It suffices to prove the following two claims:

**Claim 1** $\gamma_Y(T) \in \mathrm{Ker}(\alpha)$ *for all $T$ and $Y$.*

**Claim 2** $\pi_{j,k}(T) - [T] = \sum_Y (-1)^{\#Y} \gamma_Y(T)$, *the sum over all nonempty subsets $Y$ of the top $k$ entries of the $(j+1)^{\mathrm{st}}$ column of $T$.*

To prove the first claim, let $X$ be the set of entries in the $j^{\mathrm{th}}$ column of $T$, and define subgroups $K \subset H \subset S_n$, where $H$ consists of those permutations that are the identity outside the elements in the union $X \cup Y$, and $K$ is the subgroup of $H$ that maps each of $X$ and $Y$ to itself. Note first that

$$\sum_{k \in K} \mathrm{sgn}(k) k \cdot [T] = (\#X)! \cdot (\#Y)! \cdot [T],$$

since $K \subset C(T)$, so $k \cdot [T] = \mathrm{sgn}(k) \cdot [T]$ for all $k$ in $K$. The element $\gamma_Y(T)$ is the sum $\sum \mathrm{sgn}(\sigma)[\sigma T]$, the sum over a set of permutations $\sigma$ that make up a set of coset representatives for $H/K$. It follows that

$$(\#X)! \cdot (\#Y)! \cdot \gamma_Y(T) = \sum_{h \in H} \mathrm{sgn}(h) h \cdot [T].$$

It therefore suffices to show that this last sum is in the kernel of $\alpha$, i.e., that

$$\sum_{h \in H} \mathrm{sgn}(h) h \, v_T = \sum_{q \in C(T)} \mathrm{sgn}(q) \left( \sum_{h \in H} \mathrm{sgn}(h) h \{qT\} \right)$$

vanishes. For any $q$ in $C(T)$, $qT$ must have at least two elements of $X \cup Y$ that are in the same row. Let $t$ be the transposition of two such elements. Then

$$\sum_{h \in H} \mathrm{sgn}(h) h \{qT\} = \sum_g \mathrm{sgn}(g) \, g (1 - t) \{qT\},$$

where the second sum is over a set of coset representatives $g$ for $H/\{1, t\}$; and this vanishes since $t\{qT\} = \{qT\}$. This finishes the proof of Claim 1.

The proof of Claim 2 uses a counting argument. Let $W$ be the set of the first $k$ elements of the $(j+1)^{\mathrm{st}}$ column of $T$. For each $S$ that is obtained from $T$ by interchanging some subset $Z$ of $W$ with some set in the $j^{\mathrm{th}}$ column, consider the coefficient of $[S]$ in the sum

$$\sum_Y (-1)^{\#Y} \gamma_Y(T) = \sum_Y (-1)^{\#Y} \left( \sum \varepsilon_{(S,T)}[S] \right).$$

When $Z = W$, each such $S$ occurs just once, and $\varepsilon_{(S,T)} = (-1)^k$, so these terms sum to $\pi_{j,k}(T)$. When $Z$ is empty, $S = T$ occurs once for each nonempty $Y$, so the coefficient is $\sum_{\ell=1}^{k}(-1)^{\ell}\binom{k}{\ell} = -1$, which gives the contribution $-[T]$. For $Z$ of cardinality $m$, $0 < m < k$, $S$ will occur for each $Y$ that contains $Z$; such $Y$ has the form $Z \cup V$, for any subset $V$ of $W \setminus Z$, and the coefficient is $(-1)^{\#Y}(-1)^m = (-1)^{\#V}$. So the coefficient of $[S]$ is $\sum_V (-1)^{\#V} = \sum_{\ell=0}^{k}(-1)^{\ell}\binom{k}{\ell} = 0$. □

**Corollary** *The space $S^{\lambda}$ is the vector space with generators $v_T$, as $T$ varies over numberings of $\lambda$, and relations of the form $v_T - \sum v_S$, where the sum is over all $S$ obtained from $T$ by exchanging the top $k$ elements of one column with any $k$ elements of the preceding column, maintaining the vertical orders of each set exchanged. (There is one such relation for each numbering $T$, each choice of adjacent columns, and each $k$ at most equal to the length of the shorter column.)*

The relations introduced in the preceding proof are special cases of "Garnir elements"; cf. Sagan (1991) and James and Kerber (1981). The "quadratic relations" of the proposition seem simpler to use, having fewer terms – and no signs to remember. We will see that they are better related to representations of the general linear group and the geometry of flag varieties.

It should also be remarked that, although our basic relations interchange the *top* $k$ elements in one column with all subsets of $k$ elements in the preceding column, one gets the same relations by interchanging *any* given subset of $k$ elements in one column with all subsets of $k$ elements in the preceding column (preserving as always the orders in the sets being exchanged). Indeed, given any such subset, and any numbering $T$, let $T'$ be the numbering obtained by interchanging this set with the top $k$ elements of the column, preserving the order of the sets. The relation $[T] - \sum[S]$, where $S$ is obtained by interchanging the top $k$ elements, is the same as the relation $\pm([T'] - \sum[S'])$, where the sign is the sign of the permutation that interchanges the given $k$ elements with the top $k$ elements.

**Exercise 14** Show dually that the kernel of $\beta: M^{\lambda} \to \widetilde{S}^{\lambda}$ is generated by elements $\{T\} + (-1)^k \tilde{\pi}_{j,k}(T)$, where $\tilde{\pi}_{j,k}(T) = \sum\{S\}$, with the sum over all $S$ that are obtained from $T$ by interchanging the first $k$ elements in the $(j+1)^{\text{st}}$ row with $k$ elements in the $j^{\text{th}}$ row, maintaining the orders of the two sets.

The representations $M^\lambda$ and $\widetilde{M}^\lambda$ can be defined over the integers, and they are free with the same bases of row or (oriented) column tabloids as in the complex case. The proofs given show that the submodule $S^\lambda$ has a $\mathbb{Z}$-basis of the $v_T$, as $T$ varies over the standard tableaux on $\lambda$, and that its description as a quotient of $\widetilde{M}^\lambda$ by the quadratic relations is also valid over the integers. The same is true for $\widetilde{S}^\lambda$, which is free on the $\widetilde{v}_T$ for $T$ standard. The isomorphism of $S^\lambda$ with $\widetilde{S}^\lambda$, however, is only valid over the rationals, as seen in the following exercise. In addition, the Specht modules may fail to be irreducible in positive characteristic.

**Exercise 15**  Show that the representations $S^{(2,1)}$ and $\widetilde{S}^{(2,1)}$ are not isomorphic over $\mathbb{Z}$, or over any field of characteristic 3.

It is a fact that, over the rationals (but not over the integers), the kernel of $\alpha$ is generated by the relations using only interchanges of one entry, i.e., by all $[T] - \pi_{j,1}(T)$, and similarly for the kernel of $\beta$. In fact:

**Exercise 16**  Fix $\lambda$ and $j$, and for $k$ between 1 and the length of the $(j+1)^{\text{st}}$ column of $\lambda$, let $N_k$ be the subgroup of $\widetilde{M}^\lambda$ (constructed now as a $\mathbb{Z}$-module) generated by all $[T] - \pi_{j,k}(T)$ for numberings $T$ on $\lambda$. (a) Let $m$ be the length of the $j^{\text{th}}$ column of $\lambda$, and for $1 \le i \le m$ let $T_{i,k}$ be obtained from $T$ by interchanging the $i^{\text{th}}$ entry of the $j^{\text{th}}$ column of $T$ with the $k^{\text{th}}$ entry of the $(j+1)^{\text{st}}$ column of $T$. Show that $[T] - \sum_{i=1}^{m}[T_{i,k}]$ is in $N_1$, and that for $k > 1$,

$$\sum_{i=1}^{m} \pi_{j,k-1}(T_{i,k}) = k \cdot \pi_{j,k}(T) - (k-1) \cdot \pi_{j,k-1}(T).$$

(b) Deduce that $k \cdot N_k \subset N_1 + N_{k-1}$, so $k! \cdot N_k \subset N_1$.

The following exercise gives an equivalent but more classical construction of the Specht module.

**Exercise 17**  The symmetric group $S_n$ acts on the ring of polynomials $\mathbb{C}[x_1, \ldots, x_n]$ in $n$ variables by $(\sigma \cdot f)(x_1, \ldots, x_n) = f(x_{\sigma(1)}, \ldots, x_{\sigma(n)})$. Given a partition $\lambda$ of $n$, for each (distinct) numbering $T$ of $\lambda$, let

$$F_T = \prod_{j} \prod_{i < i'} (x_{T(i',j)} - x_{T(i,j)}),$$

where $T(i,j)$ is the entry of $T$ in the $(i,j)$ position. (a) Show that $\sigma \cdot F_T = F_{\sigma T}$, so the space spanned by these polynomials is preserved by $S_n$. (b) Show that there is an isomorphism of $S^\lambda$ with this space, with $v_T$

mapping to $F_T$. In particular, the $F_T$, as $T$ varies over standard tableaux on $\lambda$, forms a basis.

**Exercise 18** (a) Show that, if $T$ is a numbering of $\lambda$, then $c_T \cdot c_T = n_\lambda c_T$, where $n_\lambda$ is the number defined in the proof of Lemma 5. (b) Show that $c_{T'} \cdot c_T = 0$ if and only if there are two entries in the same row of $T'$ and the same column of $T$. (c) Show that $A$ is a direct sum of the ideals $A \cdot c_T$, as $T$ varies over the standard tableaux with $n$ boxes. (d) Find two distinct standard tableaux $T$ and $T'$ of the same shape such that $c_{T'} \cdot c_T \neq 0$.

**Exercise 19** Show that the integer $n_\lambda$ occurring in the proof of Lemma 5 and the preceding exercise is the product of the hook lengths of $\lambda$, i.e., $n_\lambda = n!/f^\lambda$.

**Exercise 20** Give an alternative proof of the fact that $S^\lambda$ and $\widetilde{S}^\lambda$ are isomorphic by constructing an isomorphism of $A \cdot b_T \cdot a_T$ with $A \cdot a_T \cdot b_T$.

**Exercise 21** (for those who know what Hopf algebras are) The ring $\Lambda$ has a "coproduct" $\Lambda \to \Lambda \otimes \Lambda$, $\Lambda_n \to \oplus_{p+q=n} \Lambda_p \otimes \Lambda_q$, that takes $s_\nu$ to $\sum_{\lambda,\nu} c_{\lambda\mu}^\nu s_\lambda \otimes s_\mu$. (a) Show that this makes $\Lambda$ into a Hopf algebra, and that the involution $\omega$ is compatible with the coproduct. (b) Describe the corresponding maps on $R$.

# 8

# Representations of the general linear group

The main object of this chapter is to construct and study the irreducible polynomial representations of the general linear group $GL_m\mathbb{C} = GL(E)$, where $E$ is a complex vector space of dimension $m$. These can be formed by a basic construction in linear algebra that generalizes a well known construction of symmetric and exterior products; they make sense for any module over a commutative ring. These representations are parametrized by Young diagrams $\lambda$ with at most $m$ rows, and have bases corresponding to Young tableaux on $\lambda$ with entries from $[m]$. They can also be constructed from representations of symmetric groups. Like the latter, these have useful realizations both as subspaces and as quotient spaces of naturally occurring representations, with relations given by quadratic equations. The characters of the representations are given in §8.3. To prove that these give all the irreducible representations we use a bit of the Lie group–Lie algebra story, which is sketched in this setting in §8.2. In the last section we describe some variations on the quadratic equations. In particular, we identify the sum of all polynomial representations with a ring constructed by Deruyts a century ago.

## 8.1 A construction in linear algebra

For any commutative ring $R$ and any $R$-module $E$, and any partition $\lambda$, we will construct an $R$-module denoted $E^\lambda$. (For applications in these notes, the case where $R = \mathbb{C}$, so $E$ is a complex vector space, will suffice.) When $\lambda = (n)$, $E^\lambda$ will be the symmetric product $\mathrm{Sym}^n E$, and when $\lambda = (1^n)$, $E^\lambda$ will be the exterior product $\wedge^n E$. Like these modules, the general $E^\lambda$ can be described as the solution of a universal problem.

104

The cartesian product of $n$ copies of a set $E$ is usually written in the form $E^{\times n} = E \times \ldots \times E$, which implies an ordering of the index set. However, the construction does not depend on any such ordering, and makes sense for any index set. We will write $E^{\times \lambda}$ for the cartesian product of $n = |\lambda|$ copies of $E$, but labelled by the $n$ boxes of (the Young diagram of) $\lambda$. So an element $\mathbf{v}$ of $E^{\times \lambda}$ is given by specifying an element of $E$ for each box in $\lambda$.

Consider maps $\varphi: E^{\times \lambda} \to F$ from $E^{\times \lambda}$ to an $R$-module $F$, satisfying the following three properties:

(1)      $\varphi$ is $R$-multilinear.

This means that if all the entries but one are fixed, then $\varphi$ is $R$-linear in that entry.

(2)      $\varphi$ is alternating in the entries of any column of $\lambda$.

That is, $\varphi$ vanishes whenever two entries in the same column are equal. Together with (1), this implies that $\varphi(\mathbf{v}) = -\varphi(\mathbf{v}')$ if $\mathbf{v}'$ is obtained from $\mathbf{v}$ by interchanging two entries in a column.

(3)      For any $\mathbf{v}$ in $E^{\times \lambda}$, $\varphi(\mathbf{v}) = \sum \varphi(\mathbf{w})$, where the sum is over all $\mathbf{w}$ obtained from $\mathbf{v}$ by an exchange between two given columns, with a given subset of boxes in the right chosen column.

Here an exchange is as defined in the introductory notation to Part II. Note that the two columns, and the set of boxes in the right column, are fixed. If the number of boxes chosen is $k$, and the left chosen column has length $c$, then there are $\binom{c}{k}$ such $\mathbf{w}$ for a given $\mathbf{v}$. In the presence of (1) and (2), one need only include such exchanges when the boxes are chosen from the top of the column. For example, for $\lambda = (2,2,2)$, and choosing the top box in the second column, we have the equation

$$
\varphi\left(\begin{array}{|c|c|}\hline x & u \\\hline y & v \\\hline z & w \\\hline\end{array}\right) = \varphi\left(\begin{array}{|c|c|}\hline u & x \\\hline y & v \\\hline z & w \\\hline\end{array}\right) + \varphi\left(\begin{array}{|c|c|}\hline x & y \\\hline u & v \\\hline z & w \\\hline\end{array}\right) + \varphi\left(\begin{array}{|c|c|}\hline x & z \\\hline y & v \\\hline u & w \\\hline\end{array}\right),
$$

for all $x$, $y$, $z$, $u$, $v$, $w$ in $E$. Using the top two boxes in the second column, we have the equation

$$\varphi\left(\begin{array}{cc} x & u \\ y & v \\ z & w \end{array}\right) = \varphi\left(\begin{array}{cc} u & x \\ v & y \\ z & w \end{array}\right) + \varphi\left(\begin{array}{cc} u & x \\ y & z \\ v & w \end{array}\right) + \varphi\left(\begin{array}{cc} x & y \\ u & z \\ v & w \end{array}\right).$$

Using all three boxes, we have

$$\varphi\left(\begin{array}{cc} x & u \\ y & v \\ z & w \end{array}\right) = \varphi\left(\begin{array}{cc} u & x \\ v & y \\ w & z \end{array}\right).$$

We define the **Schur module** $E^\lambda$ to be the universal target module for such maps $\varphi$. This means that $E^\lambda$ is an $R$-module, and we have a map $E^{\times\lambda} \to E^\lambda$, that we denote $\mathbf{v} \mapsto \mathbf{v}^\lambda$, satisfying (1)–(3), and such that for any $\varphi: E^{\times\lambda} \to F$ satisfying (1)–(3), there is a unique homomorphism $\widetilde{\varphi}: E^\lambda \to F$ of $R$-modules such that $\varphi(\mathbf{v}) = \widetilde{\varphi}(\mathbf{v}^\lambda)$ for all $\mathbf{v}$ in $E^{\times\lambda}$.

Consider first the two extreme cases. When $\lambda = (n)$, property (2) is empty, and (3) says that all entries commute. We see that $E^{(n)}$ is the symmetric power $\mathrm{Sym}^n(E)$, which can be constructed as usual as the quotient of $E^{\otimes n}$ by the submodule generated by all $v_1 \otimes \ldots \otimes v_n - v_{\sigma(1)} \otimes \ldots \otimes v_{\sigma(n)}$ for all $v_i$ in $E$ and $\sigma$ in $S_n$. Similarly, if $\lambda = (1^n)$, property (3) is empty, and (2) says that the entries are alternating. Therefore $E^{(1^n)}$ is the exterior power $\bigwedge^n(E)$, constructed as the quotient of $E^{\otimes n}$ by the submodule generated by all $v_1 \otimes \ldots \otimes v_n$ where any two $v_i$'s are equal.

The fact that $E^\lambda$ is unique up to canonical isomorphism follows immediately from its description as the solution to a universal problem. To construct it, note first that the universal module with property (1) is the tensor product of $n$ copies of $E$. We denote this tensor product by $E^{\otimes\lambda}$ to emphasize that the factors are indexed by the boxes in $\lambda$. The universal module with properties (1) and (2) is the quotient of $E^{\otimes\lambda}$ by the submodule generated by tensors of elements of $E$ that have two entries in the same column equal. If we number $\lambda$ down the columns from left to right, this identifies this module with the module

$$\wedge^{\mu_1} E \otimes_R \cdots \otimes_R \wedge^{\mu_\ell} E,$$

where $\mu_i$ is the length of the $i^{\text{th}}$ column of $\lambda$, i.e., $\mu = \tilde{\lambda}$. The map from $E^{\times \lambda}$ to $\otimes \wedge^{\mu_i} E$ is the obvious one: given a vector in $E^{\times \lambda}$, take the wedge product of the entries in each column, from top to bottom, and take the tensor product of these classes. For example,

$$\begin{array}{|c|c|} \hline x & u \\ \hline y & v \\ \hline z & w \\ \hline \end{array} \quad \mapsto (x \wedge y \wedge z) \otimes (u \wedge v \wedge w) \ \text{ in } \ \wedge^3 E \otimes \wedge^3 E.$$

We write this map from $E^{\times \lambda}$ to $\otimes \wedge^{\mu_i} E$ simply $\mathbf{v} \mapsto \wedge \mathbf{v}$. Then

$$(4) \qquad E^\lambda = \wedge^{\mu_1} E \otimes_R \cdots \otimes_R \wedge^{\mu_\ell} E / Q^\lambda(E),$$

where $Q^\lambda(E)$ is the submodule generated by all elements of the form $\wedge \mathbf{v} - \sum \wedge \mathbf{w}$, the sum over all $\mathbf{w}$ obtained from $\mathbf{v}$ by the exchange procedure described in (3), for some choice of columns and boxes. Indeed, it is straightforward to verify that the right side of this display satisfies the universal property to be $E^\lambda$. For example, $E^{(2,1)}$ is the quotient of $\wedge^2 E \otimes E$ by the submodule generated by all $u \wedge v \otimes w - w \wedge v \otimes u - u \wedge w \otimes v$.

By its definition, the construction of $E^\lambda$ is functorial in $E$: any homomorphism $E \to F$ determines a homomorphism $E^\lambda \to F^\lambda$. It also follows directly from this definition that the construction is **compatible with base change:** if $R \to R'$ is a homomorphism of commutative rings, then there is a canonical isomorphism $(E \otimes_R R')^\lambda \cong E^\lambda \otimes_R R'$.

Suppose now we have an ordered set $e_1, \ldots, e_m$ of elements of $E$. Then for any filling $T$ of $\lambda$ with elements in $[m]$, we get an element of $E^{\times \lambda}$ by replacing any $i$ in a box of $T$ by the element $e_i$. The image of this element in $E^\lambda$ we denote by $e_T$. First we have the following simple lemma:

**Lemma 1** *If $E$ is free on $e_1, \ldots, e_m$, then $E^\lambda \cong F/Q$, where $F$ is free on elements $e_T$ for all fillings $T$ of $\lambda$ with entries in $[m]$, and $Q$ is generated by the elements*

(i) *$e_T$ if $T$ has two equal entries in any column;*

(ii) *$e_T + e_{T'}$ where $T'$ is obtained from $T$ by interchanging two entries in a column;*

(iii)  $e_T - \sum e_S$,  *where the sum is over all  S  obtained from  T  by an exchange as in* (3).

**Proof** It follows from the multilinearity that the elements  $e_T$  generate  $E^\lambda$, so we have a surjection  $F \to E^\lambda$. Properties (2) and (3) imply that the generators of  $Q$  map to zero, so  $F/Q \twoheadrightarrow E^\lambda$. It is routine to verify that this is an isomorphism. To start, the vectors  $e_T$,  as  $T$  varies over all such fillings of  $\lambda$,  give a basis of the tensor product  $E^{\otimes\lambda}$.  The module obtained by using the relations in (i) and (ii) is exactly the tensor product  $\wedge^{\mu_1}E \otimes_R \ldots \otimes_R \wedge^{\mu_\ell}E$  (and the  $e_T$  with all columns strictly increasing form a basis for this module). The relations (iii) then generate the module of relations  $Q^\lambda(E)$,  as follows from  $R$-multilinearity and the fact that the  $e_i$  generate  $E$.  The lemma therefore follows from (4).  □

Perhaps the simplest and earliest (1851) appearance of the relations (1)–(3) is in the following basic identity in linear algebra:

**Lemma 2** (Sylvester) *For any*  $p \times p$  *matrices  M  and  N, and*  $1 \le k \le p$,

$$\det(M) \cdot \det(N) = \sum \det(M') \cdot \det(N'),$$

*where the sum is over all pairs*  $(M', N')$  *of matrices obtained from  M  and N  by interchanging a fixed set of  k  columns of  N  with any  k  columns of M,  preserving the ordering of the columns.*

**Proof** By the alternating property of determinants, there is no loss in generality in assuming that the fixed set of columns of  $N$  consists of its first  $k$  columns. For vectors  $v_1, \ldots, v_p$  in  $R^p$,  write  $|v_1 \ldots v_p|$  for the determinant of the matrix with these vectors as columns. The identity to be proved is

(5)          $|v_1 \ldots v_p| \cdot |w_1 \ldots w_p| =$

$$\sum_{i_1 < \ldots < i_k} |v_1 \ldots w_1 \ldots w_k \ldots v_p| \cdot |v_{i_1} \ldots v_{i_k} w_{k+1} \ldots w_p|,$$

where, in the sum, the vectors  $w_1, \ldots, w_k$  are interchanged with the vectors  $v_{i_1}, \ldots, v_{i_k}$.  It suffices to show that the difference of the two sides is an alternating function of the  $p+1$  vectors  $v_1, \ldots, v_p, w_1$,  since any such function must vanish  $(\wedge^{p+1}(R^p) = 0)$.  For this it suffices to show that the two sides are equal when two successive vectors  $v_i$  and  $v_{i+1}$  are equal (which is immediate), and when  $v_p = w_1$.  In the latter case, fixing  $v_p = w_1$,  it suffices to show that the difference of the two sides is an alternating function of $v_1, \ldots, v_p, w_2$. Again the case when $v_i = v_{i+1}$ is immediate, and this time the case $v_p = w_2$ is also easy.    □

Let $Z_{i,j}$ be indeterminates, $1 \leq i \leq n$, $1 \leq j \leq m$, and let $R[Z] = R[Z_{1,1}, Z_{1,2}, \ldots, Z_{n,m}]$ be the polynomial ring in these variables. For each $p$-tuple $i_1, \ldots, i_p$ of integers from $[m]$, with $p \leq n$, set

$$(6) \qquad D_{i_1, \ldots, i_p} = \det \begin{bmatrix} Z_{1,i_1} & \cdots & Z_{1,i_p} \\ \vdots & & \vdots \\ Z_{p,i_1} & \cdots & Z_{p,i_p} \end{bmatrix}.$$

This is an alternating function of the subscripts $i_1, \ldots, i_p$.

For a Young diagram $\lambda$ with at most $n$ rows, and an arbitrary filling $T$ of $\lambda$ with numbers from $[m]$, let $D_T$ be the product of the determinants corresponding to the columns of $T$, i.e.,

$$(7) \qquad D_T = \prod_{j=1}^{\ell} D_{T(1,j), T(2,j), \ldots, T(\mu_j, j)},$$

where $\mu_j$ is the length of the $j^{\text{th}}$ column of $\lambda$, $\ell = \lambda_1$, and $T(i,j)$ is the entry of $T$ in the $i^{\text{th}}$ row and $j^{\text{th}}$ column.

**Lemma 3** *If $E$ is free with basis $e_1, \ldots, e_m$, then there is a canonical homomorphism from $E^\lambda$ to $R[Z]$ that maps $e_T$ to $D_T$ for all $T$.*

**Proof** Using Lemma 1, it suffices to show that the elements $D_T$ satisfy the properties corresponding to (i)–(iii) of that lemma. Properties (i) and (ii) follow from the alternating property of determinants. Property (iii) follows from Sylvester's lemma, applied to appropriate matrices. For this, suppose the two columns of $T$ in which the exchange takes place have entries $i_1, \ldots, i_p$ in the first, and $j_1, \ldots, j_q$ in the second. Set

$$M = \begin{bmatrix} Z_{1,i_1} & \cdots & Z_{1,i_p} \\ \vdots & & \vdots \\ Z_{p,i_1} & \cdots & Z_{p,i_p} \end{bmatrix}, \quad N = \begin{bmatrix} Z_{1,j_1} & \cdots & Z_{1,j_q} & & 0 \\ \vdots & & \vdots & & \\ Z_{p,j_1} & \cdots & Z_{p,j_q} & & I_{p-q} \end{bmatrix}.$$

Here the matrix $N$ has a lower right identity matrix of size $p-q$, and an upper right $q \times (p-q)$ block of zeros. Sylvester's lemma, applied to these two matrices and the subset of columns specified by the subset of the right column of $T$ being exchanged, translates precisely to the required equation.

□

**Theorem 1** *If $E$ is free on generators $e_1, \ldots, e_m$, then $E^\lambda$ is free on generators $e_T$, as $T$ varies over the tableaux on $\lambda$ with entries in $[m]$.*

**Proof** The proof that the $e_T$ generate $E^\lambda$ is similar to the proof of Lemma 6 in §7.4, and we use the same ordering of the fillings: $T' \succ T$ if, in the right-most column which is different, the lowest box where they differ has a larger entry in $T'$. We also use the presentation $E^\lambda = F/Q$ of Lemma 1. We must show that, given any $T$ that is not a tableau, we can write $e_T$ as a linear combination of elements $e_S$, with $S \succ T$, and elements in $Q$. We may assume the entries in the columns of $T$ are strictly increasing, by using relations (i) and (ii); note that making the columns strictly increasing in $T$ replaces $T$ by a $T'$ that is larger than $T$ in the ordering. If the columns are strictly increasing but $T$ is not a tableau, suppose the $k^{\text{th}}$ entry of the $j^{\text{th}}$ column is strictly larger than the $k^{\text{th}}$ entry of the $(j+1)^{\text{st}}$ column. Then we have a relation $e_T \equiv \sum e_S$, the sum over all $S$ obtained from $T$ by exchanging the top $k$ entries of the $(j+1)^{\text{st}}$ column of $T$ with $k$ entries of the $j^{\text{th}}$ column. Since each such $S$ is larger than $T$ in the ordering, the proof is complete.

To prove that the $e_T$'s are linearly independent, we use Lemma 3, so it suffices to prove that the $D_T$ are linearly independent as $T$ varies over tableaux. For this we order the variables $Z_{i,j}$ in the order: $Z_{i,j} < Z_{i',j'}$ if $i < i'$ or $i = i'$ and $j < j'$. We order monomials in these variables lexicographically: $M_1 < M_2$ if the smallest $Z_{i,j}$ that occurs to a different power occurs to a smaller power in $M_1$ than in $M_2$. Note that if $M_1 < M_2$ and $N_1 \leq N_2$ then $M_1 N_1 < M_2 N_2$. It follows immediately from this definition that the largest monomial that appears in a determinant $D_{i_1, \ldots, i_p}$, if $i_1 < \ldots < i_p$, is the diagonal term $Z_{1,i_1} \cdot Z_{2,i_2} \cdot \ldots \cdot Z_{p,i_p}$. Therefore the largest monomial occurring in $D_T$, if $T$ has increasing columns, is $\prod (Z_{i,j})^{m_T(i,j)}$, where $m_T(i,j)$ is the number of times $j$ occurs in the $i^{\text{th}}$ row of $T$. This monomial occurs with coefficient 1.

Now order the tableaux by saying that $T < T'$ if the first row where they differ, and the first entry where they differ in that row, is smaller in $T$ than in $T'$. Equivalently, the smallest $i$ for which there is a $j$ with $m_T(i, j) \neq m_{T'}(i,j)$, and the smallest such $j$, has $m_T(i,j) > m_{T'}(i,j)$. It follows that if $T < T'$, then the largest monomial occurring in $D_T$ is larger than any monomial occurring in $D_{T'}$. From this the linear independence follows: if $\sum r_T D_T = 0$, take $T$ minimal such that $r_T \neq 0$, and then the coefficient of $\prod (Z_{i,j})^{m_T(i,j)}$ in $\sum r_T D_T$ is $r_T$.          □

**Corollary of proof** *The map from $E^\lambda$ to $R[Z]$ is injective, and its image $D^\lambda$ is free on the polynomials $D_T$, as $T$ varies over the tableaux on $\lambda$ with entries in $[m]$.*

We will need the following variation on this construction:

**Exercise 1** Show that one obtains the same module $E^\lambda$ if, in relation (3), one allows interchanges only between two adjacent columns.

If $E \to F$ is a surjection of $R$-modules, it follows immediately from the definition or construction that $E^\lambda \to F^\lambda$ is surjective. The corresponding result for injective maps is not true in general. The following exercise is fairly well known (but not so obvious) in the case of exterior products. We won't have any need for it, but it may provide an interesting challenge to those with some commutative algebra background.

**Exercise 2** Let $\varphi: E \to F$ be a homomorphism of finitely generated free $R$-modules. Show that the following are equivalent: (i) $\varphi$ is a monomorphism; (ii) $\varphi^\lambda: E^\lambda \to F^\lambda$ is a monomorphism for all $\lambda$; (iii) $\varphi^\lambda$ is a monomorphism for some $\lambda$ with at most $m$ rows, $m = \text{rank}(E)$.

By the functoriality of the construction of $E^\lambda$, any endomorphism of $E$ determines an endomorphism of $E^\lambda$. This gives a left action of the algebra $\text{End}_R(E)$ on $E^\lambda$. In particular, the group $GL(E)$ of automorphisms of $E$ acts on the left on $E^\lambda$. If $E$ is free with a given basis, thus identifying $E$ with $R^m$, then $\text{End}_R(E) = M_m R$ is the algebra of $m \times m$ matrices. Therefore $M_m R$ acts on $E^\lambda$, as does the subgroup $GL_m R$. We will need the following exercise.

**Exercise 3** If $g = (g_{i,j}) \in M_m R$, show that if $T$ has entries $j_1, \ldots, j_n$ in its $n$ boxes (ordered arbitrarily), then $g \cdot e_T = \sum g_{i_1, j_1} \cdot \ldots \cdot g_{i_n, j_n} e_{T'}$, the sum over the $m^n$ fillings $T'$ obtained from $T$ by replacing the entries $(j_1, \ldots, j_n)$ by $(i_1, \ldots, i_n)$.

The algebra $M_m R$ also acts on the left on the $R$-algebra $R[Z]$ by the formula

$$(8) \qquad g \cdot Z_{i,j} = \sum_{k=1}^{m} Z_{i,k} g_{k,j}, \quad g = (g_{i,j}) \in M_m R.$$

Regarding $R[Z]$ as the polynomial functions on the space of $n \times m$ matrices, with $Z_{i,j}$ a coordinate function, this is the action of $M_m R$ on functions by

$(g \cdot f)(A) = f(A \cdot g)$ for $g \in M_m R$, $A$ a matrix, and $f$ a function on matrices.

**Exercise 4** Show that $g \cdot D_{j_1, \ldots, j_p} = \sum g_{i_1, j_1} \cdot \ldots \cdot g_{i_p, j_p} D_{i_1, \ldots, i_p}$, the sum over all $p$-tuples $i_1, \ldots, i_p$ from $[m]$.

It follows from this exercise that the left action of $M_m R$ on $R[Z]$ maps the module $D^\lambda$ to itself.

**Exercise 5** Show that when $E = R^m$, the isomorphism from $E^\lambda$ to $D^\lambda$ is an isomorphism of $M_m R$-modules.

## 8.2 Representations of $GL(E)$

Now we specialize to the case where $R = \mathbb{C}$, so $E$ is a finite dimensional complex vector space. In this case $E^\lambda$ is a finite dimensional representation of $GL(E)$. Our object is to show that these are irreducible representations, and that all finite dimensional representations of $GL(E)$ can be described in terms of these representations.

A representation $V$ (always assumed to be a finite dimensional complex vector space) of $G = GL(E)$ is called **polynomial** if the corresponding mapping $\rho : GL(E) \to GL(V)$ is given by polynomials, i.e., after choosing bases of $E$ and $V$, so $GL(E) = GL_m \mathbb{C} \subset \mathbb{C}^{m^2}$ and $GL(V) = GL_N \mathbb{C} \subset \mathbb{C}^{N^2}$, the $N^2$ coordinate functions are polynomial functions of the $m^2$ variables. Similarly, the representation is **rational**, or **holomorphic**, if the corresponding functions are rational, or holomorphic. These notions are easily checked to be independent of bases. All representations will be assumed to be at least holomorphic. The representations $E^\lambda$ are polynomial. Our goal here is to show that these representations $E^\lambda$ are exactly the irreducible polynomial representations of $GL(E)$, where $\lambda$ varies over all Young diagrams with at most $m$ rows. (The representation $E^\lambda$ is 0 if $\lambda$ has more than $m$ rows.) They will determine all holomorphic representations of $GL(E)$ by tensoring with suitable negative powers of the **determinant representation** $D = \bigwedge^m E$. We denote by $D^{\otimes k}$ the one-dimensional representation $GL(E) \to \mathbb{C}^*$ given by $g \mapsto \det(g)^k$; this is a polynomial representation only when $k \geq 0$.

We choose a basis for $E$, which identifies $G = GL(E)$ with $GL_m \mathbb{C}$. We let $H \subset G$ denote the subgroup of diagonal matrices; write $x = \mathrm{diag}(x_1, \ldots, x_m)$ in $H$ for the diagonal matrix with these entries. A vector $v$ in a representation $V$ is called an **weight vector** with **weight**

$\alpha = (\alpha_1, \ldots, \alpha_m)$,  with  $\alpha_i$  integers, if

$$x \cdot v = x_1{}^{\alpha_1} \ldots x_m{}^{\alpha_m} v \qquad \text{for all } x \text{ in } H.$$

It is a general fact, following from the fact that the action of $H$ on $V$ is by commuting (diagonal) matrices, that any $V$ is a direct sum of its *weight spaces:*

$$V = \oplus V_\alpha, \quad V_\alpha = \{v \in V : x \cdot v = \left(\textstyle\prod x_i{}^{\alpha_i}\right) v \quad \forall x \in H\}.$$

We will see this decomposition explicitly in all the examples. For example, if $V = E^\lambda$, we see immediately from the definition (see Exercise 3) that each $e_T$ is a weight vector, with weight $\alpha$, where $\alpha_i$ is the number of times the integer $i$ occurs in $T$.

Let $B \subset G$ be the Borel group of all upper triangular matrices. A weight vector $v$ in a representation $V$ is called a *highest weight vector* if $B \cdot v = \mathbb{C}^* \cdot v$.

**Lemma 4** *Up to multiplication by a nonzero scalar, the only highest weight vector in $E^\lambda$ is the vector $e_T$, where $T = U(\lambda)$ is the tableau on $\lambda$ whose $i^{\text{th}}$ row contains only the integer $i$.*

**Proof** We use the formula $g \cdot e_T = \sum g_{i_1, j_1} \cdot \ldots \cdot g_{i_n, j_n} e_{T'}$ for multiplying $e_T$ by a matrix $g$ from Exercise 3. It follows immediately from this formula that, if $T = U(\lambda)$, and $g_{i,j} = 0$ for $i > j$, then the only nonzero $e_{T'}$ that can occur in $g \cdot e_T$ is $e_T$ itself. Similarly, suppose $T \neq U(\lambda)$, and the $p^{\text{th}}$ row is the first row that contains an element larger than $p$, and this smallest element is $q$. Define $g$ in $B$ to be the elementary matrix with $g_{i,j} = 1$ if $i = j$ or if $i = p$ and $j = q$, and $g_{i,j} = 0$ otherwise. We see that $g \cdot e_T = \sum e_{T'}$, where the sum is over all fillings $T'$ obtained from $T$ by exchanging some set (possibly empty) of the $q$'s appearing in $T$ to $p$'s. In particular, if $T'$ is the tableau obtained from $T$ by changing all of the $q$'s in its $p^{\text{th}}$ row to $p$'s, we see that $e_{T'}$ occurs in $g \cdot e_T$ with coefficient 1, which means that $e_T$ is not a highest weight vector.    □

Now we appeal to a basic fact of representation theory, which we will discuss briefly at the end of this section. A (finite dimensional, holomorphic) representation $V$ of $GL_m\mathbb{C}$ is irreducible if and only if it has a unique highest weight vector, up to multiplication by a scalar. In addition, two representations are isomorphic if and only if their highest weight vectors have the same weight. The possible highest weights are those $\alpha$ with $\alpha_1 \geq \alpha_2 \geq \cdots \geq \alpha_m$. Using these facts, we have:

**Theorem 2** (1) *If* $\lambda$ *has at most* $m$ *rows, then the representation* $E^\lambda$ *of* $GL_m\mathbb{C}$ *is an irreducible representation with highest weight* $\lambda = (\lambda_1, \ldots, \lambda_m)$. *These are all of the irreducible polynomial representations of* $GL_m\mathbb{C}$.

(2) *For any* $\alpha = (\alpha_1, \ldots, \alpha_m)$ *with* $\alpha_1 \geq \ldots \geq \alpha_m$ *integers, there is a unique irreducible representation of* $GL_m\mathbb{C}$ *with highest weight* $\alpha$, *which can be realized as* $E^\lambda \otimes D^{\otimes k}$, *for any* $k \in \mathbb{Z}$ *with* $\lambda_i = \alpha_i - k \geq 0$ *for all* $i$.

**Proof** Since $E^\lambda \otimes D^{\otimes k}$ is an irreducible representation with highest weight $\alpha$, where $\alpha_i = \lambda_i + k$, and this is polynomial exactly when each $\alpha_i$ is nonnegative, the conclusions follow from the preceding discussion.  □

It follows in particular that all (finite dimensional) holomorphic representations of $GL_m\mathbb{C}$ are actually rational. Note that $E^\lambda \otimes D^{\otimes k}$ is isomorphic to $E^{\lambda'} \otimes D^{\otimes k'}$ if and only if $\lambda_i + k = \lambda_i' + k'$ for all $i$. [1]

One can also describe all (holomorphic) representations of the subgroup $SL(E) = SL_m\mathbb{C}$ of automorphisms of determinant 1. The story is the same as for $GL_m\mathbb{C}$ except that the group $H$ now consists of diagonal matrices whose product is 1, so the weights $\alpha$ all lie in the hyperplane $\alpha_1 + \ldots + \alpha_m = 0$, and the determinant representation $D$ is trivial. The irreducible representations are precisely the $E^\lambda$, but with $E^\lambda \cong E^{\lambda'}$ if and only if $\lambda_i - \lambda_i'$ is constant. One therefore gets a unique irreducible representation for each $\lambda$ if one allows only those $\lambda$ with $\lambda_m = 0$.

**Exercise 6** Prove these assertions.

We conclude this section by sketching the ideas for proving these basic facts in representation theory. One reference for details is Fulton and Harris (1991), where one can find several other constructions and proofs that these representations are irreducible. One uses the Lie algebra $\mathfrak{g} = \mathfrak{gl}_m\mathbb{C} = M_m\mathbb{C}$ of matrices, which can be identified with the tangent space to manifold $G = GL_m\mathbb{C}$ at the identity element $I$, with its bracket $[X, Y] = X \cdot Y - Y \cdot X$. A representation of $\mathfrak{g}$ is a vector space $V$ together with an action $\mathfrak{g} \otimes V \to V$ such that $[X, Y] \cdot v = X \cdot (Y \cdot v) - Y \cdot (X \cdot v)$ for $X, Y$ in $\mathfrak{g}$ and $v$ in $V$; equivalently, one has a homomorphism of Lie algebras from $\mathfrak{g}$ to $\mathfrak{gl}(V)$. Any holomorphic representation $\rho : GL(E) \to GL(V)$ determines a homomorphism $d\rho : \mathfrak{gl}(E) \to \mathfrak{gl}(V)$ on tangent spaces at the identity, which can be seen to

[1] This suggests that there should be a theory of "rational tableaux" corresponding to $\alpha$ with possibly negative entries, allowing boxes to extend to the left. This has been carried out by Stembridge (1987).

be compatible with the bracket, and therefore determines a representation $V$ of $\mathfrak{g}$. Using the exponential map from $\mathfrak{g} = \mathfrak{gl}_m\mathbb{C}$ to $G = GL_m\mathbb{C}$, one sees that a subspace $W$ of a representation $V$ is a subrepresentation of $G$ if and only if it is preserved by $\mathfrak{g}$.

The weight space $V_\alpha$ can be described in terms of the action of $\mathfrak{g}$ by the equation $V_\alpha = \{v \in V : X \cdot v = (\sum \alpha_i x_i)v \; \forall \, X \in \mathfrak{h}\}$, where $\mathfrak{h}$ consists of the diagonal elements $X = \mathrm{diag}(x_1, \ldots, x_m)$ in $\mathfrak{g}$. For the action of $\mathfrak{g}$ on itself by the left multiplication via the bracket, we have the decomposition $\mathfrak{g} = \mathfrak{h} \oplus \bigoplus \mathfrak{g}_\alpha$, the sum over $\alpha = \alpha(i,j)$ with a 1 in the $i^{\text{th}}$ place, a $-1$ in the $j^{\text{th}}$ place, $i \neq j$. These $\mathfrak{g}_\alpha$ are called *root spaces*, and these $\alpha$ are the roots. In fact, if $E_{i,j}$ is the elementary matrix with a 1 in the $i^{\text{th}}$ row and $j^{\text{th}}$ column, with other entries 0, then $E_{i,j}$ is a basis for the corresponding root space. Those roots $\alpha(i, j)$ with $i < j$ (corresponding to upper triangular elementary matrices) will be called *positive*, those with $i > j$ *negative*. We put the corresponding partial ordering on weights by saying that

$$\alpha \geq \beta \quad \text{if} \quad \alpha_1 + \ldots + \alpha_p \geq \beta_1 + \ldots + \beta_p \quad \text{for} \quad 1 \leq p \leq m.$$

The sum of $\mathfrak{h}$ and the positive root spaces is the Lie algebra of $B$. A weight vector $v$ in $V$ is a highest weight vector exactly when $X \cdot v = 0$ for all $X$ in positive root spaces, i.e., $E_{i,j} \cdot v = 0$ for all $i < j$. An advantage of working with the Lie algebra is that if $V = \oplus V_\alpha$ is the decomposition into weight spaces, then

$$\mathfrak{h} \cdot V_\beta \subset V_\beta \quad \text{and} \quad \mathfrak{g}_\alpha \cdot V_\beta \subset V_{\alpha+\beta}.$$

If $V$ is an irreducible representation, then a highest weight vector must span its root space. The reason for this is that if $v$ is a highest weight vector, then space consisting of $\mathbb{C} \cdot v$ and the sum of all translates $\mathfrak{g}_\alpha \cdot v$, as $\alpha$ varies over the negative weights, can be seen to be a subrepresentation of $\mathfrak{g}$. It follows from this that an irreducible representation can have only one highest weight vector, up to multiplication by a nonzero scalar. Moreover, two irreducible representations are isomorphic if and only if they have the same highest weight. This is seen by looking in the direct sum of the representations, and showing that the direct sum of two highest weight vectors generates a subrepresentation that is the graph of an isomorphism between them. That the highest weights are weakly decreasing sequences can be seen directly when $m = 2$, and by restricting to appropriate subgroups isomorphic to $GL_2(\mathbb{C})$ when $m > 2$.

Another basic fact, that we can see explicitly in the representations we construct here, is the *semisimplicity* of holomorphic representations of $GL_m\mathbb{C}$; that is, that for any subrepresentation $W$ of a representation $V$ there is a complementary subrepresentation $W'$ of $V$ so that $V = W \oplus W'$. A quick proof of this is by Weyl's unitary trick, using the unitary subgroup $U(m) \subset GL_m\mathbb{C}$, as follows. Choose any linear projection of $V$ onto $W$, and by averaging (integrating) over the compact group $U(m)$, one gets a $U(m)$-linear projection onto $W$, whose kernel is a complementary subspace $W'$ that is preserved by $U(m)$. On the Lie algebra level, $W'$ is preserved by its (real) Lie algebra $\mathfrak{u}(m)$, and since $\mathfrak{u}(m) \otimes_\mathbb{R} \mathbb{C} = \mathfrak{gl}_m\mathbb{C}$, it follows that $W'$ is preserved by $\mathfrak{gl}_m\mathbb{C}$, so it is preserved by $GL_m\mathbb{C}$. From this semisimplicity it follows that every holomorphic representation is a direct sum of irreducible representations. (Note that the same argument, applied to the subgroup $H$ and its compact subgroup $(S^1)^n$, verifies that any holomorphic $V$ is a direct sum of its weight spaces.)

**Exercise 7** Prove Schur's lemma: Any homomorphism between irreducible representations must be zero if they are not isomorphic, and any homomorphism from an irreducible representation to itself is multiplication by a scalar.

## 8.3 Characters and representation rings

We begin by giving an alternative presentation of these representations, by constructing them from representations of symmetric groups.

Let $E$ be a complex vector space of dimension $m$. The symmetric group $S_n$ acts on the *right* on the $n$-fold tensor product

$$E^{\otimes n} = E \otimes_\mathbb{C} E \otimes_\mathbb{C} \ldots_\mathbb{C} \otimes E,$$

$$(u_1 \otimes \ldots \otimes u_n) \cdot \sigma = u_{\sigma(1)} \otimes \ldots \otimes u_{\sigma(n)},$$

for $u_i \in E$ and $\sigma \in S_n$. For any representation $M$ of $S_n$, we have a vector space $E(M)$ defined by

$$(9) \qquad E(M) = E^{\otimes n} \otimes_{\mathbb{C}[S_n]} M;$$

that is, $E(M)$ is the quotient space of $E^{\otimes n} \otimes_\mathbb{C} M$ by the subspace generated by all

$$(w \cdot \sigma) \otimes v - w \otimes (\sigma \cdot v), \qquad w \in E^{\otimes n},\ v \in M,\ \sigma \in S_n.$$

The general linear group $GL(E)$ of automorphisms of $E$ acts on the left on $E$, so it acts on the left on $E^{\otimes n}$ by $g \cdot (u_1 \otimes \ldots \otimes u_n) = g \cdot u_1 \otimes \ldots \otimes g \cdot u_n$. Since this action commutes with the right action by $S_n$, this determines a left action of $GL(E)$ on $E(M)$: $g \cdot (w \otimes v) = (g \cdot w) \otimes v$. All of these representations are easily seen to be polynomial representations.

For example, if $M$ is the trivial representation, then $E(M)$ is the symmetric power $\text{Sym}^n(E)$; if $M$ is the alternating representation, then $E(M)$ is the exterior power $\bigwedge^n E$. If $M = \mathbb{C}[S_n]$ is the regular representation, then $E(M) = E^{\otimes n}$, since $P \otimes_A A = P$ for any $A$-module $P$. The construction is functorial: a homomorphism $\varphi \colon M \to N$ of $S_n$-modules determines a homomorphism $E(\varphi) \colon E(M) \to E(N)$ of $GL(E)$-modules. A direct sum decomposition $M = \oplus M_i$ determines a direct sum decomposition $E(M) = \oplus E(M_i)$.

**Exercise 8**  Show that if $\varphi$ is surjective (resp. injective), then $E(\varphi)$ is surjective (resp. injective).

Two more of the representations $E(M)$ are easy to describe. If $M^\lambda$ is the representation described in §7.2, then

$$(10) \qquad E(M^\lambda) \cong \text{Sym}^{\lambda_1}(E) \otimes \ldots \otimes \text{Sym}^{\lambda_k}(E),$$
$$\lambda = (\lambda_1 \geq \ldots \geq \lambda_k > 0).$$

One can deduce this from Exercise 10 of §7.3, where we saw that choosing a numbering $U$ of $\lambda$ (with distinct numbers from 1 to $n$) determines a surjection $\mathbb{C}[S_n] \to M^\lambda$, $\sigma \mapsto \sigma\{U\}$, with kernel generated by all elements $p - 1$, as $p$ varies among all elements (or transpositions) in the row group of $U$. By the functoriality of the map from $S_n$-modules to $GL(E)$-modules, this determines a surjection $E^{\otimes n} \to E(M^\lambda)$, with kernel generated by all

$$u_{p(1)} \otimes \ldots \otimes u_{p(n)} \; - \; u_1 \otimes \ldots \otimes u_n.$$

This is the realization of the tensor product of the symmetric powers $\text{Sym}^{\lambda_i} E$ of $E$ as a quotient of $E^{\otimes n}$, by symmetrizing each set of factors that are in the same row of $U$. (Usually for this one takes $U$ to be the standard tableau that numbers the rows of $\lambda$ in order.)

Similarly, by Exercise 13(c) of §7.4, realizing $\widetilde{M}^\lambda$ as a quotient of $\mathbb{C}[S_n]$ by the ideal generated by all $q - \text{sgn}(q) \cdot 1$ for $q$ in the column group of a numbering $U$ (say by columns from left to right) determines an isomorphism of $E(\widetilde{M}^\lambda)$ with a tensor product of exterior products

$$(11) \qquad E(\widetilde{M}^\lambda) \cong \bigwedge^{\mu_1}(E) \otimes \ldots \otimes \bigwedge^{\mu_\ell}(E),$$
$$\mu = \tilde{\lambda} = (\mu_1 \geq \ldots \geq \mu_\ell > 0).$$

**Exercise 9** If $N$ is a representation of $S_n$ and $M$ is a representation of $S_m$, show that $E(N \circ M) \cong E(N) \otimes E(M)$, where $N \circ M$ is the representation of $S_{n+m}$ defined in §7.3. Use this to give another proof of (10) and (11).

**Proposition 1** *There is a canonical isomorphism* $E^\lambda \cong E(S^\lambda)$.

**Proof** Given $\mathbf{v}$ in $E^{\times \lambda}$, and a numbering $U$ of $\lambda$ with distinct numbers from 1 to $n$, we have an element $\mathbf{v}(U) = v_1 \otimes \ldots \otimes v_n$ in $E^{\otimes n}$, where $v_i$ is the element of $\mathbf{v}$ in the box where $U$ has entry $i$. We map $E^{\times \lambda}$ to $E(S^\lambda) = E^{\otimes n} \otimes_{\mathbb{C}[S_n]} S^\lambda$ by the formula

$$\mathbf{v} \mapsto \mathbf{v}(U) \otimes v_U,$$

where $v_U$ is the generator of $S^\lambda$ defined in §7.2. This is independent of the choice of $U$, for if $\sigma U$ is another, with $\sigma \in S_n$, then

$$\mathbf{v}(\sigma U) \otimes v_{\sigma U} = \mathbf{v}(\sigma U) \otimes \sigma \cdot v_U = \mathbf{v}(\sigma U) \cdot \sigma \otimes v_U = \mathbf{v}(U) \otimes v_U,$$

since $\mathbf{v}(\sigma U) \cdot \sigma = \mathbf{v}(U)$, as follows from the definition. To show that this determines a map from $E^\lambda$ to $E(S_\lambda)$, we must show that properties (1)–(3) of §8.1 are valid. The multilinearity (1) is obvious. For (2), if $\mathbf{v}$ has two equal entries in a column, and $t$ permutes these entries in $U$, then

$$\mathbf{v}(U) \otimes v_U = \mathbf{v}(tU) \otimes v_{tU} = -\mathbf{v}(U) \otimes v_U,$$

since $\mathbf{v}(tU) = \mathbf{v}(U)$ and $v_{tU} = -v_U$. For (3), if we start with $\mathbf{v}$, and let $\mathbf{w}$ denote a result of making an exchange as in (3), and we let $W$ denote the corresponding exchange carried out on $U$, then $\mathbf{w}(U) \otimes v_U = \mathbf{v}(U) \otimes v_W$, so

$$\mathbf{v} - \sum \mathbf{w} \mapsto \mathbf{v}(U) \otimes v_U - \sum \mathbf{w}(U) \otimes v_U$$

$$= \mathbf{v}(U) \otimes v_U - \sum \mathbf{v}(U) \otimes v_W$$

$$= \mathbf{v}(U) \otimes (v_U - \sum v_W),$$

and we know from Proposition 4 of §7.4 that $v_U - \sum v_W = 0$ in $S^\lambda$.

These calculations amount to showing that $E^\lambda \cong E(\widetilde{M}^\lambda)/E(Q^\lambda)$. The fact that $E(\widetilde{M}^\lambda)/E(Q^\lambda) \cong E(S^\lambda)$ follows from the isomorphism $S^\lambda = \widetilde{M}^\lambda/Q^\lambda$. (See equations (11) and (4), and note as in the preceding paragraph that the image of $E(Q^\lambda)$ in $E(\widetilde{M}^\lambda)$ is the subspace denoted $Q^\lambda(E)$ in (4).) We will soon see several other proofs of this fact.    □

This construction gives another way to prove some of the basic facts about representations of $GL(E)$. Using the obvious fact that the map $E^\lambda \to E(S^\lambda)$

constructed in the preceding proof is surjective, it follows from the first part of Theorem 1 that $\dim(E(S^\lambda))$ is at most the number $d_\lambda(m)$ of tableaux on $\lambda$ with entries in $[m]$. The fact that the regular representation $\mathbb{C}[S_n]$ is isomorphic to the direct sum of copies of $S^\lambda$, each occurring $f^\lambda$ times, implies that there is a decomposition

$$E^{\otimes n} \;=\; E(\mathbb{C}[S_n]) \;\cong\; \bigoplus_{\lambda \vdash n} (E(S^\lambda))^{\oplus f^\lambda}.$$

Therefore $m^n = \dim(E^{\otimes n}) = \sum f^\lambda \dim(E^\lambda) \leq \sum f^\lambda d_\lambda(m)$. Since $\sum f^\lambda d_\lambda(m) = m^n$ by equation (5) in §4.3, each $E^\lambda$ must have dimension $d_\lambda(m)$, and the map from each $E^\lambda$ to $E(S^\lambda)$ must therefore be an isomorphism. This also gives another proof of the second fact proved in the theorem in §8.1, that the $e_T$ are linearly independent in $E^\lambda$. (To be precise, this proves this when $E$ is free over $\mathbb{C}$, from which it follows for $E$ free over $\mathbb{Z}$, from which it follows for all $E$ free over all $R$ by base change.)

**Corollary 1** $E^{\otimes n} \cong \bigoplus (E^\lambda)^{\oplus f^\lambda}$, *the sum over partitions* $\lambda$ *of* $n$.

**Exercise 10** Show similarly that the realization $S^\lambda \cong \widetilde{S}^\lambda$ as a quotient space of $M^\lambda$ in Exercise 14 of §7.4 realizes $E^\lambda = E(S^\lambda)$ as a quotient

(12)      $E^\lambda \;\cong\; \mathrm{Sym}^{\lambda_1}(E) \otimes \ldots \otimes \mathrm{Sym}^{\lambda_k}(E) \,/\, \widetilde{Q}^\lambda(E),$

where $\widetilde{Q}^\lambda(E)$ is the subspace spanned by all relations $\xi + (-1)^k \widetilde{\pi}_{j,k}(\xi)$, where $\xi = w_1 \otimes \ldots \otimes w_\ell$, $w_i \in \mathrm{Sym}^{\lambda_i}(E)$, $w_i = x_{i,1} \cdot \ldots \cdot x_{i,\lambda_i}$, $x_{i,r} \in E$, and $\widetilde{\pi}_{j,k}(\xi)$ is the sum of all $\xi'$ obtained from $\xi$ by interchanging the first $k$ vectors $x_{j+1,1}, \ldots, x_{j+1,k}$ in $w_{j+1}$ with $k$ of the vectors in $w_j$, preserving the order in each.

**Exercise 11** Use the relations of the preceding exercise to show that $E^\lambda$ is spanned by elements $\widetilde{e}_T$, where $T$ varies over all tableaux on $\lambda$ with entries in $[m]$, and $\widetilde{e}_T$ is the image of $\mathbf{v}(U) \otimes \widetilde{v}_U$, with $U$ a numbering of $\lambda$ and $\widetilde{v}_U$ as in §7.4.

Other realizations of the representations $S^\lambda$ of $S_n$ lead to other realizations of the representations $E^\lambda$ of $GL(E)$. For example, since $S^\lambda$ is isomorphic to the image of the endomorphism of $A = \mathbb{C}[S_n]$ that is right multiplication by the Young symmetrizer $c_U = b_U \cdot a_U$, for any (distinct) numbering $U$ of $\lambda$, it follows that $E^\lambda$ is isomorphic to the image of the map $E^{\otimes n} \to E^{\otimes n}$ that is right multiplication by $c_U$. Similarly, the description of $S^\lambda$ as the image of a homomorphism $\widetilde{M}^\lambda \to M^\lambda$ gives a realization of $E^\lambda$ as the image of a homomorphism

$$\wedge^{\mu_1}(E) \otimes \ldots \otimes \wedge^{\mu_\ell}(E) \;\to\; \mathrm{Sym}^{\lambda_1}(E) \otimes \ldots \otimes \mathrm{Sym}^{\lambda_k}(E),$$

where $\mu$ is the conjugate of $\lambda$.

The **character** of a (finite dimensional holomorphic) representation $V$ of $GL_m\mathbb{C}$, denoted $\mathrm{Char}(V)$ or $\chi_V$, is the function of $m$ (nonzero) complex variables defined by

$$(13) \qquad \chi_V(x) \;=\; \chi_V(x_1, \ldots, x_m) \;=\; \text{Trace of } \mathrm{diag}(x) \text{ on } V.$$

Decomposing $V$ into weight spaces $V_\alpha$, we see that

$$\chi_V(x) \;=\; \sum_\alpha \dim(V_\alpha) x^\alpha \;=\; \sum_\alpha \dim(V_\alpha) x_1^{\alpha_1} \cdot \ldots \cdot x_m^{\alpha_m}.$$

In particular, for $E^\lambda$, where there is one weight vector $e_T$ for each tableau $T$ with entries in $[m]$, we see that

$$(14) \qquad \mathrm{Char}(E^\lambda) \;=\; \sum x^T \;=\; s_\lambda(x_1, \ldots, x_m)$$

is the Schur polynomial corresponding to $\lambda$. (In this context, the Jacobi–Trudi formula (7) of §6.1 becomes a special case of the Weyl character formula.) In general it follows immediately from the definitions that

$$(15) \qquad \mathrm{Char}(V \oplus W) \;=\; \mathrm{Char}(V) \;+\; \mathrm{Char}(W);$$

$$(16) \qquad \mathrm{Char}(V \otimes W) \;=\; \mathrm{Char}(V) \cdot \mathrm{Char}(W).$$

For example, the decomposition of Corollary 1 determines the identity

$$(17) \qquad (x_1 + \ldots + x_m)^n \;=\; \sum_{\lambda \vdash n} f^\lambda s_\lambda(x_1, \ldots, x_m).$$

Another general fact from representation theory is the fact that a representation is uniquely determined by its character. This follows from the fact that every representation is a direct sum of irreducible representations, together with the fact that an irreducible representation is determined by its highest weight. Note that the highest weight can be read off the character. In our case we can see this explicitly, since the character of a direct sum $\oplus(E^\lambda)^{\oplus m(\lambda)}$ is $\sum m(\lambda) s_\lambda(x_1, \ldots, x_m)$, and we know that the Schur polynomials are linearly independent. It follows that one can find the decomposition of any polynomial representation into its irreducible components by writing its character as a sum of corresponding Schur polynomials. From §2.2 (8), (9), and §5.2 (4) we deduce

**Corollary 2**

(a) $\mathrm{Sym}^{\lambda_1} E \otimes \ldots \otimes \mathrm{Sym}^{\lambda_n} E \cong \bigoplus (E^{\nu})^{\oplus K_{\nu\lambda}} \cong E^{\lambda} \oplus \bigoplus_{\nu \rhd \lambda} (E^{\nu})^{\oplus K_{\nu\lambda}}$,

where $K_{\nu\lambda}$ is the Kostka number.

(b) $\bigwedge^{\mu_1} E \otimes \ldots \otimes \bigwedge^{\mu_m} E \cong \bigoplus (E^{\nu})^{\oplus K_{\tilde{\nu}\mu}} \cong E^{\tilde{\mu}} \oplus \bigoplus_{\tilde{\nu} \rhd \mu} (E^{\nu})^{\oplus K_{\tilde{\nu}\mu}}$.

(c) $E^{\lambda} \otimes E^{\mu} \cong \bigoplus_{\nu} (E^{\nu})^{\oplus c_{\lambda\mu}^{\nu}}$, where $c_{\lambda\mu}^{\nu}$ is the Littlewood–Richardson number.

Alternatively, these decompositions can be deduced as in Corollary 1 from the corresponding decompositions of $M^{\lambda}$, $\tilde{M}^{\lambda}$, and $S^{\lambda} \circ S^{\mu}$ as representations of the symmetric group. Part (c), which is the original Littlewood–Richardson rule, contains the special "Pieri" cases of decomposing $E^{\lambda} \otimes \mathrm{Sym}^{p} E$ (resp. $E^{\lambda} \otimes \bigwedge^{p} E$) as the sum of those $E^{\mu}$ for which $\mu$ is obtained from $\lambda$ by adding $p$ boxes, with no two in the same column (resp. row).

**Corollary 3**

(a) $\mathrm{Sym}^{p}(E^{\oplus n}) \cong \bigoplus_{\lambda \vdash p} (E^{\lambda})^{\oplus d_{\lambda}(n)}$.

(b) $\bigwedge^{p}(E^{\oplus n}) \cong \bigoplus_{\lambda \vdash p} (E^{\lambda})^{\oplus d_{\tilde{\lambda}}(n)}$.

**Proof** For (a), $\mathrm{Sym}^{p}(E^{\oplus n}) = \bigoplus \mathrm{Sym}^{p_1}(E) \otimes \ldots \otimes \mathrm{Sym}^{p_n}(E)$, the sum over all nonnegative integers $p_1, \ldots, p_n$ that add to $p$. By (a) of Corollary 2, the number of times $E^{\lambda}$ occurs in this is the number of tableaux on $\lambda$ whose entries are $p_1$ 1's, ..., $p_n$ n's. The total number of such tableaux is the number $d_{\lambda}(n)$ of tableaux on $\lambda$ with entries from $[n]$. The proof of (b) is similar, using (b) of Corollary 2. □

There are useful generalizations of Corollary 3 that can be obtained by applying the same principle to representations of $GL(E) \times GL(F)$, for finite-dimensional vector spaces $E$ and $F$. (We won't need these generalizations here.) The Cauchy–Littlewood formula (3) of §4.3 implies that

(18) $\qquad \mathrm{Sym}^{p}(E \otimes F) \cong \bigoplus_{\lambda \vdash p} E^{\lambda} \otimes F^{\lambda}$.

A dual formula (§A.4.3, Corollary to Proposition 3) gives

(19) $\qquad \bigwedge^{p}(E \otimes F) \cong \bigoplus_{\lambda \vdash p} E^{\lambda} \otimes F^{\tilde{\lambda}}$.

Restricting these isomorphisms from $GL(E) \times GL(F)$ to $GL(E) \times \{1\}$ yields Corollary 3. Similarly, from Exercise 4 of §5.2 we deduce decompositions

(20)          $(E \oplus F)^\nu \cong \oplus (E^\lambda \otimes F^\mu)^{\oplus c_{\lambda\mu}^\nu}$

**Exercise 12** (a) Show that, for $p \geq q \geq 1$, $E^{(p,q)}$ is isomorphic to the kernel of the linear map

$$\text{Sym}^p E \otimes \text{Sym}^q E \twoheadrightarrow \text{Sym}^{p+1} E \otimes \text{Sym}^{q-1} E,$$

$$(u_1 \cdot \ldots \cdot u_p) \otimes (v_1 \cdot \ldots \cdot v_q) \mapsto \sum_{i=1}^{q} (u_1 \cdot \ldots \cdot u_p \cdot v_i) \otimes (v_1 \cdot \ldots \cdot \widehat{v_i} \cdot \ldots \cdot v_q).$$

(b) Show that, for $p \geq q \geq 1$, $E^{(2^q 1^{p-q})}$ is isomorphic to the kernel of the linear mapping

$$\wedge^p E \otimes \wedge^q E \twoheadrightarrow \wedge^{p+1} E \otimes \wedge^{q-1} E,$$

$$(u_1 \wedge \ldots \wedge u_p) \otimes (v_1 \wedge \ldots \wedge v_q) \mapsto$$

$$\sum_{i=1}^{q} (-1)^i (u_1 \wedge \ldots \wedge u_p \wedge v_i) \otimes (v_1 \wedge \ldots \wedge \widehat{v_i} \wedge \ldots \wedge v_q).$$

**Exercise 13** Show that the subspace $Q^\lambda(E) \subset \otimes \wedge^{\mu_i} E$ of quadratic relations is the sum of the images of maps

$$\wedge^{\mu_1} E \otimes \ldots \otimes \wedge^{\mu_j + 1} E \otimes \wedge^{\mu_{j+1} - 1} E \otimes \ldots \otimes \wedge^{\mu_\ell} E \to$$

$$\wedge^{\mu_1} E \otimes \ldots \otimes \wedge^{\mu_\ell} E,$$

for $1 \leq j \leq \ell - 1$. Deduce another proof of the fact that the quadratic equations $\xi = \pi_{j,k}(\xi)$ for $k > 1$ follow from those for $k = 1$.

Define the **representation ring** of $GL_m\mathbb{C}$, denoted $\mathcal{R}(m)$, to be the Grothendieck ring of polynomial representations. This is defined to be the free abelian group on the isomorphism classes $[V]$ of all polynomial representations, modulo the subgroup generated by all $[V \oplus W] - [V] - [W]$. Since every such representation is a direct sum of irreducible representations, each occurring with a well-defined multiplicity, $\mathcal{R}(m)$ is the free abelian group on the isomorphism classes of the irreducible representations. It is given a commutative ring structure from the tensor product of representations: $[V] \cdot [W] = [V \otimes_\mathbb{C} W]$.

The mapping that to a representation $M$ of a symmetric group $S_n$ assigns the representation $E(M)$ of $GL_m\mathbb{C}$ determines an additive homomorphism from the Grothendieck group $R_n$ of such representations to $\mathcal{R}(m)$, and hence, by adding over $n$, a homomorphism from $R = \bigoplus R_n$ to $\mathcal{R}(m)$. The character Char determines a homomorphism from $\mathcal{R}(m)$ to the ring $\Lambda(m)$ of symmetric polynomials in the variables $x_1, \ldots, x_m$, which is a ring homomorphism by (15) and (16), and an injection since a representation is determined by its character. We therefore have maps

$$(21) \qquad \Lambda \to R \to \mathcal{R}(m) \to \Lambda(m).$$

The first of these takes the Schur function $s_\lambda$ to the class $[S^\lambda]$ of the representation $S^\lambda$ of $S_n$, $n = |\lambda|$; the second takes $[S^\lambda]$ to $[E^\lambda]$; and the third takes $E^\lambda$ to $s_\lambda(x_1, \ldots, x_m)$. Since the Schur functions are a basis for $\Lambda$, it follows that the composite $\Lambda \to \Lambda(m)$ is simply the homomorphism that takes a function $f$ to $f(x_1, \ldots, x_m, 0, \ldots, 0)$. In particular, this composite is surjective. It follows that $\mathcal{R}(m) \to \Lambda(m)$ is an isomorphism, and that $R \to \mathcal{R}(m)$ is a surjective ring homomorphism, a result that can also be proved directly (see Exercise 9). In addition, since the kernel of the map from $\Lambda$ to $\Lambda(m)$ is generated by the Schur functions $s_\lambda$ for those $\lambda$ that have more than $m$ rows, it follows that the map from $R$ to $\mathcal{R}(m)$ determines an isomorphism

$$(22) \qquad R/(\text{Subgroup spanned by } [S^\lambda], \lambda_{m+1} \neq 0) \xrightarrow{\cong} \mathcal{R}(m).$$

It is possible to describe the map backwards, going from representations of $GL_m\mathbb{C}$ to representations of symmetric groups, without going through the ring of symmetric polynomials via characters. In simple cases, this can be done as follows. A polynomial representation $V$ of $GL_m\mathbb{C}$ is **homogeneous** of degree $n$ if its weights $\alpha$ all have $\alpha_1 + \ldots + \alpha_m = n$. By what we have seen, such a representation is a direct sum of copies of representations $E^\lambda$ as $\lambda$ varies over partitions of $n$ in at most $m$ parts. It follows in particular that such a representation has the form $E(M)$ for some presentation $M$ of $S_n$. If $n \leq m$, we have a natural inclusion

$$S_n \subset S_m \subset GL_m\mathbb{C},$$

with $\sigma \in S_m$ acting on the basis for $E$ by $\sigma(e_i) = e_{\sigma(i)}$. Let $\alpha(n)$ be the weight $(1, \ldots, 1, 0, \ldots, 0)$, with $n$ 1's. For any representation $M$ of $S_n$, consider the composite

$$M \cong (e_1 \otimes \ldots \otimes e_n) \otimes {}_\mathbb{C} M \subset E^{\otimes n} \otimes {}_\mathbb{C} M$$

$$\twoheadrightarrow E^{\otimes n} \otimes {}_{\mathbb{C}[S_n]} M = E(M).$$

**Exercise 14** Show that this composite maps $M$ isomorphically onto the weight space $E(M)_{\alpha(n)}$, determining an isomorphism $M \cong E(M)_{\alpha(n)}$ of $S_n$-modules.

This shows how to recover $M$ from $E(M)$ when $E(M)$ is homogeneous of degree $n \leq m$. The general case requires more sophisticated techniques, and can be found in Green (1980).

There is a general formula called the *Weyl character formula*, that writes the character of a representation as a ratio of two determinants. For $GL_m\mathbb{C}$ this gives exactly the Jacobi–Trudi formula for the Schur polynomials (see Fulton and Harris [1991]).

**Exercise 15** (a)  Show that each polynomial representation of $GL(E)$ occurs exactly once in $\bigoplus_k \operatorname{Sym}^k(E \oplus \wedge^2 E)$. (b)  Show that $E^\lambda$ occurs in $\operatorname{Sym}^k(E \oplus \wedge^2 E)$ exactly when $k$ is half the sum of the number of boxes in $\lambda$ and the number of odd columns of $\lambda$.

**Exercise 16** (For those who know what $\lambda$-rings are) The ring $\Lambda$ has the structure of a $\lambda$-ring, determined by the property that $\lambda^r(e_1) = e_r$ for all $r \geq 1$. The ring $\mathcal{R}(m)$ has the structure of a $\lambda$-ring, determined by setting $\lambda^r[V] = [\wedge^r V]$ for representations $V$ of $GL_m\mathbb{C}$. Show that the homomorphism $\Lambda \to \mathcal{R}(m)$ is a homomorphism of $\lambda$-rings.

## 8.4 The ideal of quadratic relations

The main results of this chapter can be rewritten in terms of symmetric algebras. Recall that for a complex vector space $V$, the symmetric algebra $\operatorname{Sym}^\bullet V$ is the direct sum of all the symmetric powers of $V$:

$$\operatorname{Sym}^\bullet V \;=\; \bigoplus_{n=0}^{\infty} \operatorname{Sym}^n V,$$

with $\operatorname{Sym}^0 V = \mathbb{C}$. The natural map $\operatorname{Sym}^n V \otimes \operatorname{Sym}^m V \to \operatorname{Sym}^{n+m}(V)$, $(v_1 \cdot \ldots \cdot v_n) \otimes (w_1 \cdot \ldots \cdot w_m) \mapsto v_1 \cdot \ldots \cdot v_n \cdot w_1 \cdot \ldots \cdot w_m$, makes $\operatorname{Sym}^\bullet V$ into a graded, commutative $\mathbb{C}$-algebra. If $V = V_1 \oplus \ldots \oplus V_r$ is a direct sum of $r$ vector spaces, there is a canonical isomorphism of algebras

$$\operatorname{Sym}^\bullet V \;=\; \operatorname{Sym}^\bullet(V_1) \otimes \operatorname{Sym}^\bullet(V_2) \otimes \ldots \otimes \operatorname{Sym}^\bullet(V_r).$$

(This follows readily from the universal property of symmetric powers.) In particular, taking a basis $X_1, \ldots, X_r$ for $V$ gives an identification of $\operatorname{Sym}^\bullet V$ with the polynomial ring $\mathbb{C}[X_1, \ldots, X_r]$.

Fix integers $m \geq d_1 > \ldots > d_s > 0$, and let $E$ be a vector space of dimension $m$. Define an algebra $S^{\cdot}(E; d_1, \ldots, d_s)$ to be the symmetric algebra on the vector spaces $\wedge^{d_1} E \oplus \ldots \oplus \wedge^{d_s} E$, modulo the ideal generated by all quadratic relations:

(23) $\qquad S^{\cdot}(E; d_1, \ldots, d_s) =$

$$\oplus \operatorname{Sym}^{a_1}(\wedge^{d_1} E) \otimes \ldots \otimes \operatorname{Sym}^{a_s}(\wedge^{d_s} E) \, / \, Q,$$

the sum over all $s$-tuples $(a_1, \ldots, a_s)$ of nonnegative integers, where $Q = Q(E{:}d_1, \ldots, d_s)$ is the two-sided ideal generated by all the quadratic relations. These relations are obtained as follows: for any pair $p \geq q$ in $\{d_1, \ldots, d_s\}$, and any $v_1, \ldots, v_p$ and $w_1, \ldots, w_q$ in $E$, a generator of $Q$ is

$$(v_1 \wedge \ldots \wedge v_p)(w_1 \wedge \ldots \wedge w_p) \, -$$

$$\sum_{i_1 < \ldots < i_k} (v_1 \ldots w_1 \ldots w_k \ldots v_p)(v_{i_1} \ldots v_{i_k} w_{k+1} \ldots w_p),$$

where, in the sum, the vectors $w_1, \ldots, w_k$ are interchanged with the vectors $v_{i_1}, \ldots, v_{i_k}$. Note that if $p > q$, this generator is in $\wedge^p E \otimes \wedge^q E$, while if $p = q$, it is in $\operatorname{Sym}^2(\wedge^p E)$.

Taking a basis $e_1, \ldots, e_m$ for $E$, i.e., identifying $E$ with $\mathbb{C}^m$, this algebra $S^{\cdot}(E; d_1, \ldots, d_s)$ can be identified with the quotient of a polynomial ring modulo an ideal. The symmetric algebra on $\oplus \wedge^{d_i} E$ is the polynomial ring with variables $X_{i_1, \ldots, i_p}$, for subsets of $p$ elements $i_1, \ldots, i_p$ of $[m]$, with $p \in \{d_1, \ldots, d_s\}$; $X_{i_1, \ldots, i_p}$ corresponds to $e_{i_1} \wedge \ldots \wedge e_{i_p}$ in $\wedge^p E$, so these variables are regarded as alternating functions of the subscripts. The ideal is generated by all quadratic relations

(24) $\qquad X_{i_1, \ldots, i_p} X_{j_1, \ldots, j_q} \, - \, \sum X_{i'_1, \ldots, i'_p} X_{j'_1, \ldots, j'_q},$

the sum over all exchanges of $j_1, \ldots, j_k$ with $k$ of the indices $i_1, \ldots, i_p$, with $p \geq q \geq k \geq 1$, $p, q \in \{d_1, \ldots, d_s\}$. We denote this ring by $S^{\cdot}(m; d_1, \ldots, d_s)$:

(25) $\qquad S^{\cdot}(m; d_1, \ldots, d_s) = \mathbb{C}[X_{i_1, \ldots, i_p}, p \in \{d_1, \ldots, d_s\}] \, / \, Q,$

where $Q$ is generated by the quadratic relations (24).

Let $\lambda$ be a partition whose columns have lengths among the set $\{d_1, \ldots, d_s\}$. That is, the conjugate $\tilde{\lambda}$ has the form $(d_1^{a_1} \ldots d_s^{a_s})$ for some nonnegative integers $a_1, \ldots, a_s$. We have seen that the representation $E^{\lambda}$ is the quotient of $\operatorname{Sym}^{a_1}(\wedge^{d_1} E) \otimes \ldots \otimes \operatorname{Sym}^{a_s}(\wedge^{d_s} E)$ by the subspace spanned by the quadratic relations. In particular, it follows that the algebra $S^{\bullet}(E; d_1, \ldots, d_s)$ is the direct sum of copies of $E^{\lambda}$, one for each such $\lambda$.

If $n \geq d_1$, the corollary to Theorem 1 in §8.1 gives a canonical isomorphism from $S^{\bullet}(m; d_1, \ldots, d_s)$ to the subalgebra of $\mathbb{C}[Z]$ generated by all $D_T$, where $T$ varies over all tableaux on Young diagrams whose columns have lengths among the numbers $d_1, \ldots, d_s$ and whose entries are in $[m]$; this isomorphism takes $X_{i_1, \ldots, i_p}$ to $D_{i_1, \ldots, i_p}$. Indeed, we have seen that each piece $E^{\lambda}$ maps isomorphically to $D^{\lambda}$, so it suffices to show that the sum of the $D^{\lambda}$ in $\mathbb{C}[Z]$ is direct. This is an immediate consequence of the fact that the $D^{\lambda}$ are nonisomorphic irreducible representations:

**Exercise 17** If a representation $V$ of $GL_m\mathbb{C}$ is a sum of subrepresentations $V_1, \ldots, V_r$, with each $V_i$ irreducible and nonzero, and $V_i$ and $V_j$ nonisomorphic if $i \neq j$, show that $V$ is the direct sum $V_1 \oplus \ldots \oplus V_r$.

In particular, the ring $S^{\bullet}(m; d_1, \ldots, d_s)$ *is an integral domain*, since it is isomorphic to a subring of the polynomial ring $\mathbb{C}[Z]$. Equivalently:

**Proposition 2** *The ideal in* $\mathrm{Sym}^{\bullet}(\wedge^{d_1}E) \otimes \ldots \otimes \mathrm{Sym}^{\bullet}(\wedge^{d_s}E)$ *generated by the quadratic relations is a prime ideal.*

The same is true when $\mathbb{C}$ is replaced by any integral domain $R$, and $E$ is a free $R$-module with $m$ generators. The ring $S^{\bullet}(m; d_1, \ldots, d_s)$ defined to be the quotient of a polynomial ring $R[X_{i_1, \ldots, i_p}]$ by the ideal generated by the quadratic relations (24) is isomorphic to the subring of $R[Z]$ generated by the polynomials $D_T$, and the $D_T$, for $T$ a tableau on a shape with column lengths in $\{d_1, \ldots, d_s\}$ and entries in $[m]$, form a basis. Indeed, the proof is the same once one knows the linear independence of these $D_T$. This is true over $\mathbb{C}$ by the above proof using representation theory; from this it follows when $R = \mathbb{Z}$, and then for any $R$ by base change from $\mathbb{Z}$ to $R$. We will prove more about these rings in §9.2.

# Part III

## Geometry

In this part we apply the results of the first two parts to study the geometry of Grassmannians and flag manifolds. In this introduction we set up the basic notation and describe some important examples.

If $E$ is any finite dimensional vector space, we denote by $\mathbb{P}(E)$ the projective space of lines through the origin in $E$. Such a line is determined by any nonzero vector $v$ in $E$, and such a vector $v$ is determined by the line up to multiplying by a nonzero vector. In other words,

$$\mathbb{P}(E) = E \smallsetminus \{0\} / \mathbb{C}^*.$$

The point in $\mathbb{P}(E)$ determined by $v$ in $E \smallsetminus \{0\}$ is often denoted $[v]$.

We will usually work with the dual projective space $\mathbb{P}^*(E)$ consisting of all hyperplanes $H \subset E$, or equivalently of all one-dimensional quotient spaces $E \twoheadrightarrow L$; two quotient maps $E \twoheadrightarrow L$ and $E \twoheadrightarrow L'$ are identified if there is an isomorphism of $L$ with $L'$ that commutes with the maps from $E$. Equivalently, $\mathbb{P}^*(E) = \mathbb{P}(E^*)$ is the set of lines in the dual space $E^*$; the line in $E^*$ corresponding to the quotient $E \twoheadrightarrow L$ is the dual line $L^* \subset E^*$. The main reason for this "dual" notation is so that $E$ will be the space of linear forms on $\mathbb{P}^*(E)$. In fact, the *symmetric algebra*

$$\mathrm{Sym}^{\bullet} E = \bigoplus_{n=0}^{\infty} \mathrm{Sym}^{n} E$$

is the algebra of polynomial forms on $\mathbb{P}^*(E)$, also called the ***homogeneous coordinate ring of*** $\mathbb{P}^*(E)$. Elements $f$ in $\mathrm{Sym}^n E$ define functions on $E^*$ that are homogeneous of degree $n$: $f(\lambda \cdot v) = \lambda^n \cdot f(v)$. The value of $f$ on a line $L^*$ in $E^*$ is therefore defined only up to a nonzero scalar, which means that we can say only whether $f(L^*) = 0$ or $f(L^*) \neq 0$. A ratio $f/g$ with $f$ and $g$ in $\mathrm{Sym}^n E$ will define a function on the open set in $\mathbb{P}^*(E)$ on which $g$ does not vanish. A homogeneous form on projective space is often called, with some abuse of terminology, a homogeneous function.

When $E$ has a basis $e_1, \ldots, e_m$, so $E^*$ has the dual basis, we write $\mathbb{P}^{m-1}$ for $\mathbb{P}^*(E) = \mathbb{P}(\mathbb{C}^m)$. The point of $\mathbb{P}^{m-1}$ determined by a nonzero vector $(x_1, \ldots, x_m)$ in $\mathbb{C}^m$ is usually denoted $[x_1 : \ldots : x_m]$. The numbers $x_i$ are called **homogeneous coordinates** of the point. The ring $\text{Sym}^{\cdot}E$ can be identified with the polynomial ring $\mathbb{C}[X_1, \ldots, X_m]$. An element of $\text{Sym}^n E$ is a homogeneous polynomial $F$ of degree $n$ in these variables, and its zeros are the points $[x_1 : \ldots : x_m]$ in $\mathbb{P}^{m-1}$ such that $F(x_1, \ldots, x_m) = 0$.

If $W$ is any representation of $GL(E)$, then $GL(E)$ acts on the projective space $\mathbb{P}^*(W)$, since any automorphism of $W$ takes a hyperplane in $W$ to another hyperplane in $W$.

In particular, $GL(E)$ acts on $\mathbb{P}^*(E^\lambda)$. We will find a closed orbit of $GL(E)$ on $\mathbb{P}^*(E^\lambda)$ that can be identified with a flag variety. In fact, if $d_1 > \ldots > d_s$ are the positive numbers that are the lengths of the columns of $\lambda$, we will see that this closed orbit is the **partial flag variety** or **manifold**

$$F\ell^{d_1, \ldots, d_s}(E) =$$

$$\{E_1 \subset \ldots \subset E_s \subset E : \text{codim}(E_i, E) = d_i, 1 \le i \le s\}$$

consisting of chains of linear subspaces of $E$ of the indicated codimensions.

The two extremes are classical: (1) The action of $GL(E)$ on $\mathbb{P}^*(\text{Sym}^n E)$ has an orbit that can be identified with $\mathbb{P}^*(E)$, embedded in $\mathbb{P}^*(\text{Sym}^n E)$ by the **Veronese embedding**

$$\mathbb{P}^*(E) \hookrightarrow \mathbb{P}^*(\text{Sym}^n E), \quad E \twoheadrightarrow L \mapsto \text{Sym}^n E \twoheadrightarrow \text{Sym}^n L.$$

(2) The action of $GL(E)$ on $\mathbb{P}^*(\wedge^n E)$ has an orbit that can be identified with the Grassmannian $Gr^n E = Gr_{m-n}E$ of $n$-dimensional quotient spaces (or $(m-n)$-dimensional subspaces) of $E$, embedded in $\mathbb{P}^*(\wedge^n E)$ by the **Plücker embedding**

$$Gr^n E \hookrightarrow \mathbb{P}^*(\wedge^n E), \quad E \twoheadrightarrow W \mapsto \wedge^n E \twoheadrightarrow \wedge^n W.$$

This uses that fact that if $W$ is an $n$-dimensional quotient space of $E$, then $\wedge^n W$ is a one-dimensional quotient space of $\wedge^n E$ (see §9.1).

**Exercise** Identify $F\ell^{2,1}(E)$ with a closed orbit of $GL(E)$ on $\mathbb{P}^*(E^{(2,1)})$.

In this part we will use a few basic notions from algebraic geometry. An **algebraic subset** of projective space $\mathbb{P}^*(E) = \mathbb{P}^{m-1}$ is a subset that is the set of zeros of a collection of homogeneous forms. For such an algebraic subset $X$, its **ideal** $I(X) = \oplus I(X)_n$ is a homogeneous ideal in $\text{Sym}^{\bullet}E = \mathbb{C}[X_1, \ldots, X_m]$, where $I(X)_n$ consists of the forms of degree $n$ that vanish on $X$; the set $X$ is then the set of zeros of a set of homogeneous generators of $I(X)$. An algebraic subset is **irreducible** if it is not the union of two proper algebraic subsets; such is an (embedded) **projective variety**. Any algebraic set is a union of a finite number of irreducible algebraic subsets; if this is done with a minimum number of irreducible subsets, they are called its **irreducible components**. If $X \subset \mathbb{P}^*(E)$ is irreducible, its ideal is a prime ideal. The graded ring $\text{Sym}^{\bullet}E/I(X)$ is called the **homogeneous coordinate ring** of $X$.

The **Nullstellensatz** states that if $I$ is any homogeneous ideal in $\text{Sym}^{\bullet}E$, and $X$ is the set of zeros of $I$, then $I(X)$ consists of all polynomials $F$ such that some power of $F$ is contained in $I$; in particular, if $I$ is a prime ideal, then $I(X) = I$. If $X$ is a projective variety in $\mathbb{P}^*(E)$, an **algebraic subset of** $X$ is the locus in $X$ defined by a collection of forms in $\text{Sym}^{\bullet}E$, or, equivalently, by a homogeneous ideal in the homogeneous coordinate ring of $X$.

We will also meet subvarieties of products of projective spaces. An algebraic subset of $\mathbb{P}^*(E_1) \times \mathbb{P}^*(E_1) \times \ldots \times \mathbb{P}^*(E_s)$ is the set of zeros of a collection of multihomogeneous polynomials, each in some $\text{Sym}^{a_1}(E_1) \otimes \text{Sym}^{a_2}(E_2) \otimes \ldots \otimes \text{Sym}^{a_s}(E_s)$. When bases are chosen for each of these vector spaces $E_i$, this tensor product is identified with the polynomial ring in the corresponding variables. There is the same notion of irreducibility, of the (multihomogeneous) ideal of a subvariety, and the **multihomogeneous coordinate ring**

$$\text{Sym}^{\bullet}(E_1) \otimes \text{Sym}^{\bullet}(E_2) \otimes \ldots \otimes \text{Sym}^{\bullet}(E_s)/I(X)$$
$$= \text{Sym}^{\bullet}(E_1 \oplus \ldots \oplus E_s)/I(X)$$

of a subvariety $X$.

We will occasionally refer to the **Zariski topology** on projective space, or a product of projective spaces, or an algebraic subvariety. This is the topology whose closed sets are just the algebraic subsets, so the open sets are defined by the nonvanishing of a finite number of homogeneous or multihomogeneous polynomials. There are many more closed or open sets in the usual "classical" topology on the complex manifold $\mathbb{P}^{m-1}$, but these are the only ones needed here. A **closed embedding** of a variety $X$ in a variety $Y$ is an isomorphism of $X$ with a closed subvariety of $Y$.

A finite-dimensional vector space $E$ determines a **_trivial vector bundle_** $E_X = X \times E$ on any variety $X$; we often abuse notation, when the variety $X$ is evident, and denote this bundle simply by $E$.

We will need the notion of the **_dimension_** of an algebraic variety. Any variety has an open subset that is a complex manifold, and its (complex) dimension can be taken to be the dimension of the variety. All the examples we will see, in fact, have an open subset isomorphic to an affine space $\mathbb{C}^r$. If $Z$ is a proper algebraic subset of a variety $Y$, then all the irreducible components of $Z$ have dimension strictly less than the dimension of $Y$.

These facts can be found in any text on algebraic geometry, such as Harris (1992) and Shafarevich (1977) or Hartshorne (1977). A few other basic facts from algebraic geometry will be quoted as needed, but mainly in the exercises. We hope the main discussion will be accessible to those without much background in algebraic geometry.

# 9

# Flag varieties

The rings constructed from representations in Chapter 8 will here be identified with the multihomogeneous coordinate rings of flag varieties for natural embeddings of these varieties in products of projective spaces. They are also rings of invariants of linear groups acting on the ring of polynomial functions on the space of $n \times m$ matrices; these basic invariant theory facts follow easily from what we have proved in representation theory. From this it follows that these rings are unique factorization domains, a fact which has useful applications in algebraic geometry. In §9.3 a link is made to the realization of the representations as sections of line bundles on homogeneous spaces. The last section presents the basic facts about intersection theory on Grassmannians. (The main results proved in this section will also be deduced from more general results on flag manifolds in Chapter 10.)

## 9.1 Projective embeddings of flag varieties

Let $E$ be a vector space of dimension $m$. For $0 < d \leq m$, $Gr^d E$ denotes the Grassmannian of subspaces of $E$ of codimension $d$. In particular, $Gr^1 E = \mathbb{P}^*(E)$ and $Gr^{m-1} E = \mathbb{P}(E)$. If $F$ is a subspace of $E$ of codimension $d$, then the kernel of the map from $\bigwedge^d(E)$ to $\bigwedge^d(E/F)$ is a hyperplane in $\bigwedge^d(E)$. Assigning this hyperplane to $F$ gives a mapping

$$Gr^d E \quad \to \quad \mathbb{P}^*(\textstyle\bigwedge^d E),$$

called the **Plücker embedding.** Note that $\operatorname{Sym}^{\cdot}(\bigwedge^d E)$ is the ring of polynomial functions on $\mathbb{P}^*(\bigwedge^d E)$ (see the introduction to Part III). This means that for any $v_1, \ldots, v_d$ in $E$, $v_1 \wedge \ldots \wedge v_d$ is a linear form on $\mathbb{P}^*(\bigwedge^d E)$, and products of such linear forms are homogeneous forms on $\mathbb{P}^*(\bigwedge^d E)$.

**Lemma 1** *The Plücker embedding is a bijection from $Gr^d E$ to the subvariety of $\mathbb{P}^*(\bigwedge^d E)$ defined by the quadratic equations*

$$(v_1 \wedge \ldots \wedge v_d) \cdot (w_1 \wedge \ldots \wedge w_d) \;-$$

$$\sum_{i_1 < \ldots < i_k} (v_1 \wedge \ldots \wedge w_1 \wedge \ldots \wedge w_k \wedge \ldots \wedge v_d) \cdot$$

$$(v_{i_1} \wedge \ldots \wedge v_{i_k} \wedge w_{k+1} \wedge \ldots \wedge w_d) \;=\; 0,$$

*for $v_1, \ldots, v_d, w_1, \ldots, w_d$, in $E$. Any polynomial vanishing on the image of $Gr^d E$ is in the ideal generated by these quadratic equations.*

Before giving the proof, we reinterpret this result in coordinates. Let $e_1, \ldots, e_m$ be a basis for $E$, thus identifying $E$ with $\mathbb{C}^m$. Then $X_{i_1, \ldots, i_d} = e_{i_1} \wedge \ldots \wedge e_{i_d}$ in $\bigwedge^d E$ is a linear form on $\mathbb{P}^*(\bigwedge^d E)$. These are skew-commutative in the subscripts. Every point of $\mathbb{P}^*(\bigwedge^d E)$ has homogeneous coordinates $x_{i_1, \ldots, i_d}$ as the subscripts vary over all $1 \leq i_1 < \ldots < i_d \leq m$; such coordinates will also be regarded as skew-commutative in the subscripts. For a subspace $V$ of $E = \mathbb{C}^m$, homogeneous coordinates of the corresponding point of $\mathbb{P}^*(\bigwedge^d E)$, called **Plücker coordinates,** are determined as follows. Write $V$ as the kernel of a $d \times m$ matrix $A: \mathbb{C}^m \to \mathbb{C}^d$ of rank $d$. The map $\bigwedge^d A$ from $\bigwedge^d(\mathbb{C}^m)$ to $\bigwedge^d(\mathbb{C}^d) = \mathbb{C}$ takes $e_{i_1} \wedge \ldots \wedge e_{i_d}$ to the determinant of the maximal minor of $A$ using the columns numbered $i_1, \ldots, i_d$. The Plücker coordinate $x_{i_1, \ldots, i_d}$ of the corresponding point in $\mathbb{P}^*(\bigwedge^d E)$ is therefore this determinant.

The relations in Lemma 1 can be written in terms of these coordinates in the form

$$(1) \qquad X_{i_1, \ldots, i_d} \cdot X_{j_1, \ldots, j_d} \;-\; \sum X_{i_1', \ldots, i_{d'}} \cdot X_{j_1', \ldots, j_{d'}},$$

the sum over all pairs obtained by interchanging a fixed set of $k$ of the subscripts $j_1, \ldots, j_d$ with $k$ of the subscripts $i_1, \ldots, i_d$, maintaining the order in each; as usual, it suffices to use the first $k$ subscripts $j_1, \ldots, j_k$ in such an exchange. Rearranging the subscripts in increasing order introduces signs. For example, the lemma asserts that the Grassmannian $Gr^2 \mathbb{C}^4 \subset \mathbb{P}^5$ is defined by the one quadratic relation $X_{1,2} \cdot X_{3,4} = X_{3,2} \cdot X_{1,4} + X_{1,3} \cdot X_{2,4}$, or

$$X_{1,2} \cdot X_{3,4} \;-\; X_{1,3} \cdot X_{2,4} \;+\; X_{2,3} \cdot X_{1,4}, \;=\; 0.$$

The equations (1) are the same as the equations of Lemma 1 when the $v$'s and $w$'s are taken from the basis elements of $E$. Conversely, if these equations hold for basis elements, they hold in general by multilinearity.

**Exercise 1** Show that the equations for $k = 1$ are equivalent to the "classical" equations $\sum_{s=1}^{d+1} (-1)^s X_{i_1, \ldots, i_{d-1}, j_s} \cdot X_{j_1, \ldots \widehat{j_s} \ldots, j_{d+1}} = 0$, for all sequences $i_1, \ldots, i_{d-1}$ and $j_1, \ldots, j_{d+1}$.

**Proof of Lemma 1** It follows from Sylvester's lemma that the coordinates arising from any linear subspace satisfy the quadratic equations (1). Indeed, if the subspace is the kernel of a matrix $A$ as above, apply Lemma 2 of §8.1 with $p = d$, and $M$ and $N$ the minors of $A$ using the columns numbered $i_1, \ldots, i_d$ and $j_1, \ldots, j_d$ respectively. Conversely, suppose we have a point in $\mathbb{P}^*(\wedge^d E)$ whose homogeneous coordinates $x_{i_1, \ldots, i_d}$ satisfy the quadratic equations (1). Fix some $i_i, \ldots, i_d$ with $x_{i_1, \ldots, i_d} \neq 0$. Since we can multiply by a nonzero scalar without changing the point, we may assume that $x_{i_1, \ldots, i_d} = 1$. Define a $d \times m$ matrix $A = (a_{s,t})$ by the formula

$$(2) \qquad a_{s,t} = x_{i_1, \ldots, i_{s-1}, t, i_{s+1}, \ldots, i_d}, \quad 1 \le s \le d, \quad 1 \le t \le m.$$

The claim is that the kernel of $A: \mathbb{C}^m \to \mathbb{C}^d$ is a subspace of codimension $d$ whose Plücker coordinates are the given $x_{j_1, \ldots, j_d}$. To see this, let $I = (i_1, \ldots, i_d)$, and consider the determinants of the minors corresponding to all $J = (j_1, \ldots, j_d)$. For $J = I$ the corresponding submatrix is the identity matrix, which shows in particular that the rank of $A$ is $d$, and that the corresponding determinant is 1. When $I$ and $J$ have $d-1$ entries in common, the determinants of the corresponding minors are seen to be the other entries of $A$, which is the correct answer in this case; indeed, if $J$ is obtained from $I$ by replacing $i_s$ by an integer $t$, then the corresponding minor looks like the identity matrix except in the $s^{\text{th}}$ column, whose entry on the diagonal is $a_{s,t}$. For other $J$ we argue by descending induction on the number of common entries in $I$ and $J$. Suppose $j_r$ does not occur in $I$. The quadratic relation (1) for $k = 1$ and this $I$ and $J$, making the exchanges with $j_r$, then writes $x_{j_1, \ldots, j_d}$ as a linear combination of products of coordinates that are known, since their subscripts have larger intersection with $I$ than $J$ does. By what we saw at the beginning of the proof, the same identity holds for the corresponding determinants of minors of $A$. It therefore follows that $x_{j_1, \ldots, j_d}$ is the corresponding minor determinant of $A$, as asserted.

To see that $Gr^d E \to \mathbb{P}^*(\wedge^d E)$ is injective, since the map does not depend on a choice of a basis of $E$, it suffices to observe that the spaces $\langle e_{p+1}, \ldots, e_m \rangle$ and $\langle e_1, \ldots, e_r, e_{p+r+1}, \ldots, e_m \rangle$ have different Plücker coordinates, if $r \ge 1$.

To prove the last assertion in the lemma, we use the Nullstellensatz, which says that if $\mathfrak{p}$ is any prime ideal in a polynomial ring such as $\mathrm{Sym}^{\cdot}(\bigwedge^d E) = \mathbb{C}[X_{i_1,\ldots,i_d}]$, then the ideal of polynomials vanishing on the set of zeros of polynomials in $\mathfrak{p}$ is $\mathfrak{p}$ itself. We have seen in §8.4 that the quadratic relations generate a prime ideal $\mathfrak{p}$. Since we have just realized the Grassmannian as the zeros of the ideal generated by the quadratic relations, this finishes the proof.     □

We have therefore identified the homogeneous coordinate ring of $Gr^d E \subset \mathbb{P}^*(\bigwedge^d E)$ with the ring

$$S^{\cdot}(m;d) \;=\; \mathrm{Sym}^{\cdot}(\textstyle\bigwedge^d E)/Q \;=\; \mathbb{C}[X_{i_1,\ldots,i_d}]/Q,$$

where $Q$ is the ideal generated by the quadratic relations.

**Exercise 2** Find the subspace of $\mathbb{C}^4$ with Plücker coordinates $x_{1,2} = 1$, $x_{1,3} = 2$, $x_{1,4} = 1$, $x_{2,3} = 1$, $x_{2,4} = 2$, and $x_{3,4} = 3$.

Consider next pairs $(V, W)$ of subspaces of $E$ of codimensions $p$ and $q$, with $p \geq q$. These are parametrized by the product of the Grassmannians $Gr^p E$ and $Gr^q E$, which is a subvariety of the product of projective spaces $\mathbb{P}^*(\bigwedge^p E) \times \mathbb{P}^*(\bigwedge^q E)$. We next ask what equations on the Plücker coordinates of $V$ and $W$ correspond to the condition that $V$ be contained in $W$. The answer yet again will be the quadratic equations.

Let $F\ell^{p,q}(E) \subset Gr^p E \times Gr^q E$ be the ***incidence variety*** of spaces $(V, W)$ of codimensions $p$ and $q$ with $V \subset W$. Note that for vectors $v_1, \ldots, v_p$ in $E$, $v_1 \wedge \ldots \wedge v_p \in \bigwedge^p E$ is a linear form on $Gr^p E \subset \mathbb{P}^*(\bigwedge^p E)$, and similarly, for any $w_1, \ldots, w_q$ in $E$, $w_1 \wedge \ldots \wedge w_q$ is a linear form on $Gr^q E \subset \mathbb{P}^*(\bigwedge^q E)$. Therefore products such as $(v_1 \wedge \ldots \wedge v_p) \cdot (w_1 \wedge \ldots \wedge w_q)$ are bihomogeneous on $\mathbb{P}^*(\bigwedge^p E) \times \mathbb{P}^*(\bigwedge^q E)$.

**Lemma 2** *The incidence variety $F\ell^{p,q}(E)$ is defined in $Gr^p E \times Gr^q E$ by the quadratic equations*

$$(v_1 \wedge \ldots \wedge v_p) \cdot (w_1 \wedge \ldots \wedge w_q) \;-$$

$$\sum_{i_1 < \ldots < i_k} (v_1 \wedge \ldots \wedge w_1 \wedge \ldots \wedge w_k \wedge \ldots \wedge v_p) \cdot$$

$$(v_{i_1} \wedge \ldots \wedge v_{i_k} \wedge w_{k+1} \wedge \ldots \wedge w_q) \;=\; 0,$$

*for $v_1, \ldots, v_p$ in $E$ and $w_1, \ldots, w_q$ in $E$, $1 \leq k \leq q$.*

As usual, the displayed sum is over all exchanges of the first $k$ of the $w_j$'s with $k$ of the $v_i$'s, preserving the order in each. In terms of the homogeneous coordinates, these can be written in the form

$$(3) \qquad X_{i_1,\dots,i_p} {\cdot} X_{j_1,\dots,j_q} - \sum X_{i_1',\dots,i_p'} {\cdot} X_{j_1',\dots,j_q'} = 0,$$

the sum over all pairs obtained by interchanging the first $k$ of the $j$ subscripts with $k$ of the $i$ subscripts, maintaining the order in each.

**Proof**  Both the incidence variety and the zeros of the quadratic equations are preserved by the action of $GL(E)$; for the zeros of the quadratic equations, this is evident from the description in the statement of Lemma 2. We may therefore take any convenient basis for $E$. For example if we are given subspaces $V \subset W$, we can take a basis so that $V = \langle e_{p+1}, \dots, e_m \rangle$ is spanned by the last $m - p$ basis vectors, and $W = \langle e_{q+1}, \dots, e_m \rangle$ by the last $m - q$ basis vectors. In this case each has only one nonzero Plücker coordinate, namely $x_{1,\dots,p}$ and $x_{1,\dots,q}$, respectively, and the validity of the relations (3) is obvious. Conversely, if $V \not\subset W$, we may take

$$V = \langle e_1, \dots, e_r, e_{p+r+1}, \dots, e_m \rangle, \quad \text{and} \quad W = \langle e_{q+1}, \dots, e_m \rangle$$

for some $r \geq 1$. Then a calculation shows that the quadratic equation (3) fails, with $k = 1$, for $I = (r+1, \dots, r+p)$ and $J = (1, \dots, q)$.    $\square$

Now fix a sequence of integers $m \geq d_1 > \dots > d_s \geq 0$. The **partial flag variety** $F\ell^{d_1,\dots,d_s}(E)$ is the set of flags (i.e., nested subspaces)

$$\{E_1 \subset E_2 \subset \dots \subset E_s \subset E : \operatorname{codim}(E_i) = d_i, \ 1 \leq i \leq s\}.$$

This is a subset of the product $Gr^{d_1}E \times \dots \times Gr^{d_s}E$ of Grassmannians, and hence, via the Plücker embeddings, of the product of projective spaces

$$\prod_{i=1}^{s} \mathbb{P}^*(\wedge^{d_i}E) = \mathbb{P}^*(\wedge^{d_1}E) \times \dots \times \mathbb{P}^*(\wedge^{d_s}E).$$

**Proposition 1**  *The flag variety $F\ell^{d_1,\dots,d_s}(E) \subset \prod_{i=1}^{s} \mathbb{P}^*(\wedge^{d_i}E)$ is the locus of zeros of the quadratic equations* (3), *for $p \geq q$ in $\{d_1, \dots, d_s\}$. These equations generate the prime ideal of all polynomials vanishing on the flag variety.*

**Proof**  It follows from the two lemmas that the flag variety is set-theoretically defined by the quadratic equations. But we have seen in §8.4 that these relations generate a prime ideal in the polynomial ring

$$\text{Sym}^{\cdot}(\wedge^{d_1}E) \otimes \ldots \otimes \text{Sym}^{\cdot}(\wedge^{d_s}E) = \text{Sym}^{\cdot}(\wedge^{d_1}E \oplus \ldots \oplus \wedge^{d_s}E)$$

$$= \mathbb{C}[X_{i_1,\ldots,i_p}], \quad 1 \le i_1 < \ldots < i_p \le m, \quad p \in \{d_1, \ldots, d_s\}.$$

The last assertion therefore follows from the Nullstellensatz.                          □

This identifies the multihomogeneous coordinate ring of the flag variety with the ring denoted $S^{\cdot}(m; d_1, \ldots, d_s)$ in §8.4. The proof shows that the flag varieties are set-theoretically defined by the quadratic relations for $k = 1$, this implies by the Nullstellensatz that the prime ideal generated by all quadratic relations is the radical of the ideal generated by those for $k = 1$. These ideals are equal in characteristic zero (see Exercise 13 of §8.3), but not always in positive characteristic (Towber [1979]; Abeasis [1980]).

We have seen that, if $\lambda$ is a partition whose conjugate has the form $(d_1{}^{a_1} \ldots d_s{}^{a_s})$ for some positive integers $a_1, \ldots, a_s$, then the kernel of the surjection $\otimes_{i=1}^{s} \text{Sym}^{a_i}(\wedge^{d_i}E) \twoheadrightarrow E^{\lambda}$ is defined by the same quadratic relations that cut out the above partial flag variety. To see what this says geometrically we need three basic constructions from projective geometry:

(i) A surjection $V \twoheadrightarrow W$ of vector spaces determines an embedding $\mathbb{P}^*(W) \subset \mathbb{P}^*(V)$, taking a hyperplane in $W$ to its inverse image in $V$; equivalently, a surjection $W \twoheadrightarrow L$ to a line determines the surjection $V \twoheadrightarrow W \twoheadrightarrow L$.

(ii) The $a$-fold **Veronese embedding** $\mathbb{P}^*(V) \subset \mathbb{P}^*(\text{Sym}^a V)$, which takes the hyperplane that is the kernel of a surjection $V \twoheadrightarrow L$ to the kernel of the induced surjection $\text{Sym}^a V \twoheadrightarrow \text{Sym}^a L$.

(iii) The **Segre embedding** $\mathbb{P}^*(V_1) \times \ldots \times \mathbb{P}^*(V_s) \subset \mathbb{P}^*(V_1 \otimes \ldots \otimes V_s)$, which takes the kernels of surjections $V_i \twoheadrightarrow L_i$ to the kernel of the induced surjection $V_1 \otimes \ldots \otimes V_s \twoheadrightarrow L_1 \otimes \ldots \otimes L_s$.

**Exercise 3** Show that each of these is a closed embedding. Find equations for the images.

The product of the $a_i$-fold Veronese embeddings gives an embedding

$$\mathbb{P}^*(\wedge^{d_1}E) \times \ldots \times \mathbb{P}^*(\wedge^{d_s}E) \subset$$
$$\mathbb{P}^*(\text{Sym}^{a_1}(\wedge^{d_1}E)) \times \ldots \times \mathbb{P}^*(\text{Sym}^{a_s}(\wedge^{d_s}E)).$$

This can be followed by the Segre embedding

$$\mathbb{P}^*(\text{Sym}^{a_1}(\wedge^{d_1}E)) \times \ldots \times \mathbb{P}^*(\text{Sym}^{a_s}(\wedge^{d_s}E)) \subset \mathbb{P}^*\left(\bigotimes_{i=1}^{s} \text{Sym}^{a_i}(\wedge^{d_i}E)\right).$$

The surjection $\otimes_{i=1}^{s} \mathrm{Sym}^{a_i}(\wedge^{d_i}E) \to E^{\lambda}$ determines an embedding

$$\mathbb{P}^*(E^{\lambda}) \subset \mathbb{P}^*\left(\overset{s}{\underset{i=1}{\otimes}} \mathrm{Sym}^{a_i}(\wedge^{d_i}E)\right).$$

The fact that the same equations define the flag manifold $F\ell^{d_1,\ldots,d_s}(E)$ in $\prod_{i=1}^{s} \mathbb{P}^*(\mathrm{Sym}^{a_i}(\wedge^{d_i}E))$ as define $\mathbb{P}^*(E^{\lambda})$ in $\mathbb{P}^*(\otimes_{i=1}^{s} \mathrm{Sym}^{a_i}(\wedge^{d_i}E))$ means first that we have a commutative diagram

(4)

$$
\begin{array}{ccc}
F\ell^{d_1,\ldots,d_s}(E) \subset \prod\limits_{i=1}^{s} Gr^{d_i}(E) & \subset & \prod\limits_{i=1}^{s} \mathbb{P}^*(\wedge^{d_i}(E)) \\
& & \cap \\
\cap & & \prod\limits_{i=1}^{s} \mathbb{P}^*(\mathrm{Sym}^{a_i}(\wedge^{d_i}E)) \\
& & \cap \\
\mathbb{P}^*(E^{\lambda}) & \subset & \mathbb{P}^*\left(\overset{s}{\underset{i=1}{\otimes}} \mathrm{Sym}^{a_i}(\wedge^{d_i}E)\right).
\end{array}
$$

In fact, this shows that the flag variety $F\ell^{d_1,\ldots,d_s}(E)$ is the intersection of $\mathbb{P}^*(E^{\lambda})$ and $\prod_{i=1}^{s} \mathbb{P}^*(\wedge^{d_i}E)$ inside $\mathbb{P}^*(\otimes_{i=1}^{s}\mathrm{Sym}^{a_i}(\wedge^{d_i}E))$. Moreover, this is a *scheme-theoretic* intersection, which means that the ideal defininig $F\ell^{d_1,\ldots,d_s}(E)$ is the sum of the ideals defining $\mathbb{P}^*(E^{\lambda})$ and $\prod_{i=1}^{s} \mathbb{P}^*(\wedge^{d_i}E)$.

## 9.2 Invariant theory

With $\mathbb{C}[Z] = \mathbb{C}[Z_{1,1},\ldots,Z_{n,m}]$ the polynomial functions on the space of $n \times m$ matrices, as in Chapter 8, the group $GL_n\mathbb{C}$ acts on the *right* on $\mathbb{C}[Z]$, by the formula

$$Z_{i,j} \cdot g = \sum_{k=1}^{n} g_{i,k} Z_{k,j}, \quad g = (g_{i,j}) \in GL_n\mathbb{C}.$$

This is the action of $GL_n\mathbb{C}$ on functions by $(f \cdot g)(A) = f(g \cdot A)$, for $A$ a matrix, $g$ in $GL_n\mathbb{C}$, and $f$ a function. In this setting, the basic problem of invariant theory is to describe the ring of invariants $\mathbb{C}[Z]^{SL_n\mathbb{C}}$ by the subgroup $SL_n\mathbb{C}$ of matrices of determinant 1. For any $i_1,\ldots,i_n$ from $[m]$ the determinant $D_{i_1,\ldots,i_n}$ defined in §8.1 is an invariant, and the *first fundamental theorem of invariant theory* asserts that these determinants generate the ring of invariants. Equivalently, the ring of invariants is the ring we have denoted $S^{\bullet}(m;n)$. The *second fundamental theorem* says that the quadratic relations give all the relations among these generators. In other words,

**Proposition 2** *The ring of invariants* $\mathbb{C}[Z]^{SL_n\mathbb{C}}$ *is*

$$\mathbb{C}[D_{i_1,\ldots,i_n}]_{1 \leq i_1 < \ldots < i_n \leq m} = \mathbb{C}[X_{i_1,\ldots,i_n}]/Q,$$

*where* $Q$ *is generated by the quadratic equations* (3) *with* $p = q = n$.

**Proof** From the identification of the ring generated by the $D_{i_1,\ldots,i_n}$ as the ring $S^{\bullet}(m;n)$, which is the sum of the representations $E^{\lambda}$, one for each $\lambda$ of the form $(\ell^n)$, we know that the dimension of the subspace of homogeneous polynomials of degree $a$ in the subring of $\mathbb{C}[Z]$ generated by the $D_{i_1,\ldots,i_n}$ is the dimension $d_{\lambda}(m)$ of $E^{\lambda}$, $\lambda = (\ell^n)$, with $\ell \cdot n = a$. To prove the theorem, it suffices to show that the space of invariant polynomials of degree $a$ has the same dimension, since the space generated by the specified determinants is a subspace of the invariants.

For this we will apply what we know about representations of $GL_n\mathbb{C}$. These were defined to be left actions, but one can turn the right action into a left one simply by defining $g \cdot f = f \cdot g^{\tau}$, i.e., $(g \cdot f)(A) = f(g^{\tau} \cdot A)$, where $g^{\tau}$ is the transpose of $g$, and $f$ is any function in $\mathbb{C}[Z]$. The ring of invariant functions for $SL_n\mathbb{C}$ is obviously the same for this left action. Let $V = \mathbb{C}^n$, with the standard left action of $GL_n\mathbb{C}$. Then $\mathbb{C}[Z]$ can be identified with the symmetric algebra $\text{Sym}^{\bullet}(V^{\oplus m})$, with $Z_{i,j}$ corresponding to the $i^{\text{th}}$ basis element of the $j^{\text{th}}$ copy of $V$. We know how to decompose $\text{Sym}^a(V^{\oplus m})$ into its irreducible factors over $GL_n\mathbb{C}$. The only factors that will be invariant under $SL_n\mathbb{C}$ are those corresponding to factors of the form $(\wedge^n V)^{\otimes \ell}$. There are such invariants only when $a = \ell \cdot n$, and then, by Corollary 3(a) in §8.3, the dimension of this space is $d_{\lambda}(m)$, with $\lambda = (\ell^n)$. This completes the proof.  □

**Corollary** *The ring* $S^{\bullet}(m;n)$ *is a unique factorization domain.*

**Proof** Let $G = SL_n\mathbb{C}$. Given $f \in \mathbb{C}[Z]^G$, factor it into irreducible polynomials in the polynomial ring $\mathbb{C}[Z]$: $f = \prod f_i^{m_i}$. It suffices to show that each irreducible factor $f_i$ is invariant under $G$. Since $f$ is invariant, any $g$ in $G$ must permute the factors, up to scalars. The subgroup of $G$ taking $f_i$ to a multiple of itself is therefore a closed subgroup of finite index in $G$. Since $G = SL_n\mathbb{C}$ is connected, this subgroup must be all of $G$, for otherwise $G$ would be a disjoint union of the cosets. Therefore $g \cdot f_i = \chi(g) f_i$ for some nonzero scalar $\chi(g)$. The mapping $g \mapsto \chi(g)$ is a (holomorphic) homomorphism from $G$ to $\mathbb{C}^*$. But, as we have seen, $SL_n\mathbb{C}$ has no nontrivial one-dimensional representations, i.e., no nontrivial characters, so $\chi$ must be identically 1 and $f_i$ must be an invariant. □

As sketched in the following exercise, the same is true for all the rings $S^{\cdot}(m; d_1, \ldots, d_s)$. Other properties of these rings, such as the fact that they are Cohen–Macaulay, are also proved by realizing them as rings of invariants; cf. Kraft (1984).

**Exercise 4** For any $m \geq d_1 > \ldots > d_s \geq 0$, let $n = d_1$, $V = \mathbb{C}^n$; let $V_i \subset V$ be the span of the first $d_i$ basis elements; and let $G(d_1, \ldots, d_s)$ be the subgroup of $GL(V)$ that maps each $V_i$ to itself with the determinant of each restriction $V_i \to V_i$ equal to 1. (a) Show that $S^{\cdot}(m; d_1, \ldots, d_s)$ is the ring of invariants $\mathbb{C}[Z]^{G(d_1, \ldots, d_s)}$. (b) Deduce that $S^{\cdot}(m; d_1, \ldots, d_s)$ is a unique factorization domain.

In the rest of this section we sketch some applications of these facts to algebraic geometry. For this we assume some knowledge of algebraic geometry, but these results will not be needed later.

It is a consequence of Proposition 2 that every hypersurface of $Gr^n E$ is defined by one homogeneous polynomial on the ambient projective space $\mathbb{P}^*(\bigwedge^n E)$. This is a general fact:

**Exercise 5** Suppose $X \subset \mathbb{P}^n$ is a subvariety whose homogeneous coordinate ring is a unique factorization domain. Show that every subvariety of codimension one in $X$ is cut out by a hypersurface in the ambient space. Deduce from this the fact that, if $X$ is not a point, the divisor class group of $X$ is $\mathbb{Z}$, generated by a hyperplane section.

This fact is used in one of the standard ways of parametrizing subvarieties of given dimension $k$ in a projective space $\mathbb{P}(E) = \mathbb{P}^{m-1}$, dating back to Cayley and Severi. Assume $k < m-1$, and set $n = k+1$. Given a variety $Z \subset \mathbb{P}(E)$ of dimension $k$, define a subset $H_Z$ of $Gr^n E$ by

$$H_Z = \{F \in Gr^n E : \mathbb{P}(F) \cap Z \neq \emptyset\}.$$

**Exercise 6** (a) Show that $H_Z$ is an irreducible subvariety of codimension one in $Gr^n E$. (b) Show that the degree of a hypersurface in $\mathbb{P}^*(\bigwedge^n E)$ whose intersection with $Gr^n E$ is $H_Z$ is equal to the degree of $Z$ in $\mathbb{P}(E)$.

The variety $Z$ is in fact determined by $H_Z$, so this gives an embedding

$$\{Z \subset \mathbb{P}(E) \text{ of dimension } k \text{ and degree } d\} \subset \mathbb{P}(A_d),$$

where $A_d = S^d(m;n)$ is the part of the homogeneous coordinate ring of $Gr^n E \subset \mathbb{P}^*(\wedge^n E)$ of degree $d$. We know that $A_d$ has a basis corresponding to tableaux on $(d^n)$ with entries in $[m]$. The closure of this locus is the **Chow variety** of cycles of dimension $k$ and degree $d$.

**Exercise 7** Suppose $X \subset \mathbb{P}^{n_1} \times \ldots \times \mathbb{P}^{n_r}$ is a subvariety whose multi-homogeneous coordinate ring is a unique factorization domain. (a) Show that any subvariety of codimension one in $X$ is cut out by a hypersurface in the ambient space. Assume that no projection from $X$ to a factor $\mathbb{P}^{n_i}$ is constant. (b) Show that the group of divisor classes on $X$ is isomorphic to $\mathbb{Z}^{\oplus r}$, with a basis coming from hyperplane sections from the factors. (c) Deduce that every hypersurface in $F\ell^{d_1, \ldots, d_s}(E)$ is cut out by a hypersurface in $\prod_{i=1}^s \mathbb{P}^*(\wedge^{d_i}(E))$, and the divisor class group of the flag variety $F\ell^{d_1, \ldots, d_s}(E)$ is free of rank $s$, if $m > d_1 > \ldots > d_s > 0$.

## 9.3 Representations and line bundles

There is a general procedure for producing representations as sections of a line bundle on a homogeneous space (which is a space on which a Lie group acts transitively). Our goal in this section is to see this explicitly in the case of the group $G = GL(E)$, with the homogeneous spaces being partial flag manifolds. For complete proofs some algebraic geometry is needed, but again, the results are not needed elsewhere.

For any irreducible representation $V$ of $G = GL(E)$, there is a dual action on $V^*$ by $(g \cdot \varphi)(v) = \varphi(g^{-1} \cdot v)$, which gives the induced action on $\mathbb{P}^*(V)$. Take a **lowest weight vector** $\varphi$ for $V^*$, which is a weight vector that is preserved by the group of *lower* triangular matrices; equivalently, for the Lie algebra action, $E_{i,j} \cdot \varphi = 0$ for all $i > j$, where $E_{i,j}$ is the matrix with a 1 in the $(i,j)$ position, and 0 elsewhere. Let $[\varphi]$ be the point in $\mathbb{P}^*(V)$ defined by $\varphi$. The corresponding **parabolic subgroup** $P$ is

$$P = \{g \in G : g \cdot \varphi \in \mathbb{C} \cdot \varphi\} = \{g \in G : g \cdot [\varphi] = [\varphi]\}.$$

The coset space $G/P$ is identified with the orbit $G \cdot [\varphi] \subset \mathbb{P}^*(V)$. It is a general fact that $G/P$ is compact, and this orbit is a closed subvariety of $\mathbb{P}^*(V)$. Knowing all irreducible representations of $G$ explicitly, we can see these facts directly as follows. Note first that there is a canonical isomorphism of $\mathbb{P}^*(V)$ with $\mathbb{P}^*(V \otimes M)$ for any one-dimensional representation $M$, given by the map that takes a quotient $V \twoheadrightarrow L$ of $V$ to the quotient $V \otimes M \twoheadrightarrow L \otimes M$ of $V \otimes M$. By Theorem 2 of §8.2 we may therefore assume that $V = E^\lambda$.

Taking a basis for $E$, we have a basis $\{e_T\}$ for $V$, and therefore a dual basis $\{e_T^*\}$ for $V^*$. A lowest weight vector is $e_T^*$, where $T = U(\lambda)$ is the tableau with all $i$'s in the $i^{\text{th}}$ row. Suppose the conjugate partition $\tilde\lambda$ is $(d_1{}^{a_1} \ldots d_s{}^{a_s})$, with $m \geq d_1 > \ldots > d_s \geq 1$, $a_i > 0$; that is, the $d_i$'s are the lengths of the columns of the Young diagram of $\lambda$, and $a_i$ is the number of columns of length $d_i$.

**Exercise 8** Show that $E_{i,j} \cdot (e_{U(\lambda)}^*) = 0$ if and only if $i > j$, or $i < j$ and $i$ and $j$ lie in one of the intervals $[1, d_s]$, $[d_s + 1, d_{s-1}]$, ..., $[d_2 + 1, d_1]$, $[d_1 + 1, m]$.

The Lie algebra $\mathfrak{p}$ of $P$ is the sum of the subalgebra $\mathfrak{h}$ and the one-dimensional spaces $\mathfrak{g}_{i,j} = \mathbb{C} \cdot E_{i,j}$ for those $i$ and $j$ in the preceding exercise. It follows that $P$ is the subgroup of $GL_m\mathbb{C}$ of those $g = (g_{i,j})$ such that $g_{i,j} = 0$ if $i < j$ and the interval $[i, j-1]$ contains some $d_k$; a matrix in $P$ has invertible matrices in blocks of size $d_s$, $d_{s-1} - d_s$, ..., $d_1 - d_2$, and $m - d_1$ down the diagonal, with arbitrary entries below these blocks:[1]

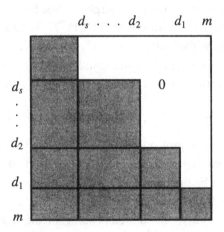

Let $Z_1 \subset Z_2 \subset \ldots \subset Z_s \subset E$ be the flag defined by

$$Z_i = \langle e_{d_i+1}, e_{d_i+2}, \ldots, e_m \rangle,$$

---

[1] The reader more comfortable with groups than Lie algebras can verify this directly by using the group elements $I + E_{i,j}$ where we used the Lie algebra elements $E_{i,j}$.

Then $P$ is exactly the subgroup fixing this flag:

$$P = \{g \in GL_m\mathbb{C} : g(Z_i) \subset Z_i \quad \text{for} \quad 1 \le i \le s\}.$$

Since $GL_m\mathbb{C}$ acts transitively on the set of all flags of fixed dimensions, the map that sends the coset of $g$ to the flag $g{\cdot}Z_1 \subset \ldots \subset g{\cdot}Z_s$ identifies $G/P$ with the flag manifold $F\ell^{d_1,\ldots,d_s}(E)$.

To see that these flag manifolds are the only closed orbits in $\mathbb{P}^*(E^\lambda)$, note first that any such orbit must contain a point fixed by the subgroup of lower triangular matrices (see Exercise 10.1). The only such point is the point $[\varphi]$ determined by a lowest weight vector $\varphi = e_{U(\lambda)}{}^*$, and since the orbit contains $[\varphi]$, it must be $G{\cdot}[\varphi]$, which we have seen is the partial flag manifold.

**Exercise 9** Show that this realization of $F\ell^{d_1,\ldots,d_s}(E)$ in $\mathbb{P}^*(E^\lambda)$ agrees with that found in §9.1.

The irreducible representation $E^\lambda$ can be realized as the space of sections of a line bundle $L^\lambda$ on the flag variety $G/P$. To see this we need some standard facts about line bundles. On any projective space $\mathbb{P}^*(V)$ there is a hyperplane line bundle $\mathcal{O}_V(1)$; its fiber over a point described by a quotient line $V \twoheadrightarrow L$ is the line $L$. The canonical map from $V$ to the space of (regular, or algebraic) sections $\Gamma(\mathbb{P}^*(V), \mathcal{O}_V(1))$ is an isomorphism (see Exercise 11 below).

Write $\mathcal{O}_V(n)$ for the tensor power $\mathcal{O}_V(1)^{\otimes n}$. For any subvariety $X$ of $\mathbb{P}^*(V)$ let $\mathcal{O}_X(n)$ be the restriction of $\mathcal{O}_V(n)$ to $X$. There is a canonical map from $V$ to $\Gamma(X, \mathcal{O}_X(1))$, and from $\operatorname{Sym}^n V$ to $\Gamma(X, \mathcal{O}_X(n))$. More generally, on a subvariety $X$ of a product $\prod_{i=1}^{s} \mathbb{P}^*(V_i)$ there are line bundles

$$\mathcal{O}_X(a_1, \ldots, a_s) = (\mathrm{pr}_1)^* \mathcal{O}_{V_1}(a_1) \otimes \ldots \otimes (\mathrm{pr}_s)^* \mathcal{O}_{V_s}(a_s),$$

where $\mathrm{pr}_i$ denotes the projection from $X$ to the $i^{\text{th}}$ factor. There are canonical maps from $\operatorname{Sym}^{a_1} V_1 \otimes \ldots \otimes \operatorname{Sym}^{a_s} V_s$ to $\Gamma(X, \mathcal{O}_X(a_1, \ldots, a_s))$.

We define $L^\lambda$ to be the bundle $\mathcal{O}_{G/P}(1)$, for the embedding of the partial flag manifold $G/P = F\ell^{d_1,\ldots,d_s}(E)$ in $\mathbb{P}^*(E^\lambda)$. We claim first that $L^\lambda = \mathcal{O}_{G/P}(a_1, \ldots, a_s)$, for the embedding of the flag variety $G/P = F\ell^{d_1,\ldots,d_s}(E)$ in $\prod_{i=1}^{s} \mathbb{P}^*(\wedge^{d_i} V_i)$. This follows from the diagram (4) at the end of §9.1 and the following exercise.

**Exercise 10** Prove this assertion by showing that, in the three canonical embeddings (i)–(iii) used in constructing the diagram (4), the hyperplane bundles restrict as follows: (i) $\mathcal{O}_V(1)$ restricts to $\mathcal{O}_W(1)$; (ii) $\mathcal{O}_{\operatorname{Sym}^a V}(1)$ restricts to $\mathcal{O}_V(a)$; (iii) $\mathcal{O}_{\otimes V_i}(1)$ restricts to the tensor product $\mathcal{O}(1, \ldots, 1)$ of the pullbacks of the bundles $\mathcal{O}_{V_i}(1)$ on the factors $\mathbb{P}^*(V_i)$.

To prove that the canonical map from $E^\lambda$ to $\Gamma(G/P, L^\lambda)$ is an isomorphism, it suffices to invoke the following general fact:

**Exercise 11** If $X \subset \mathbb{P}^*(V)$ is a subvariety whose homogeneous coordinate ring is a unique factorization domain, and $L = \mathcal{O}_X(1)$, show that the canonical map $V \to \Gamma(X, L)$ is surjective, and an isomorphism if $X$ is not contained in any hyperplane in $\mathbb{P}^*(V)$. More generally, if $X$ is a subvariety of a product $\prod_{i=1}^s \mathbb{P}^*(V_i)$, and its multihomogeneous coordinate ring is a unique factorization domain, show that the canonical maps from $\otimes \operatorname{Sym}^{a_i}(V_i)$ to $\Gamma(X, \mathcal{O}_X(a_1, \ldots, a_s))$ are surjective for all nonnegative integers $a_1, \ldots, a_s$.

The partial flag manifold $X = F\ell^{d_1, \ldots, d_s}(E)$ has a **universal**, or **tautological**, flag of subvector bundles of the trivial bundle $E_X = X \times E$:

$$U_1 \subset U_2 \subset \ldots \subset U_s \subset E_X, \quad \operatorname{rank}(U_i) = m - d_i.$$

At a point corresponding to a flag $E_1 \subset \ldots \subset E_s \subset E$, the fiber of the bundle $U_i$ is just the subspace $E_i$ of $E$. For example, on $X = \mathbb{P}^*(V)$ the bundle $\mathcal{O}(1)$ is the quotient bundle of the trivial bundle $V_X$ by the tautological subbundle of hyperplanes. On a Grassmannian $Gr^n E$, if $U$ is the universal subbundle, there is a canonical map $\wedge^n E \to \wedge^n(E/U)$, which is the pullback of the canonical map $\wedge^n E \to \mathcal{O}(1)$ via the Plücker embedding of $Gr^n E$ in $\mathbb{P}^*(\wedge^n E)$. It follows that on $X = F\ell^{d_1, \ldots, d_s}(E)$,

(5) $\qquad L^\lambda = \mathcal{O}_X(a_1, \ldots, a_s)$

$$= \wedge^{d_1}(E/U_1)^{\otimes a_1} \otimes \ldots \otimes \wedge^{d_s}(E/U_s)^{\otimes a_s}.$$

In the language of group theory, this line bundle can be constructed by the following general construction. For any character $\chi : P \to \mathbb{C}^*$ define a line bundle $L(\chi)$ over $G/P$ as a quotient space

$$L(\chi) = G \times^P \mathbb{C} = G \times \mathbb{C} / (g \cdot p \times z) \sim (g \times \chi(p)z)$$

for $g \in G$, $p \in P$, $z \in \mathbb{C}$. There is a canonical projection from $L(\chi)$ to $G/P$, taking a pair $(g \times z)$ to the left coset $gP$ of $g$. The group $G$ acts on $L(\chi)$, by the left action of $G$ on the first factor, so that the projection to $G/P$ commutes with the action of $G$. That is, $L(\chi)$ is an **equivariant line bundle**. Conversely, if $L$ is any equivariant line bundle, then $P$ acts on the left on the fiber of $L$ over the point $eP$ that is fixed by $P$. This action of an element $p$ in $P$ must be by multiplication by an element $\chi(p)$, where $\chi : P \to \mathbb{C}^*$ is a homomorphism.

**Exercise 12** With $\chi$ constructed as above from the equivariant line bundle $L$, show that $L$ is isomorphic to $L(\chi)$. Show that the character constructed in this way from $L(\chi)$ is $\chi$.

The fixed point $x$ of $P$ on $X = F\ell^{d_1, \dots, d_s}(E)$ is the given fixed flag $Z_1 \subset \dots \subset Z_s \subset E$, where $Z_i$ is spanned by the last $m - d_i$ basic vectors. The fiber of $\bigwedge^{d_i}(E/U_i)$ at $x$ is the line $\bigwedge^{d_i}(E/Z_i)$. The image of $e_1 \wedge \dots \wedge e_{d_i}$ in $\bigwedge^{d_i}(E/Z_i)$ is a generator of this line. An element $p$ in $P$ acts by multiplying this element by the determinant of the upper left $d_i \times d_i$ corner of $p$. This means that $\bigwedge^{d_i}(E/U_i) = L(\chi)$, where $\chi(g) = \det(A_i)$, and $A_i$ is the upper left $d_i \times d_i$ corner of the matrix for $g$. Therefore

$$(6) \qquad L^\lambda = L(\chi_\lambda), \qquad \chi_\lambda(g) = \det(A_1)^{a_1}\det(A_2)^{a_2} \cdot \dots \cdot \det(A_s)^{a_s}.$$

A section of a line bundle $L(\chi)$ is given by taking a coset $gP$ to a point $(g \times f(g))$, which must satisfy the property that $(g \times f(g)) \sim (g{\cdot}p \times f(g{\cdot}p))$ $\sim (g \times \chi(p)f(g{\cdot}p))$. A section is therefore given by a function $f : G \to \mathbb{C}$ satisfying the automorphic property

$$(7) \qquad \chi(p)f(g{\cdot}p) = f(g) \quad \text{for} \quad g \in G, \; p \in P;$$

equivalently $\chi(p)f(g) = f(g{\cdot}p^{-1})$. For the section to be algebraic (i.e., a morphism from $X$ to $L(\chi)$), the corresponding function $f$ must be a morphism of algebraic varieties.

Denote this space of these sections by $\Gamma(G/P, L^\lambda)$. The group $G$ acts on the left on this space by the formula $(g{\cdot}f)(g_1) = f(g^{-1}{\cdot}g_1)$ for $g, g_1 \in G$.

**Proposition 3** *The space $\Gamma(G/P, L^\lambda)$ of sections of $L^\lambda$ is isomorphic to the representation $E^\lambda$.*

**Proof** We use the general fact that the space of sections of an algebraic vector bundle on a projective variety is finite dimensional. To prove the proposition it suffices to verify that $\Gamma(G/P, L^\lambda)$ has only one highest weight vector, up to scalars, which is of weight $\lambda$. Any highest weight vector $f$ satisfies the equation $f(g{\cdot}h) = f(g)$ for all $h$ in the group $U$ of upper triangular matrices with all 1's on the diagonal. If $B'$ denotes the group of all lower triangular matrices in $G$, then $U \cdot B'$ is dense in $G$, and $B'$ is contained in $P$. From this it follows that a highest weight vector $f$ is determined by its value at the identity element $1 \in G$. There is therefore at most one highest weight vector $f$ with $f(1) = 1$. The formula $f(g) = \chi_\lambda(g^{-1})$, where $\chi_\lambda$ is defined by the formula in (6), gives such a section.[2] The weight of this section is $\lambda$ since if $x = \mathrm{diag}(x_1, \dots, x_m)$, then $(x \cdot f)(1) = f(x^{-1}) = \chi_\lambda(x) \cdot f(1) = x_1^{\lambda_1} \cdot \dots \cdot x_m^{\lambda_m} \cdot f(1)$. $\quad\square$

One can also construct an explicit isomorphism, as follows. Let $\{e_\alpha{}^*\}$ be the basis of $E^*$ dual to the basis $\{e_\alpha\}$, let $\mu$ be the conjugate partition to $\lambda$, and define a map $\otimes \wedge^{\mu_i} E \to \Gamma(G/P, L^\lambda)$ by the formula

$$\otimes (v_{i,1} \wedge \dots \wedge v_{i,\mu_i}) \mapsto f, \quad f(g) = \prod \det(e_\alpha^*(g^{-1} \cdot v_{i,\beta}))_{1 \le \alpha, \beta \le \mu_i}.$$

One verifies easily that this $f$ satisfies (7), so is a section, and that the map is a well-defined homomorphism of $GL_m\mathbb{C}$-modules. It follows from Sylvester's lemma that it passes to the quotient by the quadratic relations, thus defining a map from $E^\lambda$ to $\Gamma(G/P, L^\lambda)$. This map is seen to be nonzero by noting that taking each $v_{i,\beta} = e_\beta$ gives a function $f$ with $f(1) = 1$. (The fact just proved that the space of sections is irreducible is used to conclude that the map is surjective.) This gives another identification of $L^\lambda$ with the hyperplane bundle $\mathcal{O}_{G/P}(1)$ on $G/P \subset \mathbb{P}^*(E^\lambda)$.

The same proof shows that $E^\lambda$ is the space of holomorphic sections of $L^\lambda$, using the general fact that such a space of sections has finite dimension. Or one can use the fact that all holomorphic sections of an algebraic vector bundle on a projective manifold are algebraic.

There is no need to assume that all of the integers $a_i$ are strictly positive. The argument is valid whenever $\tilde\lambda = (d_1{}^{a_1} \dots d_s{}^{a_s})$, with $m \ge d_1 > \dots > d_s \ge 0$ and each $a_i$ is nonnegative. For example, we can fix $d_i = m - i + 1$ for $1 \le i \le s = m+1$, in which case $Fl^{(m, m-1, \dots, 1)}$ is the variety of **complete flags**. We see in particular that all the representations $E^\lambda$ can be realized as spaces of sections of line bundles on the complete flag variety.

## 9.4 Schubert calculus on Grassmannians

Tableaux also play a role in the describing the intersection theory or cohomology rings on Grassmannian varieties $Gr^n E = Gr_r E$, with $E$ an $m$-dimensional vector space, and $r = m - n$. For each Young diagram $\lambda$ with

---

[2] Note that this function $\chi_\lambda \colon G \to \mathbb{C}$ is not a homomorphism, and it may have zeros.

at most $r$ rows and $n$ columns, and a fixed complete flag

$$F_\bullet: 0 = F_0 \subset F_1 \subset F_2 \subset \ldots \subset F_m = E$$

of subspaces, with $\dim(F_i) = i$, there is a **Schubert variety** $\Omega_\lambda = \Omega_\lambda(F_\bullet)$ defined by

$$\Omega_\lambda = \Omega_\lambda(F_\bullet) = \{V \in Gr^n E: \ \dim(V \cap F_{n+i-\lambda_i}) \geq i, \ 1 \leq i \leq r\}.$$

Note that when each $\lambda_i = 0$, no conditions are put on $V$, so $\Omega_\lambda$ is the whole Grassmannian. We will see that (i) $\Omega_\lambda$ is an irreducible closed subvariety of $Gr^n(E)$ of codimension $|\lambda|$; (ii) the class $\sigma_\lambda = [\Omega_\lambda]$ of $\Omega_\lambda$ in the cohomology group $H^{2|\lambda|}(Gr^n E)$ is independent of the choice of fixed flag used in defining it; (iii) these classes $\sigma_\lambda$ give a basis over $\mathbb{Z}$ for the cohomology ring of the Grassmannian. In fact, the products of these classes satisfy the same formulas we have seen for Schur polynomials:

$$(8) \qquad \sigma_\lambda \cdot \sigma_\mu = \Sigma c_{\lambda\mu}^\nu \sigma_\nu,$$

where the coefficients $c_{\lambda\mu}^\nu$ are the Littlewood–Richardson numbers. (We define $\sigma_\lambda$ to be $0$ if $\lambda$ contains more than $r$ rows or more than $n$ columns.) When $\lambda = (k)$, $\Omega_k = \Omega_{(k)}$ is the **special Schubert variety** consisting of spaces $V$ that meet $F_{n+1-k}$ nontrivially; the corresponding classes $\sigma_k$, $1 \leq k \leq n$, are called **special** Schubert classes. A special case of (8) is **Pieri's formula:**

$$(9) \qquad \sigma_\lambda \cdot \sigma_k = \sum \sigma_{\lambda'},$$

the sum over those $\lambda'$ that are obtained from $\lambda$ by adding $k$ boxes, with no two in a column. The determinantal formula for the Schur polynomials translates to **Giambelli's formula** for the Schubert classes in terms of the special classes:

$$(10) \qquad \sigma_\lambda = \det(\sigma_{\lambda_i+j-i})_{1 \leq i,j \leq r}.$$

In this last section we sketch the proofs, assuming some general facts about cohomology that are discussed in Appendix B. In particular, we use the following facts: (i) an irreducible subvariety $Z$ of codimension $d$ in a nonsingular projective variety $Y$ determines a cohomology class $[Z]$ in $H^{2d}(Y)$; (ii) if $Y$ has dimension $N$, then $H^{2N}(Y) = \mathbb{Z}$, with the class of a point being a generator; (iii) if two varieties $Z_1$ and $Z_2$ of complementary dimension meet transversally in $t$ points, then the product of their classes is $t$ in

$H^{2N}(Y) = \mathbb{Z}$, in which case we write $\langle [Z_1], [Z_2] \rangle = t$; (iv) if $Y$ has a filtration

$$Y = Y_0 \supset Y_1 \supset \ldots \supset Y_s = \emptyset$$

by closed algebraic subsets, and $Y_i \smallsetminus Y_{i+1}$ is a disjoint union of varieties $U_{i,j}$ each isomorphic to an affine space $\mathbb{C}^{n(i,j)}$, then the classes $[\overline{U}_{i,j}]$ of the closures[3] of these varieties give an additive basis for $H^*(Y)$ over $\mathbb{Z}$.

Each Schubert variety is the closure of the locus of $r$-planes that meet the given flag with a given "attitude," i.e., in given dimensions, as follows. Define the **Schubert cell** $\Omega_\lambda^\circ$ to be the locus of $V$ in $Gr^n(E)$ satisfying the conditions

$$\dim(V \cap F_k) = i \quad \text{for} \quad n+i-\lambda_i \leq k \leq n+i-\lambda_{i+1}, \quad 0 \leq i \leq r,$$

the condition when $i = 0$ being that $V \cap F_k = 0$ for $k = n - \lambda_1$.

To make this explicit, take a basis $e_1, \ldots, e_m$ for $E$, thus identifying $E$ with $\mathbb{C}^m$, and take $F_k = \langle e_1, \ldots, e_k \rangle$ to be the subspace spanned by the first $k$ vectors in the basis. One sees that any $V$ in $\Omega_\lambda^\circ$ is spanned by the rows of a unique $r \times m$ matrix that is in a particular "reduced row echelon form": there is a 1 in the $(n+i-\lambda_i)^{\text{th}}$ position from the left in the $i^{\text{th}}$ row; all entries after this 1 in the $i^{\text{th}}$ row are 0, and all other entries in a column with such a 1 are zero. For example, if $r = 5$, $n = 7$, and $\lambda = (5,3,2,2,1)$, these matrices have the form

$$\begin{bmatrix} * & * & 1 & 0 & 0 & 0 & 0 & 0 & 0 & 0 & 0 \\ * & * & 0 & * & * & 1 & 0 & 0 & 0 & 0 & 0 \\ * & * & 0 & * & * & 0 & * & 1 & 0 & 0 & 0 & 0 \\ * & * & 0 & * & * & 0 & * & 0 & 1 & 0 & 0 & 0 \\ * & * & 0 & * & * & 0 & * & 0 & 0 & * & 1 & 0 \end{bmatrix}$$

where the entries marked with stars are arbitrary. There are $n - \lambda_i$ stars in the $i^{\text{th}}$ row, so this constructs an isomorphism of $\Omega_\lambda^\circ$ with affine space of dimension $r \cdot n - |\lambda|$. Note that when $\lambda = 0$, and $V$ is the subspace spanned by the last $r$ basic vectors, the stars give coordinates of a neighborhood of the point $V$ in $Gr^n E$. (By changing bases, these give coordinate charts on the Grassmannian, giving it its manifold structure.)

**Exercise 13** (a) Show that $Gr^n E$ is the disjoint union of these loci $\Omega_\lambda^\circ$. (b) Show that the closure of $\Omega_\lambda^\circ$ is $\Omega_\lambda$, and show that $\Omega_\lambda$ is the disjoint union

---

[3] These closures should be taken in the Zariski topology. It is a general fact, that can be seen here explicitly, that these are also the closures in the classical topology.

of all $\Omega_\mu^\circ$ with $\mu_i \geq \lambda_i$ for all $i$. In particular $\Omega_\lambda \supset \Omega_\mu \iff \lambda \subset \mu$.
(c) Show that $\Omega_\lambda \smallsetminus \Omega_\lambda^\circ$ is the union of all $\Omega_{\lambda'}$, where $\lambda'$ is obtained from
$\lambda$ by adding one box. (d)  Show that the classes $[\Omega_\lambda]$ form a basis for
$H^*(Gr^n E)$.

The classes $\sigma_\lambda = [\Omega_\lambda]$ of the Schubert varieties are independent of the
choice of fixed flag, since the group $GL(E)$ acts transitively on the flags (see
Exercise B.7). In order to intersect two Schubert varieties it is convenient to use
also the opposite fixed flag $\widetilde{F}_\bullet$, where $\widetilde{F}_k$ is spanned by the *last* $k$ vectors in
the basis for $E = \mathbb{C}^m$; we write $\widetilde{\Omega}_\lambda$ for the corresponding Schubert variety,
and $\widetilde{\Omega}_\lambda^\circ$ for the corresponding cell. These have similar parametrizations by
row echelon matrices, but with rows *beginning* with distinguished 1's, with
these 1's placed in the $(n+i-\lambda_i)^{\text{th}}$ position from the *right*, in the $i^{\text{th}}$ row
from the *bottom*. For example, with $r = 5$, $n = 7$, and $\lambda = (5,5,4,2)$, these
matrices have the form

$$\begin{bmatrix} 1 * * 0 * * 0 * 0 0 * * \\ 0 0 0 1 * * 0 * 0 0 * * \\ 0 0 0 0 0 1 * 0 0 * * \\ 0 0 0 0 0 0 0 1 0 * * \\ 0 0 0 0 0 0 0 1 * * \end{bmatrix}$$

When we are considering the intersection of $\Omega_\lambda$ and $\widetilde{\Omega}_\mu$, we will make
frequent use of the following subspaces:

$$A_i = F_{n+i-\lambda_i}, \quad B_i = \widetilde{F}_{n+i-\mu_i}, \quad C_i = A_i \cap B_{r+1-i}, \quad 1 \leq i \leq r.$$

**Exercise 14** Show that $C_i$ is spanned by those vectors $e_j$ for which

$$i + \mu_{r+1-i} \leq j \leq n + i - \lambda_i.$$

In particular, $\dim(C_i) = n + 1 - \lambda_i - \mu_{r+1-i}$ if this number is nonnegative,
and $C_i = 0$ otherwise.

**Lemma 3** *If $\Omega_\lambda$ and $\widetilde{\Omega}_\mu$ are not disjoint, then $\lambda_i + \mu_{r+1-i} \leq n$ for all*
$1 \leq i \leq r$.

**Proof** Suppose $V$ is a subspace that is in both $\Omega_\lambda$ and $\widetilde{\Omega}_\mu$. Then for any
$i$ between 1 and $r$,

$$\dim(V \cap A_i) \geq i \quad \text{and} \quad \dim(V \cap B_{r+1-i}) \geq r + 1 - i.$$

Since these two intersections take place in the $r$-dimensional vector space $V$,
and $i + (r+1-i) - r = 1$, their intersection must have dimension at least

1. In particular, the intersection of $A_i$ and $B_{r+1-i}$ must have dimension at least 1. The conclusion follows from Exercise 14.    □

**Exercise 15** Prove the converse of Lemma 3.

The numerical condition in this lemma means that, when the Young diagram of $\mu$ is rotated by $180°$ and put in the bottom right corner of the $r \times n$ rectangle, the diagram for $\lambda$ and this rotated diagram for $\mu$ fit without overlapping. If $|\lambda|+|\mu| = r \cdot n$, in particular, the intersection can be nonempty only if these diagrams exactly fit together to make up this rectangle. For example, with $r = 5$, $n = 7$, $\lambda = (5,3,2,2,1)$, and $\mu = (6,5,5,4,2)$, this is the case:

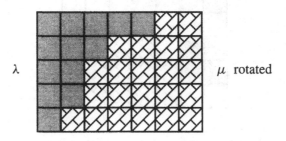

$\lambda$                                    $\mu$  rotated

Now the cells $\Omega_\lambda^\circ$ and $\widetilde{\Omega}_\mu^\circ$ can be parametrized by the stars in the corresponding matrices

$$
\begin{bmatrix}
* & * & 1 & 0 & 0 & 0 & 0 & 0 & 0 & 0 & 0 \\
* & * & 0 & * & * & 1 & 0 & 0 & 0 & 0 & 0 \\
* & * & 0 & * & * & 0 & * & 1 & 0 & 0 & 0 \\
* & * & 0 & * & * & 0 & * & 0 & 1 & 0 & 0 \\
* & * & 0 & * & * & 0 & * & 0 & 0 & * & 1 & 0
\end{bmatrix}
\quad \text{and} \quad
\begin{bmatrix}
0 & 0 & 1 & * & * & 0 & * & 0 & 0 & * & 0 & * \\
0 & 0 & 0 & 0 & 0 & 1 & * & 0 & 0 & * & 0 & * \\
0 & 0 & 0 & 0 & 0 & 0 & 0 & 1 & 0 & * & 0 & * \\
0 & 0 & 0 & 0 & 0 & 0 & 0 & 0 & 1 & * & 0 & * \\
0 & 0 & 0 & 0 & 0 & 0 & 0 & 0 & 0 & 0 & 1 & *
\end{bmatrix}
$$

In this case, from the proof of the preceding lemma, we see that $\Omega_\lambda$ and $\widetilde{\Omega}_\mu$ meet in exactly one point, which is spanned by the basis vectors corresponding to the 1's in these matrices. All the stars together give coordinates for a neighborhood of this point in the Grassmannian. The condition to be in both Schubert varieties is defined by setting all these coordinates equal to zero, from which we see that the two Schubert varieties meet transversally in one point. This proves the ***duality theorem:***

$$
(11) \qquad \sigma_\lambda \cdot \sigma_\mu \;=\; 
\begin{cases}
1 & \text{if } \lambda_i + \mu_{r+1-i} = n \quad \text{for all } 1 \le i \le r \\
0 & \text{if } \lambda_i + \mu_{r+1-i} > n \quad \text{for any } i.
\end{cases}
$$

The partition $\mu$ with $\mu_i = n - \lambda_{r+1-i}$ is sometimes called the **dual** to $\lambda$, and the class $\sigma_\mu$ the **dual class** to $\sigma_\lambda$.

We next turn to Pieri's formula (9). We must show that both sides of formula (9) have the same intersection number with all classes $\sigma_\mu$, with $|\mu| = r \cdot n - |\lambda| - k$. If the diagram of $\lambda$ is put in the top left corner of the $r \times n$ rectangle, and $\mu$ is rotated by $180°$ and put in the lower right corner, Pieri's formula is equivalent to the assertion that $\sigma_\mu \cdot \sigma_\lambda \cdot \sigma_k$ is $1$ when the two diagrams do not overlap and no two of the $k$ boxes between the two diagrams are in the same column; and that $\sigma_\mu \cdot \sigma_\lambda \cdot \sigma_k$ is $0$ otherwise. For example, with $r = 5$, $n = 7$, $\lambda = (5,3,2,2,1)$, and $\mu = (5,5,4,2,0)$ the former occurs:

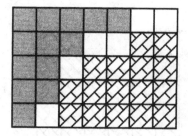

In general, this takes place exactly when

$$(12) \qquad n - \lambda_r \geq \mu_1 \geq n - \lambda_{r-1} \geq \mu_2$$

$$\geq \ \ldots \ \geq n - \lambda_1 \geq \mu_r \geq 0.$$

We use the given flag $\{F_k\}$ for the Schubert variety $\Omega_\lambda$, the dual flag $\{\widetilde{F}_k\}$ for the Schubert variety $\widetilde{\Omega}_\mu$, and we take a general linear subspace $L$ of dimension $n + 1 - k$ to define the special Schubert variety $\Omega_k(L) = \{V : \dim(V \cap L) \geq 1\}$. By Lemma 3 we may assume the two diagrams do not overlap, i.e., $\lambda_i + \mu_{r+1-i} \leq n$ for all $i$.

Pieri's formula amounts to the assertion that these three Schubert varieties meet transversally in one point when (12) is valid, and that their intersection $\Omega_\lambda \cap \widetilde{\Omega}_\mu \cap \Omega_k(L)$ is empty otherwise. This comes down to some elementary linear algebra, which we describe next. Parametrizing the Schubert varieties $\Omega_\lambda^\circ$ and $\widetilde{\Omega}_\lambda^\circ$ by row echelon matrices as above, the idea is to show that linear spaces $V$ that occur in the intersection of $\Omega_\lambda$ and $\widetilde{\Omega}_\mu$ are spanned by the rows of a matrix that has nonzero entries only between the corresponding $1$'s in echelon matrices for the two types. In the preceding example, this will be a basis taken from the rows of a matrix of the form

$$\begin{bmatrix} * & * & * & 0 & 0 & 0 & 0 & 0 & 0 & 0 & 0 & 0 \\ 0 & 0 & 0 & * & * & * & 0 & 0 & 0 & 0 & 0 & 0 \\ 0 & 0 & 0 & 0 & 0 & 0 & * & * & 0 & 0 & 0 & 0 \\ 0 & 0 & 0 & 0 & 0 & 0 & 0 & 0 & * & 0 & 0 & 0 \\ 0 & 0 & 0 & 0 & 0 & 0 & 0 & 0 & 0 & * & * & 0 \end{bmatrix}$$

The subspace represented by the stars in the $i^{\text{th}}$ row is the space $C_i$ defined before Exercise 14. The inequalities (12) say that stars occur in different columns, which will assure that nonzero vectors of this form are always independent. For spaces spanned by vectors like this, it will not be hard to see which are in $\Omega_k(L)$ for a generic linear space $L$.

To carry this out, let $C$ be the subspace of $E = \mathbb{C}^m$ spanned by the vector spaces $C_1, \ldots, C_r$. Set $A_0 = 0$ and $B_0 = 0$.

**Exercise 16** Show that: (a) $C = \bigcap_{i=0}^{r}(A_i + B_{r-i})$; (b) $\sum_{i=1}^{r} \dim(C_i) = r + k$; (c) the sum $C = C_1 + \ldots + C_r$ is a direct sum of nonempty subspaces if and only if (12) holds.

**Lemma 4** (a) *If* $V \in Gr^n E$ *is in* $\Omega_\lambda \cap \tilde{\Omega}_\mu$, *then* $V \subset C$. (b) *If, in addition,* $C_1, \ldots, C_r$ *are linearly independent, then* $\dim(V \cap C_i) = 1$ *for all* $i$, *and* $V = V \cap C_1 \oplus \ldots \oplus V \cap C_r$.

**Proof** By (a) of the exercise, we must show that $V \subset A_i + B_{r-i}$ for all $i$. This is clear if $A_i \cap B_{r-i} \neq 0$, for then $A_i + B_{r-i} = E$. So we may suppose $A_i \cap B_{r-i} = 0$. The assumption on $V$ implies that $\dim(V \cap A_i) \geq i$ and $\dim(V \cap B_{r-i}) \geq r-i$. Since $\dim(V) = r$, this means that $V$ is a direct sum of $V \cap A_i$ and $V \cap B_{r-i}$, showing in particular that $V$ is contained in $A_i + B_{r-i}$, which proves (a).

Since $\dim(V \cap A_i) \geq i$ and $\dim(V \cap B_{r+1-i}) \geq r+1-i$, as in the proof of Lemma 3 we must have $\dim(V \cap C_i) \geq i + (r+1-i) - r = 1$. If the $C_i$ are linearly independent, then $V$ contains the direct sum of the $V \cap C_i$, which has dimension at least $r$, so $V = \oplus(V \cap C_i)$ and each summand must have dimension one. $\quad \square$

If (12) fails, it follows from Exercise 16 that $C$ is not a direct sum of the spaces $C_i$, and that the dimension of $C$ is at most $r+k-1$. In this case a generic linear space $L$ of dimension $n+1-k$ will not meet $C$ except at the origin. By Lemma 4, no $V$ in $\Omega_\lambda \cap \tilde{\Omega}_\mu$ can be in $\Omega_k(L)$, so the intersection of the three Schubert varieties is empty.

If (12) is valid, then $C = \oplus C_i$, and a generic $L$ meets $C$ in a line of the form $\mathbb{C} \cdot v$, with $v = u_1 \oplus \ldots \oplus u_r$, $u_i$ a nonzero vector in $C_i$. Now the condition that $V$ meets $L$ in at least a line, together with the condition that $V$ is contained in $C$, forces $V$ to contain the vector $v$. Since $V = \oplus V \cap C_i$, we must have $u_i$ in $V$, so $V$ must be the subspace spanned by the vectors $u_1, \ldots, u_r$. This shows that the intersection of the three Schubert varieties is one point, and a local calculation, using the identifications of the open Schubert varieties with affine spaces, as before, shows that the intersection is transversal. This completes the proof of Pieri's formula. (Another proof will be given in Chapter 10.)

Now let $\Lambda$ be the ring of symmetric functions. Define an additive homomorphism $\Lambda \to H^*(Gr^n(\mathbb{C}^m))$ by sending the Schur polynomial $s_\lambda$ to $\sigma_\lambda$ if $\lambda$ has at most $r$ rows and $n$ columns, and sending $s_\lambda$ to $0$ otherwise. It is an immediate consequence of Pieri's formula, and the fact that $\Lambda$ is generated as a ring by the Schur polynomials $s_{(k)} = h_k$, that this is a homomorphism of *rings*. Formulas (8) and (10) then follow automatically, since we know that they hold in the ring $\Lambda$.

**Exercise 17** Show that the number of $r$-planes in $\mathbb{C}^m$ meeting each of $r \cdot (m-r)$ general subspaces of dimension $m-r$ nontrivially is
$$\frac{(r \cdot (m-r))! \cdot (r-1)! \cdot (r-2)! \cdot \ldots \cdot 1!}{(m-1)! \cdot (m-2)! \cdot \ldots \cdot (m-r)!}.$$

**Exercise 18** Show that the variety $\Omega_\lambda$ is defined by the conditions that $\dim(V \cap F_{n+i-\lambda_i}) \geq i$ for those $i$ such that $(i, \lambda_i)$ is an outside corner of the Young diagram $\lambda$. Show that none of these conditions can be omitted.

**Exercise 19** For $\mu = (1^k)$, $1 \leq k \leq n$, the Schubert variety $\Omega_\mu$ consists of spaces $V$ such that $\dim(V \cap F_{n+k-1}) \geq k$. Show that $\sigma_{(1^k)} \cdot \sigma_\lambda = \Sigma \sigma_{\lambda'}$, the sum over all $\lambda'$ contained from $\lambda$ by adding $k$ boxes, with no two in a row.

**Exercise 20** Show that the map that assigns to $V \subset E$ the kernel of the dual homomorphism $E^* \to V^*$ determines an isomorphism from $Gr^n(E)$ to $Gr^r(E^*)$, called the **duality isomorphism**. Show that this isomorphism takes a Schubert variety $\Omega_\lambda$ to a Schubert variety $\Omega_{\tilde{\lambda}}$.

**Exercise 21** (a) Show that $H^*(Gr^n(\mathbb{C}^m))$ is isomorphic to the polynomial ring with generators $\sigma_1, \ldots, \sigma_n$ over $\mathbb{Z}$, with $\sigma_i$ of degree $i$, modulo the ideal generated by all $p \times p$ determinants $\det(\sigma_{1+j-i})_{1 \leq i,j \leq p}$, for

$r+1 \leq p \leq m$, where in this formula $\sigma_0 = 1$ and $\sigma_k = 0$ if $i < 0$ or $i > n$. (b) Deduce that the monomials $\sigma_1^{a_1} \cdot \ldots \cdot \sigma_n^{a_n}$ are linearly independent in $H^*(Gr^n(\mathbb{C}^m))$ if $a_1 + 2a_2 + \ldots + na_n \leq r$.

In the next chapter we will also describe the intersection rings of the flag varieties. They have a similar basis consisting of closures of loci of flags with a given attitude with respect to a fixed complete flag. In general, however, one does not know formulas as explicit as (8).

# 10

# Schubert varieties and polynomials

We will describe the Schubert varieties in the complete flag manifolds $F\ell(E) = F\ell(\mathbb{C}^m) = F\ell(m) = F\ell^{(m,m-1,\ldots,1)}(\mathbb{C}^m)$, whose points consists of flags $E_{\bullet} = (E_1 \subset E_2 \subset \ldots \subset E_n = E = \mathbb{C}^m)$, $\dim(E_i) = i$. We will also define the Schubert polynomials of Lascoux and Schützenberger. Both are indexed by permutations $w$ in the symmetric group $S_m$. The Schubert polynomials, when evaluated on certain basic classes in the cohomology of the flag manifold, yield the classes of corresponding Schubert varieties. We use freely the results stated in Appendix B.

## 10.1 Fixed points of torus actions

Consider first an action of the multiplicative group $T = \mathbb{C}^*$ on projective space $\mathbb{P}^r$, given in the form

$$t \cdot [x_0 : x_1 : \ldots : x_r] = [t^{a_0} x_0 : t^{a_1} x_1 : \ldots : t^{a_r} x_r]$$

for some integers $a_0, a_1, \ldots, a_r$. For each integer $a$ that occurs among these $a_i$, there is a linear subspace $L_a$ of $\mathbb{P}^r$ defined by the equations $X_i = 0$ for all $i$ with $a_i \neq a$. It is easy to verify that the set of fixed points of this action is the disjoint union of these linear subspaces $L_a$. For example, if the $a_i$ are all distinct, the fixed point set is finite, consisting of the $r+1$ points $[1 : 0 : \ldots : 0]$, $[0 : 1 : \ldots : 0]$, $\ldots$, $[0 : 0 : \ldots : 1]$.

If $Z \subset \mathbb{P}^r$ is an algebraic subset, and the action of $T = \mathbb{C}^*$ on $\mathbb{P}^r$ maps $Z$ to itself, then the fixed points $Z^T$ of the action of $T$ on $Z$ is the intersection of $Z$ with $(\mathbb{P}^r)^T$, so

$$Z^T = \coprod X \cap L_a.$$

**Lemma 1** *If $Z$ is nonempty, then $Z^T$ is nonempty.*

**Proof** Take any point $x = [x_0: \ldots : x_r]$ in $Z$. Let $a$ be the minimum of those $a_i$ such that $x_i \neq 0$. Set $y_i = x_i$ if $a_i = a$, and $y_i = 0$ if $a_i \neq a$, and set $y = [y_0: \ldots : y_r]$. Then the points

$$t{\cdot}x = [t^{a_0}x_0: \ldots : t^{a_r}x_r] = [t^{a_0-a}x_0: \ldots : t^{a_r-a}x_r]$$

approach $y$ as $t$ approaches $0$. Since $Z$ is preserved by $T$, all these points $t{\cdot}x$ are in $Z$, and since $Z$ is closed in $\mathbb{P}^r$, the limit point $y$ is also in $Z$. And $y$ is in $Z \cap L_a \subset Z^T$.   □

This generalizes to the action of an $m$-dimensional torus $T = (\mathbb{C}^*)^m$. For $t = (t_1, \ldots, t_m) \in T$, and an $m$-tuple $\mathbf{a} = (a_1, \ldots, a_m)$ of integers, write $t^{\mathbf{a}}$ for $t_1{}^{a_1} \cdot \ldots \cdot t_m{}^{a_m}$. Suppose $T$ acts on $\mathbb{P}^r$ by a rule

$$t{\cdot}[x_0: \ldots : x_r] = [t^{\mathbf{a}(0)}x_0: \ldots : t^{\mathbf{a}(r)}x_r],$$

for some $m$-tuples $\mathbf{a}(0), \ldots, \mathbf{a}(r)$. The fixed points form a disjoint union of linear subspaces $L_{\mathbf{a}}$, as $\mathbf{a}$ varies over $m$-tuples in the set $\{\mathbf{a}(0), \ldots, \mathbf{a}(r)\}$, with $L_{\mathbf{a}}$ defined by the equations $X_i = 0$ if $\mathbf{a}(i) \neq \mathbf{a}$. This can be seen directly, or by applying the preceding case inductively to each of the $m$ factors $\mathbb{C}^* = 1 \times 1 \times \ldots \times \mathbb{C}^* \times 1 \times \ldots \times 1$ in $T$.

**Proposition 1** *If* $Z \subset \mathbb{P}^r$ *is closed and mapped to itself by* $T$, *then the fixed point set* $Z^T$ *is the disjoint union of the subsets* $Z \cap L_{\mathbf{a}}$. *If* $Z$ *is not empty, then* $Z^T$ *is not empty.*

**Proof** The last statement follows by induction on $m$, the case $m = 1$ being the lemma. Letting $T' = \mathbb{C}^* \times \ldots \times \mathbb{C}^* \times 1 \subset T$, we know that $Z^{T'}$ is not empty by induction; and if $\mathbb{C}^* = 1 \times \ldots \times 1 \times \mathbb{C}^* \subset T$, then $Z^T = (Z^{T'})^{\mathbb{C}^*}$, which is nonempty by the case $m = 1$.   □

It is a general fact, seen in Chapter 8, that for any linear algebraic action of $T$ on a vector space $V$, one can find a basis so that the action of $T$ on $\mathbb{P}(V)$ is given as above; this amounts to the simultaneous diagonalization of a collection of commuting diagonalizable matrices. The proposition is a special case of Borel's fixed point theorem: an action of a connected, solvable, linear algebraic group on a projective variety must have a fixed point. We won't need these generalizations here, but the following special case is elementary.

**Exercise 1** Let $V$ be a rational representation of $G = GL_m(\mathbb{C})$, and let $B$ be the subgroup of upper triangular matrices of $G$. (a) Show that there is a chain of subspaces $V_1 \subset V_2 \subset \ldots \subset V_r = V$, each mapped to itself by $B$, with $\dim(V_i) = i$. (b) Deduce that if $Z$ is any algebraic subset of $\mathbb{P}(V)$ that is mapped to itself by $B$, then there is a point $P$ in $Z$ that is fixed by $B$.

We apply the proposition with $T = (\mathbb{C}^*)^m$ the group of diagonal matrices in $GL_m(\mathbb{C})$, and $Z = F\ell(m)$ the flag manifold of complete flags in $\mathbb{C}^m$. We have seen how to embed $Z$ in a projective space $\mathbb{P}^r$:

$$F\ell(m) \subset \prod_{d=1}^{m} Gr^d(\mathbb{C}^m) \subset \prod_{d=1}^{m} \mathbb{P}^*(\wedge^d \mathbb{C}^m) \subset \mathbb{P}^* \left( \bigotimes_{d=1}^{m} \wedge^d \mathbb{C}^m \right) = \mathbb{P}^r.$$

The natural action of $GL_m(\mathbb{C})$ on $\mathbb{C}^m$ induces an action on each of these varieties. It is easy to see that the resulting action of $T$ on $\mathbb{P}^r$ has the form prescribed above. Indeed, if $e_1, \ldots, e_m$ is the standard basis for $\mathbb{C}^m$, then $\wedge^d \mathbb{C}^m$ has a basis of all $e_{i_1} \wedge \ldots \wedge e_{i_d}$, $1 \leq i_1 < \ldots < i_d \leq m$, and for $t = (t_1, \ldots, t_m)$,

$$t \cdot (e_{i_1} \wedge \ldots \wedge e_{i_d}) = t_{i_1} \cdot \ldots \cdot t_{i_d} e_{i_1} \wedge \ldots \wedge e_{i_d}.$$

The coordinates for $\mathbb{P}^r = \mathbb{P}^*(\bigotimes \wedge^d \mathbb{C}^m)$ correspond to products of such basis elements, from which the assertion is clear.

**Lemma 2** *The fixed points of the action of $T$ on $F\ell(m)$ are the $m!$ flags of the form*

$$\langle e_{w(1)} \rangle \subset \langle e_{w(1)}, e_{w(2)} \rangle \subset \ldots \subset \langle e_{w(1)}, e_{w(2)}, \ldots, e_{w(m)} \rangle = \mathbb{C}^m,$$

*as $w$ varies over $S_m$.*

**Proof** This is a direct calculation. Suppose a flag $E_1 \subset \ldots \subset E_m = \mathbb{C}^m$ is fixed by $T$. Suppose $E_1$ is spanned by a vector $v = \lambda_1 e_1 + \ldots + \lambda_m e_m$. Since $(t_1, \ldots, t_m) \cdot v = t_1 \lambda_1 e_1 + \ldots + t_m \lambda_m e_m$, the line $E_1$ is fixed by $T$ only if exactly one coefficient $\lambda_p$ is not zero, so $E_1 = \langle e_p \rangle$. Then $E_2$ has generators $\langle e_p, v \rangle$, with $v = \sum_{q \neq p} \lambda_q e_q$ unique up to multiplication by a nonzero scalar. Therefore $E_2$ is fixed by $T$ only if $v$ has one nonzero coefficient, say $\lambda_q$; so $E_2 = \langle e_p, e_q \rangle$ for some $q \neq p$. Continuing in this way, one sees that the flag is determined by an ordering of the basis elements. □

We write $x(w) \in F\ell(m)$ for the point corresponding to the flag of Lemma 2.

## 10.2 Schubert varieties in flag manifolds

Fix a flag $F_1 \subset F_2 \subset \ldots \subset F_m = E$. When a basis is given for $E$, identifying $E$ with $\mathbb{C}^m$, we will take $F_q = \langle e_1, \ldots, e_q \rangle$ spanned by the first $q$ elements of this basis. For each permutation $w$ in $S_m$, there is a *Schubert cell* $X_w^\circ \subset F\ell(m) = F\ell(E)$, defined as a set by the formula

$$X_w^\circ = \{E_\bullet \in F\ell(E) : \dim(E_p \cap F_q) = \#\{i \le p : w(i) \le q\}$$

$$\text{for } 1 \le p, q \le m\}.$$

Note that $X_w^\circ$ contains the point $x(w)$ defined at the end of the preceding section. We will construct an isomorphism of $X_w^\circ$ with an affine space $\mathbb{C}^{\ell(w)}$, with $x(w)$ corresponding to the origin. Here $\ell(w)$ is the number of inversions in $w$, called the *length* of $w$, i.e.,

$$\ell(w) = \#\{i < j : w(i) > w(j)\}.$$

To construct this isomorphism, note that each flag $E_\bullet$ has $E_p$ spanned by the first $p$ rows of a unique "row echelon" matrix, where the $p^{\text{th}}$ row has a 1 in the $w(p)^{\text{th}}$ column, with all 0's after this 1, and the matrix has all 0's below these 1's. For example, for $w = 4\,2\,6\,1\,3\,5$ in $S_6$, these matrices have the form

$$\begin{bmatrix} * & * & * & 1 & 0 & 0 \\ * & 1 & 0 & 0 & 0 & 0 \\ * & 0 & * & 0 & * & 1 \\ 1 & 0 & 0 & 0 & 0 & 0 \\ 0 & 0 & 1 & 0 & 0 & 0 \\ 0 & 0 & 0 & 0 & 1 & 0 \end{bmatrix}$$

where the stars denote arbitrary complex numbers. Here $\ell(w) = 7$, and $X_{426135}^\circ \cong \mathbb{C}^7$, the isomorphism given by the seven stars.

In fact, $x(w)$ has an open neighborhood $U_w$ in $F\ell(E)$ that is isomorphic with $\mathbb{C}^n$, where

$$n = m(m-1)/2 = \dim(F\ell(m))$$

is the dimension of the flag manifold. The flags in $U_w$ are spanned by rows of a matrix with 1's in the $(p, w(p))$ positions, and 0's under these 1's. In the above example, $U_{426135}$ is identified with $\mathbb{C}^{15}$ by means of the stars in the matrix

$$
\begin{bmatrix}
* & * & * & 1 & * & * \\
* & 1 & * & 0 & * & * \\
* & 0 & * & 0 & * & 1 \\
1 & 0 & * & 0 & * & 0 \\
0 & 0 & 1 & 0 & * & 0 \\
0 & 0 & 0 & 0 & 1 & 0
\end{bmatrix}
$$

**Exercise 2** Verify that $U_w$ is open in $F\ell(m)$, and that the map $\mathbb{C}^n \to U_w \subset F\ell(m)$ is an open embedding.

**Exercise 3** For $1 \le i \le m-1$, let $s_i = (i, i+1)$, so $w \cdot s_i$ is obtained from $w$ by interchanging the values in the positions $i$ and $i+1$. Show that $\ell(w \cdot s_i) = \ell(w) - 1$ if $w(i) > w(i+1)$, and $\ell(w \cdot s_i) = \ell(w) + 1$ if $w(i) < w(i+1)$. Deduce that $\ell(w)$ is the minimum $\ell$ such that $w = s_{i_1} \cdot \ldots \cdot s_{i_\ell}$.

From this description it follows that $X_w^\circ$ is a closed subvariety of $U_w$, isomorphic to an inclusion of $\mathbb{C}^{\ell(w)}$ in $\mathbb{C}^n$ as a coordinate subspace.

**Exercise 4** (a) For any flag $E.$ and $1 \le i \le m$, define a subset $\mathscr{A}_i$ of $[m]$ by the rule $\mathscr{A}_i = \{j : E_i \cap F_j \ne E_i \cap F_{j-1}\}$. Show that $\mathscr{A}_1 \subset \mathscr{A}_2 \subset \ldots \subset \mathscr{A}_m$, with $\mathscr{A}_i$ of cardinality $i$. Show in fact that $E.$ is in $X_w^\circ$ if and only if $\mathscr{A}_i = \{w(1), \ldots, w(i)\}$ for all $i$.
(b) The **diagram** $D(w)$ of a permutation $w$ in $S_m$ consists of the pairs $(i, j)$ of integers between 1 and $m$ such that $j < w(i)$ and $i < w^{-1}(j)$. Show that $X_w^\circ$ consists of flags $E.$ such that there are $v_1, \ldots, v_m$ in $E$ of the form $v_i = e_{w(i)} + \sum a_{ij} e_j$, with the sum over $(i, j) \in D(w)$, such that $E_k = \langle v_1, \ldots, v_k \rangle$ for $1 \le k \le m$.

We will also need **dual Schubert cells** $\Omega_w^\circ$, which consist of flags spanned by rows of an echelon matrix, again with 1's in the $(p, w(p))^{\text{th}}$ position, and 0's under these 1's, but this time with 0's to the *left* of these 1's. If $\widetilde{F}_q$ is the subspace of $E = \mathbb{C}^m$ spanned by the last $q$ vectors of the basis,

$$\Omega_w^\circ = \{E. \in F\ell(E) : \dim(E_p \cap \widetilde{F}_q) = \#\{i \le p : w(i) \ge m+1-q\} \ \forall \ p, q\}.$$

For example, $\Omega^{\circ}_{426135}$ is described by the matrix

$$\begin{bmatrix} 0 & 0 & 0 & 1 & * & * \\ 0 & 1 & * & 0 & * & * \\ 0 & 0 & 0 & 0 & 0 & 1 \\ 1 & 0 & * & 0 & * & 0 \\ 0 & 0 & 1 & 0 & * & 0 \\ 0 & 0 & 0 & 0 & 1 & 0 \end{bmatrix}$$

We see that $\Omega^{\circ}_w \cong \mathbb{C}^{n-\ell(w)}$ is also closed in $U_w$, and, as in §9.4, that $X^{\circ}_w$ and $\Omega^{\circ}_w$ meet transversally in $U_w$ at the point $x(w)$.

Since every flag is determined by a unique row echelon matrix, the flag manifold $F\ell(m)$ is the disjoint union of the Schubert cells $X^{\circ}_w$, one for each $w$ in the symmetric group $S_m$. These Schubert cells are exactly the orbits of the action of the group $B \subset GL_n(\mathbb{C})$ of upper triangular matrices. Similarly, the dual cells $\Omega^{\circ}_w$ are the orbits of the group $B'$ of lower triangular matrices.

The **Schubert variety** $X_w$ is defined to be the closure of the cell $X^{\circ}_w$. Similarly, define $\Omega_w$ to be the closure of $\Omega^{\circ}_w$. These are irreducible closed subvarieties of $F\ell(m)$ of dimensions $\ell(w)$ and $n - \ell(w)$, respectively. Since $B$ acts on $X^{\circ}_w$, it also acts on its closure $X_w$. In particular, $X_w$ must be a union of $X^{\circ}_w$ and some set of smaller orbits $X^{\circ}_v$ (with $\ell(v) < \ell(w)$). Similarly, $\Omega_w$ is a union of $\Omega^{\circ}_w$ and some cells $\Omega_v$, for some set of $v$ with $\ell(v) > \ell(w)$. We will describe which $v$ occur in these decompositions, and give another description of these Schubert varieties, in §10.5.

**Proposition 2** *Let $u$ and $v$ be permutations in $S_m$. If $X_u$ meets $\Omega_v$, then $\ell(v) \leq \ell(u)$, with equality holding if and only if $u = v$. The varieties $X_w$ and $\Omega_w$ meet transversally at the point $x(w)$.*

**Proof** Since $X_u$ is preserved by $B$, and $\Omega_v$ by $B'$, the intersection $Z = X_u \cap \Omega_v$ is preserved by the torus $T = B \cap B'$. By the fixed point theorem of the preceding section, $Z$ must contain some point $x(w)$. Since $x(w)$ is in $X_w$, we must have $X^{\circ}_w = B \cdot x(w) \subset X_u$, so $\ell(w) \leq \ell(u)$, with equality only if $w = u$. Similarly, $\Omega^{\circ}_w = B' \cdot x(w) \subset \Omega_v$ implies that $\ell(w) \geq \ell(v)$, with equality only if $w = v$. Therefore $\ell(v) \leq \ell(w) \leq \ell(u)$, and $\ell(v) < \ell(u)$ unless $u = w = v$, which proves the first statement.

Since $X_w \smallsetminus X^{\circ}_w$ is a union of some $X^{\circ}_v$ with $\ell(v) < \ell(w)$, it follows that $X_w \smallsetminus X^{\circ}_w$ cannot meet $\Omega_w$, and similarly, $\Omega_w \smallsetminus \Omega^{\circ}_w$ cannot meet $X_w$. Therefore $X_w \cap \Omega_w = X^{\circ}_w \cap \Omega^{\circ}_w$, and we have seen that these cells meet transversally at the point $x(w)$. $\square$

For $1 \le d \le n = m(m-1)/2$, let $Z_d \subset F\ell(m)$ be the union of those $X_w^\circ$ with $\ell(w) \le d$. By what we have seen, $Z_d$ is a closed algebraic subset of $F\ell(m)$, since it is the union of those $X_w$ with $\ell(w) \le d$. In addition, $Z_d \setminus Z_{d-1}$ is a disjoint union of the cells $X_w^\circ$, each isomorphic to $\mathbb{C}^d$. It is a general fact (see Lemma 6 of Appendix B) that the classes of the closures of such cells give an integral basis for the cohomology of the space. In this case, the fact that the classes $[X_w]$ of the varieties $X_w$ with $\ell(w) = d$ form a basis of the cohomology group $H^{2n-2d}(F\ell(m))$ over $\mathbb{Z}$ can be seen directly, by the following argument. We will see in Proposition 3 that $H^*(F\ell(m))$ is free over $\mathbb{Z}$ of rank $m!$, with all odd groups $H^{2d+1}(F\ell(m))$ vanishing. The intersection pairing is a bilinear map

$$H^{2d}(F\ell(m)) \times H^{2n-2d}(F\ell(m)) \;\to\; H^{2n}(F\ell(m)) = \mathbb{Z}, \quad \alpha \times \beta \mapsto \langle \alpha, \beta \rangle.$$

That this is a perfect pairing follows from the fact that $F\ell(m)$ is a compact oriented manifold of real dimension $2n$ whose homology is torsion free. In this case it can be seen directly as follows. The pairing has the property that the classes of two closed subvarieties of complementary dimension have intersection number $0$ if these varieties are disjoint, and intersection number $1$ if they meet transversally in one point. From the proposition it follows that as $u$ and $v$ vary over the permutations of length $d$, we have

$$(1) \qquad \langle [\Omega_v], [X_u] \rangle \;=\; \delta_{uv}.$$

From this it follows first that the classes $\{[X_u] : u \in S_m\}$ are linearly independent. Since there are $m!$ of these classes, they must give a basis for the cohomology with rational coefficients. But if a cohomology class is expanded as a rational linear combination of the classes $[X_u]$, then equation (1) implies that all these coefficients are integers. This shows that the classes $\{[X_u] : \ell(u) = d\}$ form a basis for $H^{2n-2d}(F\ell(m))$, and the classes $\{[\Omega_v] : \ell(v) = d\}$ form a dual basis for $H^{2d}(F\ell(m))$.

Denote by $w_o$ the permutation in $S_m$ that takes $i$ to $m+1-i$ for $1 \le i \le m$.

**Lemma 3** *For any $w$ in $S_m$, $[\Omega_w] = [X_{w^{\vee}}]$, where $w^{\vee} = w_o \cdot w$, i.e., $w^{\vee}(i) = m+1 - w(i)$ for $1 \le i \le m$.*

**Proof** Writing $X_w(\widetilde{F}_{\textbf{.}})$ for the Schubert variety constructed from the flag $\widetilde{F}_{\textbf{.}}$ with $\widetilde{F}_p$ spanned by the last $p$ basis elements, we see immediately from the definitions that $\Omega_w = X_{w^{\vee}}(\widetilde{F}_{\textbf{.}})$. The result therefore follows from the general fact that, for any $v$, $X_v(\widetilde{F}_{\textbf{.}})$ and $X_v(F_{\textbf{.}})$ determine the same

cohomology class. As in §9.4, using Exercise B.7, this follows from the fact
that the connected group $GL(E)$ acts transitively on the flags.    □

We define, for any $w$ in $S_m$, the **Schubert class** $\sigma_w$ in $H^{2\ell(w)}(F\ell(m))$
by the formula

$$\sigma_w \; = \; [\Omega_w] \; = \; [X_{w^{\vee}}] \; = \; [X_{w_o \cdot w}].$$

By (1) and Lemma 3, we therefore have

(2)          $\langle \sigma_u, \sigma_{v^{\vee}} \rangle \; = \; \langle \sigma_u, \sigma_{w_o \cdot v} \rangle \; = \; \delta_{uv}.$

We will need a presentation of the cohomology ring of the flag variety
$F\ell(m)$. This ring is generated by some basic classes $x_1, \ldots, x_m$ in
$H^2(F\ell(m))$, with relations generated by elementary symmetric polynomials
in these variables. This is easiest to describe in terms of projective bundles and
Chern classes. On $X = F\ell(E)$ there is a **universal** or **tautological** filtration

$$0 \; = \; U_0 \subset U_1 \subset U_2 \subset \; \ldots \; \subset U_{m-1} \subset U_m \; = \; E_X$$

of subbundles of the trivial bundle $E_X$ of rank $m$ on $X$: over a point in
$F\ell(E)$ corresponding to a flag $E.$ the fibers of these bundles are the vector
spaces $E_i$ of the flag. The first Chern classes of the line bundles $U_i/U_{i-1}$
are the generators of the cohomology ring. Precisely, we set

(3)          $L_i \; = \; U_i/U_{i-1}, \quad x_i \; = \; - \, c_1(L_i), \quad 1 \leq i \leq m.$

This line bundle $L_i$ can be identified with the bundle $L(\chi)$ constructed from
a character as in §9.3. In fact, with $B$ the group of upper triangular matrices
in $G$, $L_i = L(\chi_i)$, where $\chi_i: B \to \mathbb{C}^*$ takes an element $g$ to the $i^{\text{th}}$ entry
down the diagonal of the matrix representing $g$. Indeed, the fixed point $x$ of
$B$ in $X$ is the flag whose $i^{\text{th}}$ term is spanned by the first $i$ members of the
basis for $E$, so the fiber of $L_i$ at $x$ is spanned by the image of $e_i$, which
is multiplied by $\chi_i(g)$ for $g$ in $B$. So $x_i = -c_1(L(\chi_i)) = c_1(L(\chi_i^{-1}))$.

We will see later that these classes are closely related to the classes $\sigma_w$ that
are in $H^2(X)$; in fact, $\sigma_{s_i} = x_1 + \ldots + x_i$ for $1 \leq i \leq m-1$, where
$s_i$ is the transposition of $i$ and $i+1$. This can be used to give an alternative
definition of the basic classes: $x_i = \sigma_{s_i} - \sigma_{s_{i-1}}$ for $1 \leq i \leq m-1$, and
$x_m = -\sigma_{s_{m-1}}$.

**Proposition 3** *The cohomology ring of $X = F\ell(m)$ is generated by the
basic classes $x_1, \ldots, x_m$, subject to the relations $e_i(x_1, \ldots, x_m) = 0$,*

$1 \leq i \leq m$. *That is,* $H^*(X) = R(m)$, *where*

$$R(m) = \mathbb{Z}[X_1, \ldots, X_m]/(e_1(X_1, \ldots, X_m), \ldots, e_m(X_1, \ldots, X_m)).$$

*In addition, the classes* $x_1^{i_1} \cdot x_2^{i_2} \cdot \ldots \cdot x_m^{i_m}$, *with exponents* $i_j \leq m-j$, *form a basis for* $H^*(F\ell(m))$ *over* $\mathbb{Z}$.

**Proof** We use some basic facts about projective bundles and Chern classes (see §B.4). Let $V$ be a vector bundle of rank $r$ on a variety $Y$, and $\rho : \mathbb{P}(V) \to Y$ the corresponding projective bundle, whose fiber over a point in $Y$ is the projective space of lines through the origin in $V$. On $\mathbb{P}(V)$ there is a tautological line bundle $L \subset \rho^*(V)$. Let $\zeta = -c_1(L)$. Then

$$H^*(\mathbb{P}(V)) = H^*(Y)[\zeta]/(\zeta^r + a_1\zeta^{r-1} + \ldots + a_r),$$

for some unique elements $a_1, \ldots, a_r$ in $H^*(Y)$. In fact, $a_i$ is the $i^{\text{th}}$ Chern class $c_i(V)$ in $H^{2i}(Y)$. The classes $1, \zeta, \ldots, \zeta^{r-1}$ form a basis for $H^*(\mathbb{P}(V))$ over $H^*(Y)$.

The flag manifold $X = F\ell(E)$ can be constructed as a sequence of projective bundles. First one has the projective space $\mathbb{P}(E)$ with its tautological line bundle $U_1 \subset E$. (In this discussion we suppress the notations for pullbacks of bundles.) On $\mathbb{P}(E)$ we have a bundle $E/U_1$ of rank $m-1$, and we construct $\mathbb{P}(E/U_1) \to \mathbb{P}(E)$. The tautological line bundle on $\mathbb{P}(E/U_1)$ has the form $U_2/U_1$ for some bundle $U_2$ of rank 2 with $U_1 \subset U_2 \subset E$ on $\mathbb{P}(E/U_1)$. Then one constructs $\mathbb{P}(E/U_2)$, with its tautological bundle $U_3/U_2$, and so on, until one arrives at the flag manifold $\mathbb{P}(E/U_{m-2})$ with its tautological bundle $U_{m-1}/U_{m-2}$. From the assertions about projective bundles in the preceding paragraph, it follows that the $x_1^{i_1} \cdot x_2^{i_2} \cdot \ldots \cdot x_m^{i_m}$, for $i_j \leq m - j$, form an additive basis for $H^*(F\ell(m))$.

Since the bundle $E_X$ on $X$ has a filtration with line bundle quotients $L_i$, it follows from the Whitney sum formula that the $i^{\text{th}}$ Chern class of $E_X$ is the $i^{\text{th}}$ elementary symmetric polynomial in the Chern classes $c_1(L_1), \ldots,$ $c_1(L_m)$. Since $E_X$ is a trivial bundle, its Chern classes must vanish. This implies that $e_i(x_1, \ldots, x_m) = 0$ for $1 \leq i \leq m$.

Let $R(m)$ denote the $\mathbb{Z}$-algebra defined in the proposition. Consider the canonical surjection

$$R(m) \twoheadrightarrow H^*(F\ell(m)), \quad X_i \mapsto x_i, \quad 1 \leq i \leq m.$$

To show that this map is an isomorphism, it suffices to show that the images of the classes $X_1^{i_1} \cdot X_2^{i_2} \cdot \ldots \cdot X_m^{i_m}$, for $i_j \leq m-j$, span $R(m)$ over $\mathbb{Z}$.

This is a purely algebraic fact about symmetric polynomials; in this proof, let $x_i$ denote the image of $X_i$ in $R(m)$. In the ring $R(m)[t]$ we have an identity

$$(4) \qquad \prod_{i=1}^{p} \frac{1}{1-x_i t} \;=\; \prod_{i=p+1}^{m} (1-x_i t).$$

This follows from the fact that

$$\prod_{i=1}^{m}(1-x_i t) \;=\; 1 - e_1 t + e_2 t^2 - \ldots + (-1)^m e_m t^m \;=\; 1.$$

The left side of (4) is the sum $\sum_{i \geq 0} h_i(x_1, \ldots, x_p)t^i$. Comparing coefficients of $t^i$ we see that $h_i(x_1, \ldots, x_p) = 0$ for $i > m-p$. In particular, using the equations

$$h_{m-p+1}(x_1, \ldots, x_p) \;=\; x_p^{m-p+1} + \ldots \;=\; 0, \quad 1 \leq p \leq m,$$

by descending induction on $p$, one sees that the asserted monomials generate $R(m)$ as a $\mathbb{Z}$-module.  □

One of the goals of this chapter is to find a "Giambelli formula," to write the classes of the "geometric basis" $\sigma_w$, $w \in S_n$, as polynomials in the "algebraic" basis $x_1^{i_1} \cdot x_2^{i_2} \cdot \ldots \cdot x_m^{i_m}$, $i_j \leq m-j$. We will see that there are polynomials – the Schubert polynomials – that provide a universal solution to this problem.

## 10.3   Relations among Schubert varieties

There is a canonical embedding $\iota : F\ell(m) \hookrightarrow F\ell(m+1)$, that takes a flag $E_\bullet$ in $E = \mathbb{C}^m$ to the following flag in $E' = E \oplus \mathbb{C} = \mathbb{C}^{m+1}$:

$$E_1 \oplus 0 \subset E_2 \oplus 0 \subset \ldots \subset E_{m-1} \oplus 0 \subset E_m \oplus 0$$

$$= E \oplus 0 \subset E' = E \oplus \mathbb{C}.$$

This is a closed embedding, and identifies $F\ell(m)$ with the set of flags in $E'$ whose $m^{\text{th}}$ member is $E \oplus 0$. From this definition we see immediately that, for every $w$ in $S_m$, $\iota$ maps the Schubert cell $X_w^\circ$ in $F\ell(m)$ isomorphically onto the Schubert cell denoted $X_w^\circ$ in $F\ell(m+1)$. Here and in what follows we regard $S_m$ as usual as the subgroup of $S_{m+1}$ fixing $m+1$. Since

$F\ell(m)$ is closed in $F\ell(m+1)$, it follows that $\iota(X_w)$ is the Schubert variety corresponding to $w$ in $F\ell(m+1)$.

Now $\iota$ determines covariant homomorphisms

$$\iota_* : H_{2d}(F\ell(m)) \rightarrow H_{2d}(F\ell(m+1)),$$

with the property that $\iota_*[Z] = [\iota(Z)]$ for a closed subvariety $Z$ of $F\ell(m)$. When homology is identified with cohomology by Poincaré duality, $\iota_*$ maps $H^{2r}(F\ell(m))$ to $H^{2r+2m}(F\ell(m+1))$, since $m$ is the codimension of $F\ell(m)$ in $F\ell(m+1)$. We also have the contravariant ring homomorphisms

$$\iota^* : H^{2d}(F\ell(m+1)) \rightarrow H^{2d}(F\ell(m)).$$

These are related by the projection formula: $\iota_*(\iota^*(\alpha) \cdot \beta) = \alpha \cdot \iota_*(\beta)$ for $\alpha \in H^*(F\ell(m+1))$ and $\beta \in H^*(F\ell(m))$. For $w \in S_m$, we denote the element $\sigma_w$ in $H^{2\ell(w)}(F\ell(m))$ by $\sigma_w{}^{(m)}$, when we are considering more than one $F\ell(m)$ at a time.

**Lemma 4** *For $w \in S_m$, the homomorphism $\iota^*$ from $H^{2\ell(w)}(F\ell(m+1))$ to $H^{2\ell(w)}(F\ell(m))$ maps $\sigma_w{}^{(m+1)}$ to $\sigma_w{}^{(m)}$.*

**Proof** By the projection formula it follows that for $v \in S_m$,

$$\left\langle \iota^*(\sigma_w{}^{(m+1)}), [X_v] \right\rangle = \left\langle \sigma_w{}^{(m+1)}, \iota_*[X_v] \right\rangle = \left\langle \sigma_w{}^{(m+1)}, [\iota(X_v)] \right\rangle.$$

Since $\iota(X_v)$ is the Schubert variety in $F\ell(m+1)$ corresponding to $v$, it follows from (1) (applied to $F\ell(m+1)$) that the right side of this display is 1 if $v = w$ and 0 otherwise. But we know that $\sigma_w{}^{(m)}$ is the only element of $H^{2\ell(w)}(F\ell(m))$ such that $\langle \sigma_w{}^{(m)}, [X_v] \rangle = \delta_{wv}$ for all $v$. Hence $\iota^*(\sigma_w^{(m+1)}) = \sigma_w{}^{(m)}$.  □

The tautological flag of bundles on $F\ell(E')$ restricts by $\iota$ to the flag of bundles $U_1 \subset \ldots \subset U_m = E \subset E \oplus \mathbb{C}$ on $F\ell(E)$. If we write $x_1, \ldots, x_{m+1}$ for the basic classes on $F\ell(m+1)$, and $x_1, \ldots, x_m$ for the basic classes on $F\ell(m)$, it follows that

$$\iota^*(x_i) = x_i \quad \text{for} \quad 1 \le i \le m, \quad \text{and} \quad \iota^*(x_{m+1}) = 0.$$

In other words, with $R(m)$ as defined in the preceding proposition, if we define a map from $R(m+1)$ to $R(m)$ by the formula $X_i \mapsto X_i$ for $i \le m$

and $X_{m+1} \mapsto 0$, then the diagram

$$
\begin{array}{ccc}
R(m+1) & \to & H^*(F\ell(m+1)) \\
\downarrow & & \downarrow \iota^* \\
R(m) & \to & H^*(F\ell(m))
\end{array}
$$

commutes.

If $P$ is a polynomial in $\mathbb{Z}[X_1, \ldots, X_m]$, let $s_i(P)$ denote the result of interchanging $X_i$ and $X_{i+1}$ in $P$, for any $i$ between 1 and $m-1$. Following Bernstein, Gelfand, and Gelfand (1973) and Demazure (1974), define $\mathbb{Z}$-linear *difference operators* $\partial_i$ on the polynomial ring $\mathbb{Z}[X_1, \ldots, X_m]$ by the rule

$$
(5) \qquad \partial_i(P) = \frac{P - s_i(P)}{X_i - X_{i+1}}, \quad 1 \le i \le m-1.
$$

Note that $P - s_i(P)$ is divisible by $X_i - X_{i+1}$, so the result is always a polynomial. If $P$ is homogeneous of degree $d$, then $\partial_i(P)$ is homogeneous of degree $d - 1$, and $\partial_i(P)$ is always symmetric in the variables $X_i$ and $X_{i+1}$. It follows from the definition that $\partial_i(P) = 0$ if and only if $s_i(P) = P$, i.e., if $P$ is symmetric in the variables $X_i$ and $X_{i+1}$. In particular, $\partial_i(\partial_i(P)) = 0$ for all $P$. It also follows from the definition that for polynomials $P$ and $Q$,

$$
(6) \qquad \partial_i(P \cdot Q) = \frac{PQ - s_i(P)s_i(Q)}{X_i - X_{i+1}}
$$

$$
= \frac{(P - s_i(P))Q + s_i(P)(Q - s_i(Q))}{X_i - X_{i+1}}
$$

$$
= \partial_i(P) \cdot Q + s_i(P) \cdot \partial_i(Q).
$$

In particular, if $Q$ is symmetric in $X_i$ and $X_{i+1}$, then $\partial_i(P \cdot Q) = \partial_i(P) \cdot Q$. It follows from (6) that the operator $\partial_i$ maps the ideal generated by the elementary symmetric polynomials into itself, so $\partial_i$ induces an operator, still denoted by $\partial_i$, on the quotient ring $R(m)$.

A few elementary facts about the actions of these operators on polynomials and on the rings $R(m)$ are proved in the next three lemmas.

**Lemma 5** *Let $P$ be a polynomial in $\mathbb{Z}[X_1, \ldots, X_k]$, and suppose that $\partial_{p_r} \circ \ldots \circ \partial_{p_1}(P)$ is in $\mathbb{Z}[X_1, \ldots, X_k]$ for every choice of $p_1, \ldots, p_r$ taken from $\{1, \ldots, k\}$, and $\partial_{p_r} \circ \ldots \circ \partial_{p_1}(P) = 0$ if any $p_i = k$. Then $P = \sum a_I X^I$, the sum over $I = (i_1, \ldots, i_k)$ with $i_j \le k - j$ for all $j$.*

**Proof** From the definition of $\partial_p$ we have the basic equations

$$\partial_p(X_p^a X_{p+1}^b) =$$

$$(7) \quad \begin{cases} X_p{}^{a-1}X_{p+1}{}^b + X_p{}^{a-2}X_{p+1}{}^{b+1} + \ldots + X_p{}^b X_{p+1}{}^{a-1} & \text{if } a > b \\ 0 & \text{if } a = b \\ -X_p{}^a X_{p+1}{}^{b-1} - X_p{}^{a+1}X_{p+1}{}^{b-2} - \ldots - X_p{}^{b-1}X_{p+1}{}^a & \text{if } a < b. \end{cases}$$

Let $M$ be the $\mathbb{Z}$-submodule of $\mathbb{Z}[X_1, \ldots, X_k]$ spanned by the $X^I$, for $I = (i_1, \ldots, i_k)$, $i_j \leq k-j$, $1 \leq j \leq k$. From (7) we see that each $\partial_p$ maps $M$ to itself. By induction on the degree of $P$ it suffices to show that if $\partial_p(P)$ is in $M$ for $1 \leq p \leq k$, and $\partial_k(P) = 0$, then $P$ is in $M$. Write $P = \sum a_I X^I$, and, if $P$ is not in $M$, let $p$ be the maximal index such that some $a_I$ is nonzero, with $I = (i_1, \ldots, i_k)$ and $i_p > k-p$; let $a$ be maximal among the $i_p$'s that occur in such $I$, and let $b$ be maximal among the $i_{p+1}$'s that occur with $i_p = a$. If $p = k$, then $\partial_p(P) = 0$, and this gives a contradiction. Otherwise from (7) one sees that $\partial_p(P)$ has a nonzero term of the form $a_J X^J$ with

$$J = (i_1, \ldots, i_{p-1}, b, a-1, i_{p+2}, \ldots, i_k),$$

and this contradicts the hypothesis that $\partial_p(P)$ is in $M$.            □

**Lemma 6** *Let $P \in R(m)$, and write $P = \sum a_I x^I$, where the sum is over those $I = (i_1, \ldots, i_m)$ with $i_j \leq m-j$ for all $j$. Suppose there is an integer $k < m$ such that $\partial_i(P) = 0$ in $R(m)$ for all $k < i < m$. Then $a_I = 0$ for any $I = (i_1, \ldots, i_m)$ such that $i_j > 0$ for some $j > k$.*

**Proof** The proof is by descending induction on $k$, the case $k = m-1$ being trivial. We assume the assertion for some $k \leq m-1$, and prove it for $k-1$. By the inductive assumption, $P$ has a unique expression

$$P = \sum_{p=0}^{m-k} Q_p x_k{}^p,$$

with $Q_p$ a linear combination of monomials $x^J$ with $J$ of the form $(i_1, \ldots, i_{k-1})$, $i_j \leq m-j$. Now we have

$$0 = \partial_k(P) = \sum_{p=1}^{m-k} Q_p(x_k{}^{p-1} + x_k{}^{p-2}x_{k+1} + \ldots + x_{k+1}{}^{p-1}).$$

All the monomials $x_1{}^{i_1} \cdot x_2{}^{i_2} \cdot \ldots \cdot x_{k-1}{}^{i_{k-1}} x_k{}^s x_{k+1}{}^t$ that occur in this expression are linearly independent in $R(m)$, since $s$ and $t$ are at most $m-k-1$. It follows that $Q_p = 0$ for $p > 0$, so $P = Q_0$, as desired. $\square$

It also follows from the definitions that, for $i \leq m-1$, $\partial_i$ commutes with the homomorphism from $R(m+1)$ to $R(m)$ defined above.

**Lemma 7** *Suppose for some* $N \geq k \geq 0$ *and* $d \geq 0$, *we have elements* $P^{(m)}$ *in* $R(m)$ *for all* $m \geq N$, *each homogeneous of degree* $d$, *such that*

(1) *the canonical map from* $R(m+1)$ *to* $R(m)$ *maps* $P^{(m+1)}$ *to* $P^{(m)}$, *for all* $m \geq N$;

(2) $\partial_i(P^{(m)}) = 0$ *for all* $i > k$ *and* $m \geq N$.

*Then there is a unique polynomial* $P$ *in* $\mathbb{Z}[X_1, \ldots, X_k]$ *such that the canonical map from* $\mathbb{Z}[X_1, \ldots, X_k]$ *to* $R(m)$ *maps* $P$ *to* $P^{(m)}$ *for all* $m \geq N$.

**Proof** By Lemma 6, each $P^{(m)}$ has a unique expression $P^{(m)} = \sum a_I x^I$, $I = (i_1, \ldots, i_k)$ with $i_j \leq m-j$ for all $j$, $\sum i_j = d$. Note that, since $i_j \leq d$, the condition $i_j \leq m-j$ is vacuous if $m \geq d+k$. It follows that for $m \geq d+k$ and $m \geq N$, the condition that $P^{(m+1)}$ maps to $P^{(m)}$ implies that $P^{(m)}$ and $P^{(m+1)}$ have exactly the same expression as a polynomial in $x_1, \ldots, x_k$. The polynomial $P$ is this unique expression. $\square$

Identifying each $R(m)$ with $H^*(F\ell(m))$, the classes we denoted by $\sigma_w{}^{(m)}$ satisfy condition (1) of Lemma 7. In fact, if $w$ is in $S_k$, we may take $N = k$. The following proposition shows that they satisfy condition (2), for the same $k$. This means that there is a unique polynomial in $\mathbb{Z}[X_1, \ldots, X_k]$ that maps to $\sigma_w{}^{(m)}$ in $H^{2\ell(w)}(F\ell(m))$ for all $m \geq k$. These polynomials will be the Schubert polynomials.

**Proposition 4** *Let* $w \in S_m$, *and let* $1 \leq i \leq m-1$. *Let* $w' = w \cdot s_i$ *be the result of interchanging the values of* $w$ *in positions* $i$ *and* $i+1$.

(1) *If* $w(i) > w(i+1)$, *then* $\partial_i(\sigma_w) = \sigma_{w'}$.
(2) *If* $w(i) < w(i+1)$, *then* $\partial_i(\sigma_w) = 0$.

The proof of this proposition will be carried out by constructing an appropriate $\mathbb{P}^1$-bundle. Fix $i$, and let $Y$ denote the partial flag manifold consisting of flags

$$0 \subset E_1 \subset \ldots \subset E_{i-1} \subset E_{i+1} \subset \ldots \subset E_m = E = \mathbb{C}^m$$

of subspaces of $E$ of all dimensions except $i$. We have a projection $f$ from $X = F\ell(E)$ to $Y$ that takes a complete flag to the flag that omits its $i^{\text{th}}$ member. On $Y$ we have a tautological flag of subbundles of the trivial bundle $E = E_Y$:

$$T_1 \subset \ldots \subset T_{i-1} \subset T_{i-1} \subset \ldots \subset T_m = E.$$

The projection $f$ realizes $X$ as a $\mathbb{P}^1$-bundle over $Y$, namely, the bundle $\mathbb{P}(U)$, where $U = T_{i+1}/T_{i-1}$. Let $Z = X \times_Y X$ be the fiber product, which consists of pairs of flags $(E_\bullet, E_\bullet')$ in which $E_j = E_j'$ for all $j \neq i$. Let $p_1$ and $p_2$ denote the two projections from $Z$ to $X$, each of which realizes $Z$ as a $\mathbb{P}^1$-bundle over $X$.

The proposition will follow from the next two lemmas. For the following lemma, we need the notion of a ***birational*** morphism of algebraic varieties: this is a morphism $\pi : V \to W$ of varieties such that $W$ has a nonempty Zariski open set $U$ such that the map from $\pi^{-1}(U)$ to $U$ is an isomorphism.[1]

**Lemma 8**

(1) If $w(i) < w(i+1)$, then $p_1$ maps $p_2^{-1}(X_w)$ birationally onto $X_{w'}$, where $w' = w \cdot s_i$.

(2) If $w(i) > w(i+1)$, then $p_1$ maps $p_2^{-1}(X_w)$ into $X_w$.

**Proof** For a flag $E_\bullet'$ in $X$, $p_2^{-1}(E_\bullet')$ can be identified with the set of flags $E_\bullet$ such that $E_j = E_j'$ for all $j \neq i$. Suppose $E_\bullet'$ is in $X_w^\circ$, and let $v_1, \ldots, v_m$ be vectors such that $E_k'$ is spanned by $v_1, \ldots, v_k$; these are unique if taken as the rows of a row-reduced matrix as before. A flag $E_\bullet$ is in $p_2^{-1}(E_\bullet)$ if and only if it is either equal to $E_\bullet'$ or it is a flag $E_\bullet'(t)$ with spanning vectors

$$v_1, \ldots, v_{i-1}, \, t \cdot v_i + v_{i+1}, \, v_i, v_{i+2}, \ldots, v_n,$$

for some scalar $t$. (Indeed, these flags give an affine line of flags in the fiber, which, with the point $E_\bullet'$, make up the full projective line of the fiber.) Now if $w(i) > w(i+1)$, one sees from the definition of the Schubert cells that each of these flags $E_\bullet'(t)$ is in $X_w^\circ$. It follows that $p_1$ maps $p_2^{-1}(X_w^\circ)$ into $X_w$, and, taking closures, assertion (2) of the lemma follows.

If $w(i) < w(i+1)$, however, the above flags $E_\bullet'(t)$ are in $X_{w'}^\circ$. In fact, one sees easily that every flag in $X_{w'}^\circ$ has the form $E_\bullet'(t)$ for a unique

---

[1] In characteristic zero, as we are, it is enough to find such a $U$ such that $\pi^{-1}(U) \to U$ is bijective, for then the restriction over a smaller $U$ must be an isomorphism.

scalar $t$ and a unique $E.'$ in $X_w^\circ$. It follows that if $\Delta$ denotes the diagonal (isomorphic to $X$) in $X \times_Y X$, then $p_1$ maps $p_2^{-1}(X_w^\circ) \smallsetminus \Delta$ bijectively onto $X_{w'}^\circ$. (In fact, one can verify that this is an isomorphism, using the natural identification of $X_{w'}^\circ$ with affine space.) Assertion (1) follows by taking closures. □

We will use some general facts about pullback and pushforward maps on cohomology, as described in Appendix B, equations (1)–(8), and we will use Lemma 9 of §B.4. We have realized $X$ as a $\mathbb{P}^1$-bundle $\mathbb{P}(U)$ over $Y$, with projection $f$. If $L$ is the tautological line subbundle of the pullback of $U$ on $X$, and $x = -c_1(L)$, we know that every element of $H^*(X)$ can be written in the form $\alpha x + \beta$, for some unique $\alpha$ and $\beta$ in $H^*(Y)$. By the projection formula, we deduce from Lemma B.9 the following formula, which describes the pushforward $f_*$ entirely:

(8) $\qquad f_*(\alpha x + \beta) = \alpha.$

Now let $Z = X \times_Y X$, and let $p_1$ and $p_2$ be the projections from $Z$ to $X$. The composite $(p_1)_* \circ (p_2^*) : H^*(X) \to H^*(Z) \to H^*(X)$ is determined by the formula

(9) $\qquad (p_1)_* \circ (p_2^*)(\alpha x + \beta) = \alpha \quad$ for all $\quad \alpha, \beta \in H^*(Y).$

This follows from the fact that $p_1 : Z \to X$ is the $\mathbb{P}^1$-bundle of the vector bundle $p_2^*(U)$, so the first Chern class of its tautological line bundle is $p_2^*(-x)$. Then (9) follows from (8), together with the fact that $p_2^*(\gamma) = p_1^*(\gamma)$ for any $\gamma$ coming from $H^*(Y)$, since $f \circ p_2 = f \circ p_1$.

**Lemma 9** *With* $X = F\ell(m)$, *and the identification of* $H^*(X)$ *with* $R(m)$, *the composite* $(p_1)_* \circ (p_2^*) : H^{2d}(X) \to H^{2d}(Z) \to H^{2d-2}(X)$ *is equal to the operator* $\partial_i$.

**Proof** By Proposition 3, but applied to the variables $x_i$ in a different order so that $x_i$ and $x_{i+1}$ are taken to be the last two, we see that every class in $H^*(X) = R(m)$ has a unique expression of the form $\alpha x_i + \beta$, where $\alpha$ and $\beta$ are polynomials in the other $m-2$ variables $x_j$, for $j \neq i, i+1$. From the definition of $\partial_i$ we have $\partial_i(\alpha x_i + \beta) = \alpha$.

Since these other $x_j$ come from the classes $-c_1(T_j/T_{j-1})$ on $Y$, $\alpha$ and $\beta$ must come from classes in $H^*(Y)$. Now $x = -c_1(U_i/U_{i-1}) = x_i$, and so formula (9) implies that $(p_1)_* \circ (p_2^*)(\alpha x_i + \beta) = \alpha$. This completes the proof that $\partial_i = (p_1)_* \circ (p_2^*)$. □

**Proof of Proposition 4** Recall that $\sigma_w = [X_{w^\vee}]$, where $w^\vee = w_o{\cdot}w$.
Note that $w(i) > w(i+1)$ exactly when $w^\vee(i) < w^\vee(i+1)$. It follows that
$p_2{}^*(\sigma_w) = [p_2{}^{-1}(X_{w^\vee})]$. By Lemma 8, $p_1$ maps $p_2{}^{-1}(X_{w^\vee})$ birationally
onto $X_{w^\vee{\cdot}s_i}$ if $w(i) > w(i+1)$, and $p_1$ maps $p_2{}^{-1}(X_{w^\vee})$ into the smaller
variety $X_{w^\vee}$ if $w(i) < w(i+1)$. Hence (by (7) of §B.1)

$$(p_1)_*([p_2{}^{-1}(X_{w^\vee})]) = \begin{cases} [X_{w^\vee{\cdot}s_i}] & \text{if } w(i) > w(i+1) \\ 0 & \text{if } w(i) < w(i+1). \end{cases}$$

But since $w^\vee{\cdot}s_i = w_o{\cdot}w{\cdot}s_i = (w{\cdot}s_i)^\vee$, this means that $(p_1)_*(p_2{}^*(\sigma_w)) = \sigma_{w{\cdot}s_i}$ if $w(i) > w(i+1)$, and $(p_1)_*(p_2{}^*(\sigma_w)) = 0$ otherwise. Applying
Lemma 9, these equations become the assertions of Proposition 4.          □

## 10.4 Schubert polynomials

Let $w$ be a permutation in some $S_k$. We have a class $\sigma_w{}^{(m)}$ in
$R(m) = H^*(F\ell(m))$ for all $m \geq k$. By Lemma 4 and Proposition 4(2) these
classes satisfy the conditions of Lemma 7, for $N = k$, with $d = \ell(w)$. Hence
there is a unique homogeneous polynomial of degree $\ell(w)$ in $\mathbb{Z}[X_1, \ldots, X_k]$,
denoted $\mathfrak{S}_w = \mathfrak{S}_w(X_1, \ldots, X_k)$, that maps to $\sigma_w{}^{(m)}$ in $H^{2\ell(w)}(F\ell(m))$
for all $m \geq k$. This polynomial is called the **Schubert polynomial** corre-
sponding to $w$.

### Proposition 5

(1) *For any $i$, $\partial_i(\mathfrak{S}_w) = \mathfrak{S}_{w{\cdot}s_i}$ if $w(i) > w(i+1)$, and $\partial_i(\mathfrak{S}_w) = 0$ if*
$w(i) < w(i+1)$.
(2) *For $w \in S_k$, $\mathfrak{S}_w = \sum a_I X^I$, the sum over $I = (i_1, \ldots, i_k)$ with*
$i_j \leq k-j$ for all $j$.

**Proof** (1) For any $m \geq k$ and $m > i$, we know from the definition of $\partial_i$
that $\partial_i(\mathfrak{S}_w)$ maps to $\partial_i(\sigma_w)$ in $R(m) = H^*(F\ell(m))$. By Proposition 4,
this class is $\sigma_{w{\cdot}s_i}$ (resp. 0) if $w(i) > w(i+1)$ (resp. $w(i) < w(i+1)$). It
follows that $\partial_i(\mathfrak{S}_w)$ and $\mathfrak{S}_{w{\cdot}s_i}$ (resp. 0) map to the same class for all large
$m$, so they must be equal by Lemma 7 again.

(2) Apply part (1) and Lemma 5, noting that if $w \in S_k$, then $w{\cdot}s_i$ is in
$S_k$ for all $i < k$, and $\partial_k(\mathfrak{S}_w) = 0$ by (1).          □

So far we have not calculated a single Schubert polynomial. The only one that is immediately obvious from the definition is that corresponding to the identity permutation $w = 1\ 2\ldots m$. In this case $\mathfrak{S}_w = 1$, since its image in $H^0(F\ell(m))$ must be the class of $F\ell(m)$. Proposition 5, however, leads to an algorithm for calculating all of the Schubert polynomials. For example, for those of degree 1, the polynomial $\mathfrak{S}_{s_i}$ must have the property that $\partial_i(\mathfrak{S}_{s_i}) = 1$, and $\partial_j(\mathfrak{S}_{s_i}) = 0$ for all $j \neq i$. In fact,

$$\mathfrak{S}_{s_i} = X_1 + X_2 + \ldots + X_i.$$

Indeed, multiples of this are the only homogeneous polynomials of degree 1 that are symmetric in $X_j$ and $X_{j+1}$ for all $j \neq i$; and the coefficient of $X_i$ must be 1. From this one can similarly calculate all the Schubert polynomials of degree 2, and so on. It is more direct, however, to calculate the Schubert polynomial for some permutations of large length, and then apply the operators $\partial_i$ to derive formulas for shorter permutations. For this we need the following lemma.

**Lemma 10** *For* $w_\circ = m\ m-1\ \ldots\ 2\ 1$ *the permutation of longest length in* $S_m$,

$$\mathfrak{S}_{w_\circ} = X_1{}^{m-1} \cdot X_2{}^{m-2} \cdot \ldots \cdot X_{m-2}{}^2 \cdot X_{m-1}.$$

**Proof** There is only one monomial of the required form of length $n = \ell(w_\circ) = m(m-1)/2$, namely the one displayed in the lemma; by Proposition 5(2), $\mathfrak{S}_{w_\circ}$ must be a scalar multiple of this monomial. Now $w_\circ$ is taken to the identity permutation by multiplying on the right by the transpositions

(10) $\qquad (s_1 s_2 \cdot \ldots \cdot s_{m-1})(s_1 s_2 \cdot \ldots \cdot s_{m-2}) \cdot \ldots \cdot (s_1 s_2) s_1.$

Applying the corresponding sequence of operators to this monomial, one sees immediately that it is taken to 1. By Proposition 5(1), the same operators must take $\mathfrak{S}_{w_\circ}$ to 1, so $\mathfrak{S}_{w_\circ}$ must be equal to the required monomial. $\qquad \square$

This gives the following algorithm for calculating any Schubert polynomial. Given $w$ in $S_m$, write $w = w_\circ \cdot s_{i_1} \cdot s_{i_2} \cdot \ldots \cdot s_{i_r}$, where $\ell(w_\circ \cdot s_{i_1} \cdot \ldots \cdot s_{i_p}) = n - p$ for $1 \leq p \leq r$. Then

(11) $\qquad \mathfrak{S}_w = \partial_{i_r} \circ \ldots \circ \partial_{i_2} \circ \partial_{i_1} \left( X_1{}^{m-1} \cdot X_2{}^{m-2} \cdot \ldots \cdot X_{m-2}{}^2 \cdot X_{m-1} \right).$

By what we have proved, this is independent of the choice of sequence of elementary transpositions, and even of the choice of $m$. The general rule is that successive numerals can be interchanged when the first is larger than the second. As an example, we work out the Schubert polynomial $\mathfrak{S}_w$ for $w = 4\,1\,3\,5\,2$. Starting with $w_\circ = 5\,4\,3\,2\,1$, first interchange the first two digits to arrive at $4\,5\,3\,2\,1$:

$$\mathfrak{S}_{45321} = \partial_1(X_1{}^4 X_2{}^3 X_3{}^2 X_4) = X_1{}^3 X_2{}^3 X_3{}^2 X_4.$$

Then interchange the fourth and fifth:

$$\mathfrak{S}_{45312} = \partial_4(X_1{}^3 X_2{}^3 X_3{}^2 X_4) = X_1{}^3 X_2{}^3 X_3{}^2.$$

Then interchange the third and fourth:

$$\mathfrak{S}_{45132} = \partial_3(X_1{}^3 X_2{}^3 X_3{}^2) = X_1{}^3 X_2{}^3 X_3 + X_1{}^3 X_2{}^3 X_4.$$

Then interchange the second and third:

$$\mathfrak{S}_{41532} = \partial_2(X_1{}^3 X_2{}^3 X_3 + X_1{}^3 X_2{}^3 X_4)$$
$$= X_1{}^3 X_2{}^2 X_3 + X_1{}^3 X_2 X_3{}^2 + X_1{}^3 X_2{}^2 X_4$$
$$+ X_1{}^3 X_2 X_3 X_4 + X_1{}^3 X_3{}^2 X_4.$$

Then interchange the third and fourth again:

$$\mathfrak{S}_{41352} = \partial_3(\mathfrak{S}_{41532})$$
$$= X_1{}^3 X_2{}^2 + X_1{}^3 X_2 X_3 + X_1{}^3 X_2 X_4 - X_1{}^3 X_2{}^2 + X_1{}^3 X_3 X_4$$
$$= X_1{}^3 X_2 X_3 + X_1{}^3 X_2 X_4 + X_1{}^3 X_3 X_4.$$

One could arrive at the same result in several other ways, e.g., by applying $\partial_2 \circ \partial_3 \circ \partial_2 \circ \partial_1 \circ \partial_4$ or $\partial_2 \circ \partial_3 \circ \partial_4 \circ \partial_2 \circ \partial_1$ in place of $\partial_3 \circ \partial_2 \circ \partial_3 \circ \partial_4 \circ \partial_1$.

**Exercise 5** Calculate $\mathfrak{S}_{21543}$, and show that the coefficient of the monomial $X_1{}^2 X_2 X_3$ is 2.

It is a fact that all the coefficients of the monomials $X^I$ are nonnegative, whenever one expands the Schubert polynomials in terms of this basis (see Macdonald [1991a], 4.17); there are combinatorial formulas for the coefficients (see Billey, Jockusch, and Stanley [1993]), although mysteries about them remain.

We have a polynomial $\mathfrak{S}_w$ for every $w$ in $S_\infty = \bigcup S_m$. We have seen that $\mathfrak{S}_w$ is in $\mathbb{Z}[X_1, \ldots, X_k]$ if and only if $w(i) < w(i+1)$ for all $i \geq k$.

**Proposition 6** *The Schubert polynomials $\mathfrak{S}_w$, as $w$ varies over all permutations in $S_\infty$ such that $w(i) < w(i+1)$ for all $i \geq k$, form an additive basis for $\mathbb{Z}[X_1, \ldots, X_k]$.*

**Proof** If $P$ has degree $d$, choose $m \geq d+k$, and write $P = \sum a_w \mathfrak{S}_w$ in $R(m)$. All monomials appearing on both sides of this equation are of the form $X^I$, $I = (i_1, \ldots, i_k)$, $i_j \leq m-j$, and these monomials are independent in $R(m)$, so this is an equality of polynomials. For $i \geq k$, we have $0 = \partial_i(P) = \sum a_w \partial_i(\mathfrak{S}_w) = \sum a_w \mathfrak{S}_{w \cdot s_i}$, the latter sum over those $w$ with $w(i) > w(i+1)$. Since these $\mathfrak{S}_{w \cdot s_i}$ are linearly independent in $R(m)$, we must have $a_w = 0$ if $w(i) > w(i+1)$.  □

For any permutation $w$ in $S_\infty$ one can define an operator $\partial_w$ on the ring of polynomials $\mathbb{Z}[X_1, X_2, \ldots]$ by writing $w = s_{i_1} \cdot \ldots \cdot s_{i_\ell}$, with $\ell = \ell(w)$, and setting $\partial_w = \partial_{i_1} \cdot \ldots \cdot \partial_{i_\ell}$. It follows from Proposition 5 that $\partial_w(\mathfrak{S}_v) = \mathfrak{S}_{v \cdot w^{-1}}$ if $\ell(v \cdot w^{-1}) = \ell(v) - \ell(w)$, and $\partial_w(\mathfrak{S}_v) = 0$ otherwise. These assertions are independent of how $w$ is written as a product of these transpositions, so, by Proposition 6, the operator $\partial_w$ is independent of the representation of $w$ as a reduced word. Similarly, we see that

$$(12) \qquad \partial_{u \cdot v} = \begin{cases} \partial_u \circ \partial_v & \text{if } \ell(u \cdot v) = \ell(u) + \ell(v) \\ 0 & \text{otherwise.} \end{cases}$$

The following exercise will be used in §10.6.

**Exercise 6** If $w(1) > w(2) > \ldots > w(d)$, and $w(i) < w(i+1)$ for all $i \geq d$, show that $\mathfrak{S}_w = X_1^{w(1)-1} \cdot X_2^{w(2)-1} \cdot \ldots \cdot X_d^{w(d)-1}$.

**Exercise 7** For any subset $T$ of $[k]$, show that the Schubert polynomials $\mathfrak{S}_w$, where $w$ varies over permutations in $S_k$ such that $w(i) < w(i+1)$ if $i \notin T$, form a basis for the polynomials in $\mathbb{Z}[X_1, \ldots, X_k]$ that are symmetric in variables $X_i$ and $X_{i+1}$ for all $i$ in $T$.

It is a useful exercise to work out the Schubert polynomials for permutations in $S_3$ and $S_4$. For the answers, see Macdonald (1991a), p. 63.

## 10.5 The Bruhat order

Our object in this section is to describe for which pairs $u$ and $v$ of permutations, $X_u$ is contained in $X_v$, i.e., $X_u^\circ$ is contained in the closure of $X_v^\circ$. For $w$ in $S_m$, define

$$r_w(p, q) = \#\{i \leq p : w(i) \leq q\},$$

for all $1 \leq p, q \leq m$. Our first combinatorial definition of the Bruhat order, denoted " $\leq$ ," will be that $u \leq v$ if $r_u \geq r_v$, which means that $r_u(p,q) \geq r_v(p,q)$ for all $p$ and $q$.

**Lemma 11** *Suppose $u \leq v$, $u \neq v$. Let $j$ be the smallest integer such that $u(j) \neq v(j)$. Then $v(j) > u(j)$. Let $k$ be the smallest integer greater than $j$ such that $v(j) > v(k) \geq u(j)$. Let $v' = v \cdot (j,k)$ be the result of interchanging the values of $v$ in positions $j$ and $k$. Then $u \leq v' \leq v$.*

**Proof** That $v' \leq v$ is clear from the definition. We must show that $u \leq v'$, i.e., $r_u \geq r_{v'}$. With $j$ fixed as in the statement of the lemma, define, for any $w$ in $S_m$, and, for any $p \geq j$ and any $q$,

$$\tilde{r}_w(p,q) \;=\; \#\{i \in [j,p] : w(i) \leq q\}.$$

Since $u$ and $v'$ have the same values at all $i < j$, it suffices to show that $\tilde{r}_u \geq \tilde{r}_{v'}$.

First, for $p \notin [j,k)$, it is evident that $\tilde{r}_{v'}(p,q) = \tilde{r}_v(p,q)$, which implies that $\tilde{r}_u(p,q) \geq \tilde{r}_v(p,q) = \tilde{r}_{v'}(p,q)$. We therefore may assume that $j \leq p < k$. If $q \notin [v(k), v(j))$, it is again evident that $\tilde{r}_{v'}(p,q) = \tilde{r}_v(p,q)$, so we may assume that $v(k) \leq q < v(j)$. For such $p$ and $q$, $\tilde{r}_{v'}(p,q) = \tilde{r}_v(p,q) + 1$, so it suffices to show that $\tilde{r}_u(p,q) > \tilde{r}_v(p,q)$. Now

$$\tilde{r}_u(p,q) \;>\; \tilde{r}_u(p, u(j)-1)$$

since $q \geq u(j)$, and $j$ will count for the left side but not the right. By assumption,

$$\tilde{r}_u(p, u(j)-1) \;\geq\; \tilde{r}_v(p, u(j)-1).$$

And

$$\tilde{r}_v(p, u(j)-1) \;=\; \tilde{r}_v(p,q)$$

since none of $v(j), v(j+1), \ldots, v(p)$ are in $[u(j), q] \subset [u(j), v(j)]$. These combine to give the required inequality $\tilde{r}_u(p,q) > \tilde{r}_v(p,q)$.  □

Note that for $j < k$ and $v(j) > v(k)$, it is clear from the definition that $v' = v \cdot (j,k)$ is less than $v$ in the Bruhat order. If, in addition, $v$ and $v'$ are as in Lemma 11, then $\ell(v') = \ell(v) - 1$. The procedure of Lemma 11 gives a canonical way to construct a sequence from $v$ to $u$ if $v > u$, with each larger than the next in the Bruhat order, and each of length one larger than the next. In fact, this procedure can be carried out with any $v$ and $u$, and if the chain does not arrive at $u$, then $u$ is not less than $v$ in the Bruhat order. For example, take $u = 4\,2\,8\,3\,6\,1\,7\,9\,5$ and $v = 6\,7\,9\,2\,5\,1\,8\,3\,4$. The sequence constructed by this algorithm is as indicated, with the underlined pair switched at the next step:

$$v = 6\,7\,9\,2\,\underline{5}\,1\,8\,3\,4 \;\mapsto\; \underline{5}\,7\,9\,2\,6\,1\,8\,3\,\underline{4} \;\mapsto\; 4\,7\,\underline{9}\,2\,6\,1\,8\,3\,5$$

$$\mapsto\; 4\,2\,\underline{9}\,7\,6\,1\,\underline{8}\,3\,5 \;\mapsto\; 4\,2\,8\,\underline{7}\,6\,\underline{1}\,9\,3\,5 \;\mapsto\; 4\,2\,8\,\underline{6}\,7\,1\,9\,\underline{3}\,5$$

$$\mapsto\; 4\,2\,8\,3\,\underline{7}\,1\,9\,\underline{6}\,5 \;\mapsto\; 4\,2\,8\,3\,6\,1\,\underline{9}\,\underline{7}\,5 \;\mapsto\; 4\,2\,8\,3\,6\,1\,7\,9\,5 \;=\; u.$$

Similarly, with the same $v$ but with $u = 4\,2\,8\,6\,7\,1\,9\,5\,3$, the same chain, when it arrives at the end of the second line, shows that $u \not\le v$.

**Corollary 1** *For $u$ and $v$ in $S_m$, $u \le v$ if and only if there is a sequence $(j_1,k_1), \ldots, (j_r,k_r)$ with $j_i < k_i$ for all $i$, such that with $v_0 = v$, and $v_i = v\cdot(j_1,k_1)\cdot\ldots\cdot(j_i,k_i)$, then $v_{i-1}(j_i) > v_{i-1}(k_i)$ for $1 \le i \le r$, and $v_r = u$.* □

**Exercise 8** Show that $u \le v$ if and only if, for $1 \le p \le m$, when the sets $\{u(1), \ldots, u(p)\}$ and $\{v(1), \ldots, v(p)\}$ are each arranged in increasing order, each term of the first is less than or equal to the corresponding term of the second.

The following exercise, not needed here, shows that the definition of the Bruhat order given in this section is equivalent to the standard one of Chevalley (1994).

**Exercise 9** (a) Suppose $u(k) < u(k+1)$ and $v(k) < v(k+1)$. Show that $u \le v \iff u\cdot s_k \le v\cdot s_k$. (b) Let $\ell = \ell(v)$, $d = \ell - \ell(u)$. Show that $u \le v \iff u$ can be obtained from any (or every) representation of $v$ as a product of $\ell$ elements of $\{s_1, \ldots, s_{m-1}\}$ by removing $d$ terms.

**Proposition 7** *For $u$ and $v$ in $S_m$, the following are equivalent:*

(i)   $u \le v$

(ii)  $X_u \subset X_v$

(iii) $\Omega_u \supset \Omega_v$.

**Proof** We prove (i) $\Rightarrow$ (ii). If $u \le v$ we may assume that $u = v\cdot(j,k)$ with $j < k$ and $v(j) > v(k)$. Since $X_u$ and $X_v$ are preserved by the Borel group $B$, it suffices to show that the point $x(u)$ is in $X_v$. For $t \ne 0$, consider the flag $E_\cdot(t)$ spanned by successive vectors $f_1, \ldots, f_m$, where $f_i = e_{v(i)}$ for $i \ne j, k$, and

$$f_j \;=\; e_{v(j)} + \frac{1}{t}e_{v(k)}, \qquad f_k \;=\; e_{v(k)}.$$

Equivalently, we can take

$$f_j \;=\; e_{v(k)} + t e_{v(j)} \;=\; e_{u(j)} + t\, e_{u(k)}, \qquad f_k \;=\; e_{v(j)} \;=\; e_{u(k)}.$$

The first form shows that each $E_\bullet(t)$ is in $X_v^\circ$, and the second shows that the limit, as $t \to 0$, is $x(u)$. (Since the topology of the flag manifold is that induced by its embedding in projective space, this is verified by computing Plücker coordinates.) This completes the proof of (i) $\Rightarrow$ (ii).

Let (temporarily) $\mathfrak{X}_w$ be the set of flags $E_\bullet$ in $F\ell(m)$ such that $\dim(E_p \cap F_q) \geq r_w(p,q)$ for all $p$ and $q$. This is a closed algebraic subset of $F\ell(m)$, locally defined by setting certain minors equal to zero. Clearly $X_w^\circ \subset \mathfrak{X}_w$, since $X_w^\circ$ is the locus where equality holds in these inequalities. It follows that $X_w \subset \mathfrak{X}_w$. To prove (ii) $\Rightarrow$ (i), it suffices to show that if $u \not\leq v$, then $X_u^\circ \cap \mathfrak{X}_v = \emptyset$. In fact, if $p$ and $q$ are chosen so that $r_u(p,q) < r_v(p,q)$, then it follows from the definitions that no point of $X_u^\circ$ can be in $\mathfrak{X}_v$. This certainly implies that $X_u \not\subset X_v$. In fact, this shows in addition that $X_w = \mathfrak{X}_w$ for all $w$.

Finally we prove (ii) $\Longleftrightarrow$ (iii). Since $\Omega_w = X_{w_\circ \cdot w}(\tilde{F}_\bullet)$, this is equivalent to the assertion that $u \leq v \Longleftrightarrow w_\circ \cdot u \geq w_\circ \cdot v$. Since

$$r_{w_\circ \cdot w}(p,q) \;=\; \#\{i \leq p : m+1-w(i) \leq q\}$$

$$=\; p - \#\{i \leq p : w(i) \leq m-q\}$$

$$=\; p - r_w(p, m-q),$$

this is clear from the definition. $\square$

**Corollary of proof** *The Schubert variety $X_w$ is the locus of flags $E_\bullet$ satisfying the conditions*

$$\dim(E_p \cap F_q) \;\geq\; r_w(p,q) \quad \text{for all } p \text{ and } q.$$

**Exercise 10** Suppose $T$ is a subset of $[m]$ such that $w(p) > w(p+1)$ for all $p \notin T$. Show that the conditions that $\dim(E_p \cap F_q) \geq r_w(p,q)$ for all $p$ and $q$ follow from those for $p \in T$ and all $q$.

It is a fact that the prime ideal of $X_w$ in the multihomogeneous coordinate ring of the flag manifold is generated by the corresponding homogeneous coordinates. The multihomogeneous coordinates $X_I$ correspond to subsets $I$ of $[m]$. The generators of the prime ideal of $X_w$ are those $X_I$ for which $I$ satisfies the following condition: if $I = \{i_1 < \ldots < i_d\}$, and

$K_d(w) = \{k_1 < \ldots < k_d\}$ consists of the elements $\{w(n+1-d), \ldots, w(n)\}$, but put in increasing order, then $i_j < k_j$ for some $j$.

**Exercise 11** Let $J_w$ be the ideal generated by the $X_I$ described in the preceding paragraph. (a) Show that $X_w$ is the set of zeros of $J_w$. Let $A$ be the $m \times m$ matrix whose entry $A_{i,j}$ is 0 if $j < m+1-w(i)$ or if $i > w^{-1}(m+1-j)$, and whose other entries are indeterminates. Map the polynomials ring $\mathbb{C}[\{X_I\}]$ to $\mathbb{C}[\{A_{i,j}\}]$ by sending $X_I$, for $I = \{i_1 < \ldots < i_d\}$, to the determinant of the minor of $A$ consisting of the first $d$ rows and the columns numbered $i_1, \ldots, i_d$. (b) Show that the kernel of this homomorphism is the ideal of $X_w$.

In addition, one can give an additive basis for the multihomogeneous co-ordinate ring $\mathbb{C}[\{X_I\}]/I(X_w) = \mathbb{C}[\{X_I\}]/J_w$. These are the images of the monomials $e_T$, as $T$ varies over the Young tableaux with entries in $[m]$ such that $w_-(T) \geq w_\circ \cdot w$, where $w_-(T)$ is the permutation obtained from the left key $K_-(T)$ as at the end of §A.5 of Appendix A.

Although we were able to use some representation theory to give elementary proofs of the corresponding facts for the coordinate ring of the the flag manifold itself, we know no such proofs of the facts stated here for the Schubert subvarieties, even for the equality of $J_w$ with $I(X_w)$. They can be extracted from the "standard monomial theory" developed by Lakshmibai, Musili, and Seshadri (see Lakshmibai and Seshadri [1986], as well as Ramanathan [1987]).[2]

The Robinson correspondence of Chapter 4 can also be found in the geometry of the flag variety. If $u$ is a unipotent automorphism of a vector space $E$, then $u$ determines a partition $\lambda$ of the dimension $m$ of $E$, where the parts of $\lambda$ are the sizes of the Jordan blocks of $u$. If $E_\bullet$ is a complete flag fixed by $u$, it is not hard to see that the partitions $\lambda^{(i)}$ of the restriction of $u$ to $E_i$ are nested: $\lambda^{(1)} \subset \ldots \subset \lambda^{(m)} = \lambda$. Therefore $E_\bullet$ determines a standard tableau on $\lambda$, that places $i$ in the box of the skew diagram $\lambda^{(i)}/\lambda^{(i-1)}$. Steinberg (1988) shows that for a generic flag $E_\bullet$ determining a standard tableau $P$ of shape $\lambda$, and a generic flag $F_\bullet$ determining a standard tableau $Q$ of shape $\lambda$, then the permutation $w$ corresponding to the pair $(P,Q)$ by the Robinson correspondence is the permutation $w$ describing the attitude of $E_\bullet$ with respect to $F_\bullet$: the dimension of $E_p \cap F_q$ is the number $r_w(p,q) = \#\{i \leq p : w(i) \leq q\}$.

---

[2] V. Reiner and M. Shimozono have given elementary proofs in a new preprint: "Straightening for standard monomials on Schubert varieties."

## 10.6 Applications to the Grassmannian

In order to relate the geometry on flag manifolds to the geometry on Grassmannians, we need to calculate a less trivial class of Schubert polynomials.
For this we need the following lemma (see Macdonald [1991a]).

**Lemma 12** *For* $w_o = m \ m{-}1 \ . \ . \ 2 \ 1$ *in* $S_m$,

$$\partial_{w_o} = \frac{1}{\Delta} \sum_{w \in S_m} \mathrm{sgn}(w)w,$$

*where* $\Delta = \prod_{i<j}(X_i - X_j)$.

**Proof** Let $u = w_o$, and write $\partial_i = (X_i - X_{i+1})^{-1} \cdot (1 - s_i)$, where $s_i$
operates on polynomials by interchanging $X_i$ and $X_{i+1}$. Any composite of
such operators can be written as a linear combination of rational functions times
operators $w$, as $w$ varies over the symmetric group. Write $\partial_u = \sum R_w \cdot w$,
for some $R_w \in \mathbb{Q}(X_1, \dots, X_m)$. By (12), $\partial_v \circ \partial_u = 0$ for all $v$ in $S_m$,
so $v \cdot \partial_u = \partial_u$ for all $v$ in $S_m$. Hence $v(R_w) = R_{v \cdot w}$ for all $v$ and $w$.
So it suffices to show that $R_u = \mathrm{sgn}(u) \cdot \Delta^{-1}$. Using the factorization (10)
for $u$, we have

$$\partial_u = (\partial_1 \partial_2 \cdot \ldots \cdot \partial_{m-1})(\partial_1 \partial_2 \cdot \ldots \cdot \partial_{m-2}) \cdot \ldots \cdot (\partial_1 \partial_2) \partial_1.$$

To pick out the coefficient of $u$ in this expression is a fairly straightforward
calculation. To carry it out, one can use the following identity: with $t_i = (X_i - X_{i+1})^{-1} \cdot s_i$, and $1 \leq p \leq e < m$,

$$t_{e-p} \cdot t_{e-p+1} \cdot \ldots \cdot t_e \cdot \prod_{1 \leq i < j \leq e} (X_i - X_j)^{-1}$$

$$= \prod_{\substack{1 \leq i < j \leq e+1 \\ j \neq e-p}} (X_i - X_j)^{-1} \cdot (s_{e-p} \cdot s_{e-p+1} \cdot \ldots \cdot s_e).$$

This is easily proved by induction on $p$, and then the expression for $R_u$
follows. $\quad\square$

With this we can show that some of the Schubert polynomials coincide with
the Schur polynomials we studied in Chapters 4 and 6:

**Proposition 8** *If* $w(i) < w(i{+}1)$ *for all* $i \neq r$, *then* $\mathfrak{S}_w = s_\lambda(X_1, \dots, X_r)$,
*where* $\lambda = (w(r) - r, w(r{-}1) - (r{-}1), \dots, w(2) - 2, w(1) - 1)$.

**Proof** Let $u = w_o^{(r)} = r\ r-1\ldots2\ 1$, and let $w' = w{\cdot}u$, which rearranges the first $r$ entries of $w$ in descending order. By Exercise 6,

$$\mathfrak{S}_{w'} = X_1^{w(r)-1}X_2^{w(r-1)-1}{\cdot}\ldots{\cdot}X_r^{w(1)-1}$$
$$= X_1^{\lambda_1+r-1}X_2^{\lambda_2+r-2}{\cdot}\ldots{\cdot}X_r^{\lambda_r}.$$

Now, $\mathfrak{S}_w = \partial_u(\mathfrak{S}_{w'})$, and the conclusion follows from Lemma 12 and the Jacobi–Trudi formula for the Schur polynomials.  □

For a vector space $E$ of dimension $m$, and any $r$ between 1 and $m$, there is a canonical projection $\rho: F\ell(E) \to Gr_r E$ from the flag manifold to the Grassmann variety of subspaces of $E$ of dimension $r$; it takes a flag $E_{\scriptscriptstyle\bullet}$ to the term $E_r$ of dimension $r$. Choose a basis, identifying $E$ with $\mathbb{C}^m$. For each partition $\lambda$ of the form $m-r \geq \lambda_1 \geq \ldots \geq \lambda_r \geq 0$, we defined in §9.4 a Schubert variety $\Omega_\lambda \subset Gr_r E = Gr^{m-r} E$.

**Proposition 9** *For $\lambda$ and $w$ related as in Proposition 8,*

$$\rho^{-1}(\Omega_\lambda) = \Omega_w, \quad so \quad \rho^*(\sigma_\lambda) = \sigma_w \quad in \quad H^{2\ell(w)}(F\ell(m)).$$

**Proof** By the corollary in the preceding section, we know that $\Omega_w$ is the locus of flags $E_{\scriptscriptstyle\bullet}$ such that $\dim(E_s \cap \widetilde{F}_t) \geq \#\{i \leq s : w(i) \geq m+1-t\}$ for all $s$ and $t$. By Exercise 10, we know that this is the locus described by the condition that $\dim(E_r \cap \widetilde{F}_t) \geq \#\{i \leq r : w(i) \geq m+1-t\}$ for all $t$. But for $i \leq r$, $w(i) = \lambda_{r+1-i} + i$, so

$$\#\{i \leq r : w(i) \geq m+1-t\} = \#\{i \leq r : \lambda_{r+1-i}+i \geq m+1-t\}$$
$$= \#\{i \leq r : \lambda_i+r+1-i \geq m+1-t\}$$
$$= \#\{i \leq r : \lambda_i \geq m-r+i-t\}.$$

For $t = m-r+i-\lambda_i$, this number is $i$, so $\Omega_w$ is the locus of flags $E_{\scriptscriptstyle\bullet}$ such that $\dim(E_r \cap \widetilde{F}_{m-r+i-\lambda_i}) \geq i$ for $1 \leq i \leq r$. But $\Omega_\lambda$ is the locus of $r$-planes $E_r$ defined by the same conditions, and this means that $\rho^{-1}(\Omega_\lambda) = \Omega_w$.

We know that $\Omega_\lambda$ is an irreducible subvariety of $Gr_r E$. Since $\rho$ is a composite of a sequence of projective bundle projections, it follows that $\rho^{-1}(\Omega_\lambda)$ is an irreducible subvariety of $F\ell(m)$, and, by (8) of §B.1, that $\rho^*([\Omega_\lambda]) = [\rho^{-1}(\Omega_\lambda)]$. Since we know that $\rho^{-1}(\Omega_\lambda) = \Omega_w$, it follows that $\rho^*([\Omega_\lambda]) = [\Omega_w]$, so $\rho^*(\sigma_\lambda) = \sigma_w$.  □

Propositions 8 and 9 can be used to recover the facts we proved by other methods in §9.4. Since the pullback $\rho^*$ is one-to-one (being a composition of projective bundle projections), it follows that $\sigma_\lambda$ is the unique class in $H^{2|\lambda|}(Gr_r E)$ whose pullback to $H^{2|\lambda|}(F\ell(E))$ is the Schur polynomial. From this the Giambelli formula for $\sigma_\lambda$ follows, and the Pieri formula and the general formula for multiplying two such classes follow formally.

**Exercise 12** Show that for any irreducible subvariety $Z$ of codimension $d$ of $F\ell(m)$, one has an equation $[Z] = \sum a_w[\Omega_w]$, the sum over permutations $w$ of length $d$ in $S_m$, with coefficients $a_w$ all nonnegative integers. Deduce that for any $u$ and $v$ in $S_m$, there is an equation $[\Omega_u] \cdot [\Omega_v] = \sum c_{u,v}^w [\Omega_w]$, the sum over $w$ in $S_m$ with $\ell(w) = \ell(u) + \ell(v)$, and coefficients $c_{u,v}^w$ nonnegative.

It follows from the preceding exercise that there are identities

$$\mathfrak{S}_u \cdot \mathfrak{S}_v = \sum c_{u,v}^w \mathfrak{S}_w$$

the sum over $w$ in $S_m$ with $\ell(w) = \ell(u) + \ell(v)$, and coefficients $c_{u,v}^w$ some nonnegative integers. Although there are algorithms for calculating these coefficients, it is remarkable that there is not yet any combinatorial formula for these numbers, such as the Littlewood–Richardson rule we have for the Schur polynomials. In fact, the only known proof that the coefficients are nonnegative uses the geometry of the flag manifold.

One case where an explicit formula is known is the analogue of Pieri's formula, known as ***Monk's formula,*** which gives the product of a linear Schubert polynomial $\mathfrak{S}_{s_r} = X_1 + \ldots + X_r$ times an arbitrary Schubert polynomial $\mathfrak{S}_w$, $w \in S_m$. Monk's formula can be written

$$(13) \qquad \mathfrak{S}_{s_r} \cdot \mathfrak{S}_w = \sum \mathfrak{S}_v,$$

the sum over all $v$ that are obtained from $w$ by interchanging values of a pair $p$ and $q$, with $1 \leq p \leq r$ and $r < q \leq m$ such that $w(p) < w(q)$ and $w(i)$ is not in the interval $(w(p), w(q))$ for any $i$ in the interval $(p, q)$. (These are exactly the $v$ of the form $v = w \cdot t$, where $t$ is a transposition of some $p \leq r$ and $q > r$ with $\ell(w \cdot t) = \ell(w) + 1$.) This can be proved by a geometric argument similar to the one we gave in §9.4 (see Monk [1959]). Monk's formula is equivalent to the formula

$$(14) \qquad X_r \cdot \mathfrak{S}_w = \sum \mathfrak{S}_{w'} - \sum \mathfrak{S}_{w''},$$

the first sum over those $w'$ obtained from $w$ by interchanging the values of $w$ in positions $r$ and $q$ for those $r < q$ with $w(r) < w(q)$, and

$w(i) \notin (w(r), w(q))$ if $i \in (r,q)$, and the second sum over those $w''$ obtained from $w$ by interchanging the values of $w$ in positions $r$ and $p$ for those $p < r$ with $w(p) < w(r)$, and $w(i) \notin (w(p), w(r))$ if $i \in (p,r)$. This identity can be proved by induction on the length of $w$, by showing that the difference of the two sides is annihilated by all $\partial_i$ for $1 \leq i \leq m$. For a more elegant proof, see Macdonald (1991a), 4.15. For a generalization, see Sottile (1996).

The results of this section can be generalized from Grassmannians to general partial flag varieties $X' = F\ell_{r_1,\ldots,r_s}(E)$, consisting of flags $V_1 \subset \ldots \subset V_s \subset E$ with $\dim(V_i) = r_i$; that is, $X' = F\ell^{d_1,\ldots,d_s}(E)$, with $d_i = m - r_i$. We have a canonical projection $\rho : F\ell(E) \to X'$, and again $\rho^*$ embeds the cohomology of $X'$ in that of $F\ell(E)$. Again there are Schubert varieties $X'_w$, defined for those $w$ in $S_m$ such that $w(p) > w(p+1)$ for all $p \notin \{r_1, \ldots, r_s\}$; $X'_w$ is defined to be the set of flags of given ranks satisfying the conditions that $\dim(V_p \cap F_q) \geq r_w(r_p, q)$ for $1 \leq p \leq s$ and $1 \leq q \leq m$. It follows from Exercise 10 that $\rho^{-1}(X'_w) = X_w$. Similarly, if $w(p) < w(p+1)$ for all $p \notin \{r_1, \ldots, r_s\}$, and $\Omega'_w$ is defined by the conditions

$$\dim(V_p \cap F_q) \geq \#\{i \leq r_p : w(i) \geq m+1-q\},$$

then $\rho^{-1}(\Omega'_w) = \Omega_w$. It follows that the class $[\Omega'_w]$ of this Schubert variety in $H^{2\ell(w)}(X')$ is equal to the Schubert polynomial $\mathfrak{S}_w(x_1, \ldots, x_m)$. Note that for such $w$ this Schubert polynomial is symmetric in the variables $x_i$ and $x_{i+1}$ for all $i \notin \{r_1, \ldots, r_s\}$, so it can be expressed as a polynomial in the elementary symmetric functions of the sets of variables

$$\{x_1, \ldots, x_{r_1}\}, \{x_{r_1+1}, \ldots, x_{r_2}\}, \ldots, \{x_{r_s+1}, \ldots, x_{r_m}\}.$$

These elementary symmetric functions are the Chern classes of the duals of the corresponding bundles $U_1, U_2/U_1, \ldots, E_{X'}/U_s$, where $U_i$ is the tautological subbundle of rank $r_i$ on $X'$. In particular, this shows how to express $\mathfrak{S}_w(x_1, \ldots, x_m)$ as a cohomology class on $X'$. In fact, if $W$ is the subgroup of $S_m$ generated by those $s_i$ for $i \notin \{r_1, \ldots, r_s\}$, then $H^*(X') = H^*(F\ell(E))^W$ is the subring of elements invariant by $W$.

A generalization of the results and some of the methods of this chapter, which amounts to replacing the vector space $E$ by a vector bundle over a base variety, can be found in Fulton (1992). This version includes formulas for degeneracy loci for maps between vector bundles.

We point out again that although the representation theory must be modified if the ground field $\mathbb{C}$ is replaced by a field of positive characteristic, the geometry as we have given it is valid over an arbitrary field without change. The only modification is that one must use another theory, such as rational equivalence (cf. Fulton [1984]), in place of homology theory.

It may be worth mentioning that there is another convention (followed in Fulton [1992]) that can be used, replacing our $x_i = -c_1(U_i/U_{i-1})$ by $x_i = c_1(U_{m+1-i}/U_{m-i})$. With this convention, if $\chi_i$ is the character of the group of *lower* triangular matrices whose value on a matrix is the $i^{\text{th}}$ term down the diagonal, then $x_i = c_1(L(\chi_i))$; if $\omega_i$ is the determinant of the upper left $i \times i$ submatrix, corresponding to the $i^{\text{th}}$ fundamental weight, then $c_1(L(\omega_i)) = x_1 + \ldots + x_i$ is the Schubert polynomial $\mathfrak{S}_{s_i}$. However, with this convention, the matrix descriptions of the Schubert varieties must be reversed, so that the 1's would appear in the $p^{\text{th}}$ row $w(p)$ steps from the *right*. The fact that these two conventions are equally valid can be seen from the ***duality isomorphism*** $\varphi : F\ell(E) \to F\ell(E^*)$, that takes a flag $E.$ in $E$ to the "dual" flag $E.'$ in $E^*$, where $E_i'$ is the kernel of the canonical map from $E^*$ to $E_{m-i}^*$. The bundle $U_i/U_{i-1}$ on $F\ell(E)$ corresponds by this isomorphism to the dual of the bundle denoted by $U_{m+1-i}/U_{m-i}$ on $F\ell(E^*)$, which accounts for the above change in the $x_i$'s.

**Exercise 13** Show that a Schubert variety $X_w$ defined in $F\ell(E)$ with respect to a fixed flag corresponds by this isomorphism to the Schubert variety $X_{w_o \cdot w \cdot w_o}$ in $F\ell(E^*)$ with respect to the dual flag, and similarly for the dual Schubert varieties $\Omega_w$ and $\Omega_{w_o \cdot w \cdot w_o}$.

# APPENDIX A

# Combinatorial variations

In this appendix we discuss a few of the many variations on the themes of Part I. Several of these give alternative constructions of the product of tableaux. Others give new versions of the Littlewood–Richardson correspondences. Still others describe "dual" versions of notions from Part I. They are included to tie together a variety of approaches and results in the literature, and to illustrate the richness of the combinatorics of tableaux (or at least this author's inability to resist the temptation). The reader may use them as a source of exercises for Part I.

In this appendix we follow the "compass" conventions introduced in §4.2.

## A.1 Dual alphabets and tableaux

A first construction, which seems best qualified for the designation "duality," is one that, in the language of words, replaces each word by a word in a dual or opposite alphabet. On tableaux, this corresponds to a construction using the reverse sliding algorithm, sometimes called "evacuation."

For any alphabet, $\mathcal{A}$ we have an *opposite alphabet* $\mathcal{A}^*$, that reverses the order in $\mathcal{A}$. We will let $x^*$ denote the letter in $\mathcal{A}^*$ corresponding to $x$ in $\mathcal{A}$. So $x < y \Longleftrightarrow x^* > y^*$. (For our usual alphabet $\mathcal{A} = [m]$, one can identify $\mathcal{A}^*$ with $[m]$ by identifying $a^*$ with $m+1-a$, but this risks hiding the ideas in arithmetic, so usually we won't follow this convention.) For any word $w = x_1 x_2 \ldots x_r$ in the alphabet $\mathcal{A}$, let

$$ w^* = x_r^* \ldots x_2^* x_1^* $$

This determines an anti-isomorphism of words in $\mathcal{A}$ with words in $\mathcal{A}^*$: $(u \cdot v)^* = v^* \cdot u^*$. Identifying $(\mathcal{A}^*)^*$ with $\mathcal{A}$, we have $(w^*)^* = w$. In reversing the order, the basic Knuth equivalence of §2.1 is preserved: $w_1 \equiv w_2 \Longleftrightarrow w_1^* \equiv w_2^*$. To verify this, look at the Knuth transformations. For example, the relation $x\,z\,y \equiv z\,x\,y$, if $x \leq y < z$, maps to the relation $y^*\,z^*x^* \equiv y^*\,x^*\,z^*$, which is a Knuth transformation since $z^* < y^* \leq x^*$; and symmetrically for the other case.

183

Given a tableau $T$ in the alphabet $\mathcal{A}$, we construct, using Schützenberger's sliding algorithm, a **dual tableau** $T^*$ on the alphabet $\mathcal{A}^*$, as follows. Remove the entry, say $x$, from the upper left corner of $T$, and perform the sliding algorithm on the skew tableau that is left. This gives a tableau that we will denote by $\Delta T$, whose diagram has one box removed from the diagram of $T$. Put the letter $x^*$ in this box. For $T$ as in §1.1, this gives

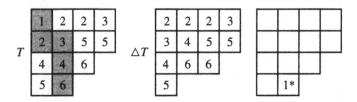

Repeat the algorithm on $\Delta T$, getting a smaller tableau $\Delta^2 T$, and putting $y^*$ in the box removed, where $y$ is the letter in the upper corner of $\Delta T$. Continue until all the entries have been removed. The Young diagram of $T$ has been filled with the duals of the letters in $T$. The result is denoted by $T^*$. For example, a short calculation gives

$$
T = \begin{array}{|c|c|c|c|}\hline 1 & 2 & 2 & 3 \\\hline 2 & 3 & 5 & 5 \\\hline 4 & 4 & 6 \\\cline{1-3} 5 & 6 \\\cline{1-2}\end{array}
\quad\leadsto\quad
T^* = \begin{array}{|c|c|c|c|}\hline 6^* & 6^* & 5^* & 4^* \\\hline 5^* & 5^* & 4^* & 2^* \\\hline 3^* & 3^* & 2^* \\\cline{1-3} 2^* & 1^* \\\cline{1-2}\end{array}
$$

This procedure of constructing $T^*$ from $T$ is often called **evacuation.**

**Duality theorem**

(1)  $T^*$ is a tableau of the same shape as $T$;

(2)  $(T^*)^* = T$;

(3)  $w(T^*) \equiv w(T)^*$;

(4)  *If an array* $\omega = \begin{pmatrix} u_1 & u_2 & \cdots & u_r \\ v_1 & v_2 & \cdots & v_r \end{pmatrix}$ *corresponds to a tableau pair* $(P,Q)$ *then*

  *the array* $\omega^* = \begin{pmatrix} u_1^* & \cdots & u_r^* \\ v_1^* & \cdots & v_r^* \end{pmatrix}$ *corresponds to the tableau-pair* $(P^*, Q^*)$.

**Proof** Given a tableau $T$, define a tableau $T^{\vee}$ on $\mathcal{A}^*$ to be the unique tableau whose word $w(T^{\vee})$ is Knuth equivalent to $w(T)^*$. The fact that $(w^*)^* = w$ implies immediately that $(T^{\vee})^{\vee} = T$. Parts (1), (2), and (3) of the theorem will therefore be proved if we show that $T^{\vee} = T^*$. Our first step toward proving this is to show that $T^{\vee}$ has the same shape as $T$. For this it suffices to observe that, with the notation of Chapter 3,

$$L(w^*, k) = L(w, k)$$

for any word $w$ and any integer $k$, since the collection of these numbers determines the shape of a tableau. This equation is immediate from the definitions, since any disjoint collection of $k$ weakly increasing sequences from $w$ will give, by reading them backwards in the dual word, $k$ weakly increasing sequences from $w^*$.

Now we prove that $T^{\vee} = T^*$ by induction on the number of boxes. Let $x$ be the entry in the upper left corner of $T$, and let $B$ be the box that is in $T$ but not in $\Delta T$. Then $T^*$ is obtained from $(\Delta T)^*$ by putting $x^*$ in $B$. By induction, $(\Delta T)^* = (\Delta T)^{\vee}$, and since $T^{\vee}$ has the same shape as $T^*$, it suffices to prove that $T^{\vee}$ is obtained from $(\Delta T)^{\vee}$ by putting $x^*$ in some box. Let

$$w(T) \ = \ \alpha \cdot x \cdot \beta,$$

where $x$ is the smallest entry in the word (with the usual convention that among equals, left is smaller); so $\alpha$ is the word of the tableau that is below the first row of $T$, and $\beta$ is the word of the part of the first row after the left corner. Then we have, from Proposition 2 of §2.1 and the definitions,

$$w(\Delta T) \ \equiv \ \alpha \cdot \beta, \quad w((\Delta T)^{\vee}) \ \equiv \ \beta^* \cdot \alpha^*, \quad w(T^{\vee}) \ \equiv \ \beta^* \cdot x^* \cdot \alpha^*.$$

Consider how $T^{\vee}$ is formed by the canonical row bumping process from the word $\beta^* \cdot x^* \cdot \alpha^*$. First a row is constructed from $\beta^*$, and $x^*$ is placed on the end of the first row of this tableau; then, since all the letters of $\alpha^*$ are strictly smaller than $x^*$, the letters of $\alpha^*$ find their places without regard to $x^*$. This means that one obtains the same tableau as the tableau $(\Delta T)^{\vee}$ that was obtained from the word $\beta^* \cdot \alpha^*$, but with $x^*$ in some box; and this is what we needed to prove.

To prove (4), we may assume the array $\omega$ is in lexicographic order. The lexicographic order for the array $\omega^*$ is then

$$\begin{pmatrix} u_r^* \ \dots \ u_1^* \\ v_r^* \ \dots \ v_1^* \end{pmatrix}.$$

The tableau of the word of the bottom row is the tableau $P^*$, so $\omega^*$ corresponds to $(P^*, Y)$ for some tableau $Y$. Applying the same to the array obtained from $\omega$ by interchanging rows, it follows that

$$\begin{pmatrix} v_1^* \ \dots \ v_r^* \\ u_1^* \ \dots \ u_r^* \end{pmatrix}$$

corresponds to $(Q^*, Z)$ for some tableau $Z$. From the Symmetry Theorem of §4.1, applied to $\omega^*$, we have $(Q^*, Z) = (Y, P^*)$, so $Y = Q^*$.   □

In terms of the tableau or plactic monoid $M(\mathcal{A})$, the map $T \mapsto T^*$ is the anti-isomorphism from $M(\mathcal{A})$ to $M(\mathcal{A}^*)$ determined by $w \mapsto w^*$. Duality can be used to prove some of the things we have seen before by other methods. For example, a bijection between the sets $\mathcal{T}(\lambda, \mu, V)$ and $\mathcal{T}(\mu, \lambda, V^*)$ described in §5.1 is given by sending the pair $[T, U]$ to the pair $[U^*, T^*]$, which shows again the identity $c_{\mu\,\lambda}^{\,\nu} = c_{\lambda\,\mu}^{\,\nu}$ of Littlewood–Richardson numbers.

**Exercise 1** Let $T$ be a tableau on $\lambda = (\lambda_1 \geq \ldots \geq \lambda_k > 0)$, and let $w(T) = v_1 \ldots v_n$. Show that the lexicographic array

$$\begin{pmatrix} 1\,1\ldots 2 \ldots k \\ v_n^* \quad \ldots \quad v_1^* \end{pmatrix},$$

where the top row of the array has $\lambda_1$ 1's, $\lambda_2$ 2's, ..., up to $\lambda_k$ $k$'s, corresponds by the R–S–K correspondence to the pair of tableaux $(T^*, U(\lambda))$, where $U(\lambda)$ is the tableau defined in §5.2.

### A.2 Column bumping

There is a construction that is "dual" to the row-insertion of Chapter 1, called ***column-insertion*** or ***column bumping***, that takes a positive integer $x$ and a tableau $T$ and puts $x$ in a new box at the bottom of the first column if possible, i.e., if it is strictly larger than all the entries of the column. If not, it bumps the highest (i.e., smallest) entry in the column that is larger than *or equal to* $x$. This bumped entry moves to the next column, going to the end if possible, and bumping an element to the next column otherwise. The process continues until the bumped entry can go at the end of the next column, or until it becomes the only entry of a new column. The resulting tableau is denoted $x \rightarrow T$. As before, the bumping takes place in a zig-zag path that moves to the right, never moving down, and the result is always another tableau. And as before, if the location of the box that is added is known, the process can be reversed. Here is an example of a column-insertion of 3 in a tableau:

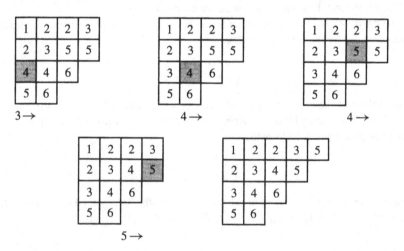

**Exercise 2** Show that column-insertion of an element $x$ in a column with word $u \cdot x' \cdot v$ can be described by the symbols

$$x \cdot (u \cdot x' \cdot v) \quad \rightsquigarrow \quad u \cdot x \cdot v \cdot x' \quad \text{if} \quad u > x' \geq x > v$$

with $u$ and $v$ words of strictly decreasing letters. Show that this transformation is a succession of Knuth transformations of type $(K'')$ and $(K')$ as in Chapter 2, so in particular the words $x \cdot (u \cdot x' \cdot v)$ and $u \cdot x \cdot v \cdot x'$ are Knuth equivalent.

It follows from this exercise that $w_{\mathrm{col}}(x \rightarrow T) \equiv x \cdot w_{\mathrm{col}}(T)$, where $w_{\mathrm{col}}(T)$ is the column word of $T$ defined in §2.3. This gives us new ways to construct the product $T \cdot U$ of two tableaux. Take any word $w = y_1 \ldots y_t$ that is Knuth equivalent to the word of $T$, for example $w = w(T)$ or $w = w_{\mathrm{col}}(T)$. Then $T \cdot U$ is obtained by successively column-inserting $y_t, \ldots, y_1$ into $U$. That is

$$T \cdot U \ = \ y_1 \ \rightarrow \ (y_2 \ \rightarrow \ (\ldots (y_{t-1} \ \rightarrow \ (y_t \ \rightarrow \ U)) \ldots )).$$

Indeed, the above shows that the word of the construction on the right is Knuth equivalent to $w(T) \cdot w(U)$, and we know that a tableau is uniquely determined by the Knuth equivalence class of its word.

In particular, when $T$ consists of one box with entry $y$, then $T \cdot U$ is the column-insertion of $y$ in $U$. The associativity of the product $(\boxed{y} \cdot T) \cdot \boxed{x} = \boxed{y} \cdot (T \cdot \boxed{x})$ implies that row and column insertion commute:

$$(y \ \rightarrow \ T) \ \leftarrow \ x \ = \ y \ \rightarrow \ (T \ \leftarrow \ x)$$

for all tableaux $T$ and all $x$ and $y$. This special case of associativity can be (and was originally) proved directly by a case by case analysis.

**Exercise 3** Prove the **Column Bumping Lemma:** Consider two successive column-insertions, first column-inserting $x$ in a tableau $T$ and then column-inserting $x'$ in the resulting tableau $x \rightarrow T$, giving rise to two bumping routes $R$ and $R'$, and two new boxes $B$ and $B'$. Show that if $x < x'$, then $R'$ lies strictly below $R$, and $B'$ is Southwest of $B$. If $x \geq x'$, show that $R'$ lies weakly above $R$, and $B'$ is northEast of $B$.

Column bumping can be used to give a dual way to construct the tableau pair $(P, Q)$ corresponding to a lexicographic array $\begin{pmatrix} u_1\ u_2\ \ldots\ u_r \\ v_1\ v_2\ \ldots\ v_r \end{pmatrix}$. We know that $P$ can be constructed by starting with $\boxed{v_r}$ and successively column-inserting $v_{r-1}, \ldots, v_1$, so $P = P_1$, where $P_r = \boxed{v_r}$ and

$$P_k \ = \ v_k \ \rightarrow \ (v_{k+1} \ \rightarrow \ \ldots \ \rightarrow \ (v_{r-1} \ \rightarrow \ \boxed{v_r}) \ldots ).$$

A new box is created at each stage. To construct $Q$, start with $Q_r = \boxed{u_r}$ and successively *slide* in the entries $u_{r-1}, \ldots, u_1$. That is, $Q_k$ is obtained from $Q_{k+1}$ by performing the reverse sliding algorithm, using the box that is in $P_k$ but not $P_{k+1}$, and then placing $u_k$ in the upper left corner of the result. For example, for the array

$\begin{pmatrix} 1\,1\,2\,2\,3 \\ 2\,2\,1\,2\,1 \end{pmatrix}$ this process gives the pairs $(P_5, Q_5), \ldots, (P_1, Q_1)$ :

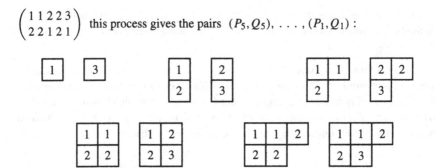

The resulting pair $(P_1, Q_1)$ is in fact the same as that of the R–S–K correspondence obtained by row-bumping the bottom row and placing the upper row, or by the matrix-ball method:

**Proposition 1** *If* $\begin{pmatrix} u_1\,u_2\,\ldots\,u_r \\ v_1\,v_2\,\ldots\,v_r \end{pmatrix}$ *is a lexicographic array, the result of column-inserting* $v_1 \rightarrow (\ \ldots\ \rightarrow (v_{r-1} \rightarrow \boxed{v_r})\ldots)$ *and successively sliding in* $u_r, \ldots, u_1$ *is the tableau pair* $(P, Q)$ *of the R–S–K correspondence.*

**Proof** Let $(P, Q)$ be the tableau pair corresponding to the array by the R–S–K correspondence, and $(P_1, Q_1)$ the last step in the above process. We have seen that $P_1 = P$, and we must show that $Q_1 = Q$. By induction on $r$, $(P_2, Q_2)$ corresponds to $\begin{pmatrix} u_2\,\ldots\,u_r \\ v_2\,\ldots\,v_r \end{pmatrix}$ by the R–S–K correspondence. In the notation of the preceding section, $Q_1$ is defined by the condition that $Q_2 = \Delta(Q_1)$, together with the fact that $Q_1$ has $u_1$ in its upper left corner, so it suffices to show that $Q$ has the same properties. Placing $u_1$ in the upper left corner of $Q$ is the first step in the R–S–K correspondence, and the fact that $\Delta(Q) = Q_2$ is a special case of Proposition 1 of Chapter 5, applied to the array $\begin{pmatrix} u_2\,\ldots\,u_r \\ v_2\,\ldots\,v_r \end{pmatrix}$ and the tableau $T = \boxed{v_1}$.     □

**Exercise 4** Show that one can start with any pair in the array, forming $(\boxed{v_k},\ \boxed{u_k})$ then successively moving to left or right until the array is exhausted; for a move to the right, use the algorithm of row-inserting the bottom element in the left tableau, and placing the top element in the right tableau, and for a move to the left, column-insert the bottom element in the left tableau, and slide the top element in the right tableau by the inverse sliding procedure just described.

**Exercise 5** For any tableau $T$ and $v_1, \ldots, v_r$, show that the succession of new boxes arrived at by the column-insertions

$$v_r \ \rightarrow \ \ldots \ \rightarrow \ v_1 \ \rightarrow \ T$$

is the same as the succession of new boxes arrived at by the row insertions

$$T^* \ \leftarrow \ v_1^* \ \leftarrow \ \ldots \ \leftarrow \ v_r^*.$$

**Exercise 6** If $w = v_1 \ldots v_r$ is a word with no two letters equal, show that $Q(w^{\text{rev}})$ and $Q(w^*)$ are conjugate tableaux, where $w^{\text{rev}} = v_r \ldots v_1$.

**Exercise 7** For any Young diagram $\lambda$, let $Q_{\text{row}}(\lambda)$ (resp. $Q_{\text{col}}(\lambda)$) be the standard tableau on $\lambda$ that numbers each row (resp. column) by consecutive integers. (a) Show that for any tableau $P$ on $\lambda$, the word $w_{\text{col}}(P)$ corresponds by the Robinson–Schensted correspondence to the pair $(P, Q_{\text{col}}(\lambda))$. For any standard tableau $Q$ with $n$ boxes, let $S(Q)$ be the result of applying the identification of $[n]^*$ with $[n]$ to the dual $Q^*$. (b) Show that the word $w_{\text{row}}(P)$ corresponds by the R–S–K correspondence to the pair $(P, S(Q_{\text{row}}(\lambda)))$.

Note that when the entries in a tableau $T$ are all distinct, then the transpose $T^\tau$ is also a tableau.

**Exercise 8** If the entries of a tableau $T$ are all distinct, show that $(T^\tau)^* = (T^*)^\tau$.

**Exercise 9** Use duality to prove the following correspondence of Thomas (1978) between $\mathcal{T}(\lambda, \mu, V_o)$ and the set of Littlewood–Richardson skew tableaux on $\nu/\mu$ of content $\lambda$. Given $[T, U]$ in $\mathcal{T}(\lambda, \mu, V_o)$, let $w(T) = v_1 \ldots v_n$, and successively column insert $v_n, \ldots, v_1$ into $U$, getting a sequence of tableaux ending with $V_o$. Number the new boxes in $\nu/\mu$ that arise with $\lambda_1$ 1's, then $\lambda_2$ 2's, and so on. Show that the result is a Littlewood–Richardson skew tableau $S$ on $\nu/\mu$ of content $\lambda$, and that each of them arises uniquely this way.

## A.3 Shape changes and Littlewood–Richardson correspondences

This section discusses a notion that, on the level of words, can be regarded as dual to Knuth equivalence. On tableaux, it describes how the shapes change when the jeu de taquin is played. This leads to explicit descriptions of the Littlewood–Richardson correspondences from Chapter 5.

If $S$ is a skew tableau on $\nu/\lambda$, there are many choices of jeu de taquin to rectify $S$. Such a jeu de taquin is a succession of $n$ choices of inside corners, where $n$ is the number of boxes in $\lambda$; if these inside corners are numbered successively from $n$ down to 1, this gives a standard tableau $J_o$ on $\lambda$, and every standard tableau on $\lambda$ gives a jeu de taquin. For each such $J_o$, at the first slide a box $B_1$ is removed from an outside corner of $\nu/\lambda$, and a box $B_2$ for the second, and so on, until at the last slide a box $B_n$ is removed. These boxes describe the changes of shapes of the skew diagrams as the successive slides are made. Let us say that two skew tableaux $S$ and $S'$ on $\nu/\lambda$ **have the same shape change by** $J_o$ if the same boxes $B_1, \ldots, B_n$ are removed in the same order. For example,

have the same shape changes by

as seen by observing the shape changes:

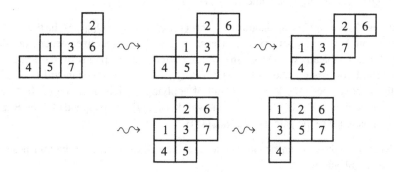

and

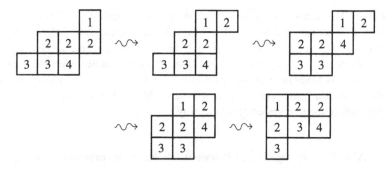

The reader may check that these skew tableaux also have the same shape changes by the other two choices of jeu de taquin, using

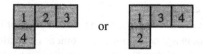     or

We will see that this is a general fact: *if two skew tableaux have the same shape changes by one jeu de taquin, then they have the same shape changes by any other.* More generally, let us say that two skew tableaux on the same shape are **shape equivalent** if any sequence of slides and reverse slides that can be applied to one of them can also be applied to the other, and the sequence of shape changes is the same for both.

In Chapter 5 we saw that the number of skew tableaux on $\nu/\lambda$ whose rectification is a given tableau $U_o$ depends only on the shape $\mu$ of $U_o$. In Proposition 2 of §5.1 we constructed for any tableau $V_o$ on $\nu$ a correspondence between the set $S(\nu/\lambda, U_o)$ of skew tableaux on $\nu/\lambda$ with rectification $U_o$ and the set $\mathcal{T}(\lambda, \mu, V_o)$ of pairs $[T, U]$ of tableaux $T$ on $\lambda$, $U$ on $\mu$, with $T \cdot U = V_o$. Let us say that two skew tableaux $S$ and $S'$ on $\nu/\lambda$ **L-R correspond by** $V_o$ if they have rectifications of the same shape $\mu$, and they determine the same pair $[T, U]$ in $\mathcal{T}(\lambda, \mu, V_o)$, i.e., if they correspond by the correspondence

$$S \in \mathcal{S}(\nu/\lambda, \text{Rect } (S)) \longleftrightarrow \mathcal{T}(\lambda, \mu, V_\circ) \longleftrightarrow \mathcal{S}(\nu/\lambda, \text{Rect } (S')) \ni S'.$$

We will see in fact that *this correspondence is always independent of choice of* $V_\circ$. Let us call two skew tableaux of the same shape *L–R equivalent* if they L–R correspond for all choices of $V_\circ$. In the above example, the reader may take any $V_\circ$ on $\nu = (4, 4, 3)$ and verify that $S$ and $S'$ L–R correspond by $V_\circ$.

A third way to compare skew tableaux is to look at their words. Recall that $w(S)$ is the row word of a skew tableau $S$, read from the entries by row, from left to right, and bottom to top. We know that words $w$ correspond by the R–S–K correspondence to pairs $(P, Q) = (P(w), Q(w))$ of tableaux, with $Q$ standard. If $w = w(S)$, then $P(w)$ is the rectification of $S$. We say that two words $w$ and $w'$ are *Q-equivalent* if $Q(w) = Q(w')$. We have seen that $P(w) = P(w')$ exactly when $w$ and $w'$ are Knuth equivalent. For this reason what we call $Q$-equivalence is sometimes called *dual Knuth equivalence*. We say that two skew tableaux are *Q-equivalent* if their row words are $Q$-equivalent. The reader may check that the two tableaux in the above example are $Q$-equivalent.

**Shape Change Theorem.** *Let $S_1$ and $S_2$ be skew tableaux on the shape $\nu/\lambda$. The following are equivalent:*

(i) $S_1$ *and* $S_2$ *have the same shape changes by some choice of jeu de taquin;*

(ii) $S_1$ *and* $S_2$ *are shape equivalent;*

(iii) $S_1$ *and* $S_2$ *L–R correspond by some tableau* $V_\circ$ *on* $\nu$;

(iv) $S_1$ *and* $S_2$ *are L–R equivalent;*

(v) $S_1$ *and* $S_2$ *are Q equivalent.*

To prove this we will need some preparation. First we analyze the notion of $Q$-equivalence of words that are permutations. Knuth equivalence of words means that one can be changed into the other by a sequence of elementary Knuth transformations. The dual notion is the following. Define an *elementary dual Knuth transformation* on a permutation $w = x_1 \ldots x_r$ to be the interchange of two letters $x_i = k$ and $x_j = k+1$, provided one of the letters $k-1$ or $k+2$ occurs between them in the word. For example, starting with $w = 3\,1\,5\,2\,4\,6$, some of these transformations are

$$3\,1\,\underline{5}\,2\,\underline{4}\,\dot{6} \,\longmapsto\, \underline{3}\,\dot{1}\,6\,\underline{2}\,4\,5 \,\longmapsto\, 2\,1\,\underline{6}\,3\,4\,\underline{5}$$

$$\longmapsto\, 2\,1\,\underline{5}\,\dot{3}\,\underline{4}\,6 \,\longmapsto\, 2\,\dot{1}\,\underline{4}\,3\,\underline{5}\,6 \,\longmapsto\, 3\,1\,4\,2\,5\,6,$$

where the underlined numbers are interchanged in the next step, and the dotted numbers satisfy the condition making the interchange permissible.

**Lemma 1** *For any permutations $w$ and $w'$, $Q(w) = Q(w')$ if and only if $w$ and $w'$ can be transformed into each other by a finite sequence of elementary dual Knuth transformations.*

**Proof** By the symmetry theorem, $Q(w) = P(w^{-1})$, and the lemma follows from the fact that an elementary dual Knuth transformation on $w$ is precisely the same as an elementary Knuth transformation on $w^{-1}$.   $\square$

**Lemma 2** *Let $S_1$ and $S_2$ be Q-equivalent skew tableaux of the same shape. Choose an inside or outside corner, and do a slide to each, obtaining skew tableaux $S_1'$ and $S_2'$. Then $S_1'$ and $S_2'$ are Q-equivalent skew tableaux of the same shape.*

**Proof** We assume first that the entries of both skew tableaux consist of the integers from 1 up to $n$, so their words are permutations. For such skew tableaux any elementary dual Knuth transformation on their words can be carried out on the skew tableaux. That is, if the letter $x-1$ or $x+2$ occurs between $x$ and $x+1$ in the row word of $S$, then $x$ and $x+1$ can be interchanged in $S$, and the result is still a skew tableau; this follows from the fact that $x$ and $x+1$ cannot be in the same row or column of $S$. By Lemma 1 we may therefore assume that $S_2$ is obtained from $S_1$ by such an elementary transformation.

Let us consider a slide given by specifying an inside corner. In most cases the routes of digging the holes through $S_1$ and $S_2$ are exactly the same, and $S_2'$ is obtained from $S_1'$ by the elementary dual Knuth transformation of exchanging the same two entries. The only way there can be two different routes is when the two entries $x$ and $y$ being exchanged are in adjacent rows and columns as indicated:

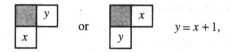

and the slide arrives at the darkened square. The slide cannot arrive at this darkened square if the letter $x-1$ is located in that square, so the only case that can arise is that in the following diagram, with $z = y+1$. The slidings in the two tableaux proceed as indicated:

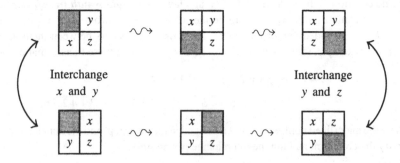

After this, the sliding is the same in each. Note that the interchange of $y$ and $z$ is allowed in the results, since the element $x$ that precedes them is now located between them in the row word. This shows that $S_1'$ and $S_2'$ have the same shape, and can be obtained from each other by a simple transposition. The same fact for reverse slides is seen by reading this proof backward.

For the general case, we resort to the method of replacing words by permutations. The letters in each word can be linearly ordered, by ordering identical letters from left to right. For a word $w$, let $w^{\#}$ be the permutation obtained by replacing each letter in $w$ by its number in this ordering. For example, if $w = 2\ 1\ 3\ 2\ 1\ 1$, then $w^{\#} = 4\ 1\ 6\ 5\ 2\ 3$. From the definition of $Q(w)$ via row bumping, one sees that $Q(w^{\#}) = Q(w)$. One can do the same for a skew tableau $S$, defining a skew tableau $S^{\#}$ of the same shape, whose entries are distinct, so that $w(S^{\#}) = w(S)^{\#}$ is a permutation. If $S'$ denotes the result of applying a slide or reverse slide to $S$, starting at a given inside or outside corner, then $(S')^{\#} = (S^{\#})'$; this follows from the fact that sliding regards the left-most of two equal entries of a skew tableau as smaller. These facts allow us to replace $S_1$ and $S_2$ by $S_1^{\#}$ and $S_2^{\#}$, reducing us to the case already proved.   □

We turn now to the proof of the theorem. The implications (ii) $\Rightarrow$ (i) and (iv) $\Rightarrow$ (iii) are trivial. It follows from Lemma 2 that (v) $\Rightarrow$ (ii). The implication (i) $\Rightarrow$ (v) is similar, once one verifies that any two tableaux on the same shape $\mu$ are $Q$-equivalent. The verification of this is immediate from the definition of $Q$ from row bumping the entries of the tableau, starting at the bottom, and observing where the new boxes arise (cf. Exercise 7 in §A.2). This proves the equivalence of conditions (i), (ii), and (v).

Recall from Chapter 5 that a skew tableau $S$ on $\nu/\lambda$ corresponds to $[T, U]$ in $\mathcal{T}(\lambda, \mu, V_o)$ by the following prescription. For any (or every) tableau $T_o$ on $\lambda$ on an alphabet whose letters come before the letters of $S$, take lexicographic arrays corresponding to the tableau pairs $(T, T_o)$ and $(U, \text{Rect}(S))$:

$$(T, T_o) \longleftrightarrow \begin{pmatrix} s_1 \ \dots \ s_n \\ t_1 \ \dots \ t_n \end{pmatrix}, \ (U, \text{Rect}(S)) \longleftrightarrow \begin{pmatrix} u_1 \ \dots \ u_m \\ v_1 \ \dots \ v_m \end{pmatrix}.$$

For $S$ and $[T, U]$ to correspond, the concatenation of these arrays must correspond to the tableau pair $(V_o, (T_o)_S)$:

$$(V_o, (T_o)_S) \longleftrightarrow \begin{pmatrix} s_1 \ \dots \ s_n \ u_1 \ \dots \ u_m \\ t_1 \ \dots \ t_n \ v_1 \ \dots \ v_m \end{pmatrix},$$

where $(T_o)_S$ is the tableau on $\nu$ that is $T_o$ on $\lambda$ and $S$ on $\nu/\lambda$. By Proposition 1 of §A.2, this means that if we do column-insertions

$$t_1 \ \rightarrow \ \dots \ \rightarrow \ t_n \ \rightarrow \ U,$$

and successively slide in $s_n, \dots, s_1$ into $\text{Rect}(S)$, the result is $(T_o)_S$. It follows that $S$ and $S'$ L–R correspond by $V_o = T \cdot U$ exactly when, doing this column bumping, sliding in $s_n, \dots, s_1$ into $\text{Rect}(S)$ gives $(T_o)_S$, and sliding in $s_n, \dots, s_1$ into $\text{Rect}(S')$ gives $(T_o)_{S'}$. It follows that if $S_1$ and $S_2$ L–R correspond by $V_o$, they must have the same shape changes for at least one jeu de taquin to their rectifications, which shows that (iii) implies (i) in the theorem. Conversely, since (i) is equivalent to (ii), if all shape changes are the same for two skew tableaux, the same argument shows that they L–R correspond by $V_o$. This completes the proof of the theorem.   □

The following exercise shows that there is a unique reverse lattice word that is $Q$-equivalent to a given word $w$; we denote it by $w^\natural$.

**Exercise 10** Show that every word is $Q$-equivalent to a unique reverse lattice word. Show in fact that $w$ is $Q$-equivalent to the word $w^\natural$ such that $U(w^\natural) = Q(w^*)$, where $U(w^\natural)$ is the standard tableau defined by a reverse lattice word in §5.3, and $w^*$ is the dual word to $w$.

The theorem shows in particular that there is a canonical correspondence between $\mathcal{S}(\nu/\lambda, U_o)$ and $\mathcal{S}(\nu/\lambda, U_o')$ for any tableaux $U_o$ and $U_o'$ on $\mu$. Taking $U_o'$ to be $U(\mu)$, this gives a canonical correspondence between skew tableaux on $\nu/\lambda$ with given rectification and Littlewood–Richardson skew tableaux of the same shape. Let $S^\natural$ denote the Littlewood–Richardson skew tableau corresponding to $S$ by this correspondence.

**Exercise 11** Show that the word of $S^\natural$ is determined by the identity $U(w(S^\natural)) = Q(w(S)^*)$, with $U(w(S^\natural))$ as in Exercise 10. Show that two tableaux $S$ and $S'$ on the same shape correspond via the theorem if and only if $S^\natural = (S')^\natural$.

**Exercise 12** Consider the map that assigns to a word $w$ the reverse lattice word $w^\natural$ defined by the condition that $U(w^\natural) = Q(w^*)$. Show that $(w^\natural)^\natural = w^\natural$, and that $w$ is a reverse lattice word if and only if $w^\natural = w$. Show that the map $w \mapsto w^\natural$ is a one-to-one correspondence between words in a given Knuth equivalence class and reverse lattice words of content $\mu$, where $\mu$ is the shape of $P(w)$. Show that $w$ and $w^\natural$ are words of skew tableaux on exactly the same shapes; that is, for any skew shape, $w$ is the word of a skew tableau on this shape if and only if $w^\natural$ is.

Robinson, augmented by Thomas (1978), has given a prescription for producing a reverse lattice word from a given word $w = v_1 \dots v_n$. For $1 \le i \le n$ define the **index** $I(i)$ to be 0 if $v_i = 1$, and otherwise

$$I(i) = \#\{j \ge i : v_j = v_i\} - \#\{j \ge i : v_j = v_i - 1\}.$$

A word is a reverse lattice word exactly when each $I(i) \le 0$. For any integer $k$ let $J(k)$ be the maximum index of any $i$ such that $v_i = k$. Define a **permissible move** to be the choice of any $k$ such that $J(k)$ is positive, then taking the largest $i$ for which $v_i = k$ and $I(i) = J(k)$, and replacing $v_i$ by $v_i - 1$ in the word. For example, if $w = 2\,3\,3\,1\,2\,2$, a sequence of permissible moves, with dots over possible choices at each stage, is

$$2\,3\,3\,1\,\dot{2}\,\dot{2} \mapsto 2\,\dot{3}\,3\,1\,1\,\dot{2} \mapsto 2\,2\,3\,1\,1\,\dot{2}$$

$$\mapsto 2\,2\,\dot{3}\,1\,1\,1 \mapsto 2\,2\,2\,1\,1\,1.$$

**Exercise 13** If $w'$ is obtained from $w$ by a permissible move, show that $Q(w') = Q(w)$, and if $w$ is the row word of a skew tableau, show that $w'$ is the row word of a skew tableau of the same shape.

Thomas shows that any sequence of permissible moves takes $w$ to a reverse lattice word, and that the resulting word and number of moves are independent of choice. In fact, one has the

**Corollary 1** *Given any word* $w = v_1 \ldots v_n$, *let* $\lambda$ *be the shape of* $P(w)$, *and let* $N = \sum v_i - \sum k\lambda_k$. *Any sequence of permissible moves ends in* $N$ *steps and takes* $w$ *to the reverse lattice word* $w^\natural$.

In particular, this gives another prescription for finding the Littlewood–Richardson skew tableau that corresponds to a given skew tableau. Equivalently, two skew tableaux on the same shape correspond exactly when this prescription on their words leads to the same reverse lattice word. For the proof of the corollary, observe that in each permissible move the sum of the letters in a word decreases by 1, and if $w^\natural$ is a reverse lattice word, with $\lambda$ the shape of $U(w^\natural)$, then the sum of the entries in $w^\natural$ is $\Sigma k\lambda_k$. The corollary then follows from the theorem and the preceding exercise.

It also follows from the theorem that there is a canonical correspondence between $\mathcal{T}(\lambda,\mu,V_\mathrm{o})$ and $\mathcal{T}(\lambda,\mu,V_\mathrm{o}')$ for any tableaux $V_\mathrm{o}$ and $V_\mathrm{o}'$ of shape $\nu$. This can be defined using Proposition 2 of §5.1 by way of $\mathcal{S}(\nu/\lambda,U_\mathrm{o})$ for any tableau $U_\mathrm{o}$ on $\mu$:

$$\mathcal{T}(\lambda, \mu, V_\mathrm{o}) \longleftrightarrow \mathcal{S}(\nu/\lambda, U_\mathrm{o}) \longleftrightarrow \mathcal{T}(\lambda, \mu, V_\mathrm{o}')$$

**Exercise 14** Show that $[T,U]$ and $[T',U']$ correspond if and only if there is a word $w = v_1 \ldots v_m$ Knuth equivalent to $w(U)$ and a word $w' = v_1' \ldots v_m'$ Knuth equivalent to $w(U')$ such that $Q(w) = Q(w')$ and the row bumpings

$$T \leftarrow v_1 \leftarrow \ldots \leftarrow v_m \quad \text{and} \quad T' \leftarrow v_1' \leftarrow \ldots \leftarrow v_m'$$

produce the same new boxes in the same order. In particular, the correspondence $\mathcal{T}(\lambda, \mu, V_\mathrm{o}) \longleftrightarrow \mathcal{T}(\lambda,\mu,V_\mathrm{o}')$ does not depend on $U_\mathrm{o}$.

**Exercise 15** Using the correspondence of the preceding exercise, show that $[T,U]$ and $[T',U']$ correspond if and only if $[U^*,T^*]$ and $[U'^*,T'^*]$ correspond.

**Exercise 16** Verify the transitivity of the Littlewood–Richardson correspondences: for three skew tableaux on the same shape, if $S$ corresponds to $S'$, and $S'$ to $S''$, then $S$ corresponds to $S''$; and similarly for pairs $[T,U]$, $[T',U']$, and $[T'',U'']$.

Let us next consider relations between skew tableaux on a shape $\nu/\lambda$ and skew tableaux on the conjugate shape $\tilde{\nu}/\tilde{\lambda}$. For skew tableaux with distinct entries, there is an obvious correspondence given by $S \mapsto S^\tau$, taking the transpose of skew tableaux. For any jeu de taquin for $S$, given by a standard tableau $J_\mathrm{o}$ on $\lambda$, there is a conjugate jeu de taquin for $S^\tau$, given by the conjugate standard tableau $J_\mathrm{o}^\tau$ on $\tilde{\lambda}$. The shape changes for this conjugate game will be exactly the conjugates of those of the original game. If entries are allowed to coincide, however, conjugates of skew tableaux may no longer be skew tableaux, but we may use the conjugate shape changes to see if skew tableaux correspond. We say that a skew tableau on $\nu/\lambda$ is ***conjugate shape equivalent*** to a skew tableau on $\tilde{\nu}/\tilde{\lambda}$ if the shape changes for any sequence of slides or reverse slides on one are conjugates of the shapes of corresponding slides or reverse slides on the other.

One can also use the Littlewood–Richardson correspondence to construct a correspondence between skew tableaux on $\nu/\lambda$ and skew tableaux on $\nu/\lambda$. For this, choose a standard tableau $T_{\circ}$ on $\lambda$. Then for any tableau $U_{\circ}$ on a shape $\mu$, and $U_{\circ}'$ on the conjugate shape $\tilde{\mu}$, we have one-to-one correspondences

$$\mathcal{S}(\nu/\lambda, U_{\circ}) \longleftrightarrow \mathcal{S}(\nu/\lambda, T_{\circ}) \longleftrightarrow \mathcal{S}(\tilde{\nu}/\tilde{\lambda}, T_{\circ}^{\tau}) \longleftrightarrow \mathcal{S}(\tilde{\nu}/\tilde{\lambda}, U_{\circ}'),$$

the first and last by the correspondences constructed in the Shape Change Theorem, and the correspondence in the middle given by conjugation of skew tableaux with distinct entries. We say that skew tableaux on $\nu/\lambda$ and $\tilde{\nu}/\tilde{\lambda}$ ***L–R correspond by*** $T_{\circ}$ if they correspond in this bijection, and that they are ***conjugate L–R equivalent*** if they L–R correspond for all such $T_{\circ}$.

We also want a computational criterion for skew tableaux on conjugate shapes to correspond, in terms of insertion tableaux of words. For this we need a permutation $\sigma = \sigma_{\nu/\lambda}$ in $S_m$, depending on the skew diagram $\nu/\lambda$ with $m$ boxes. This is obtained by numbering the skew diagram in its row numbering (from left to right in rows, then bottom to top) and its column numbering (from bottom to top in columns, then left to right). Define $\sigma_{\nu/\lambda}(j) = k$ if the box numbered $j$ in the row numbering is numbered $k$ in the column numbering. For example, for $\nu = (5,5,4,1)$ and $\lambda = (3,2,1)$,

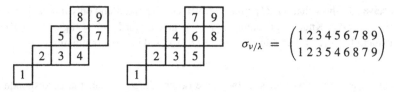

$$\sigma_{\nu/\lambda} \;=\; \begin{pmatrix} 1\,2\,3\,4\,5\,6\,7\,8\,9 \\ 1\,2\,3\,5\,4\,6\,8\,7\,9 \end{pmatrix}$$

If $T$ is a numbering of a diagram with distinct entries $1, \dots, m$, then, for any $\sigma \in S_m$, $\sigma(T)$ denotes the numbering obtained by replacing each entry $i$ by $\sigma(i)$.

**Corollary 2** *Let $S$ be a skew tableau on $\nu/\lambda$, and $S'$ a skew tableau on $\tilde{\nu}/\tilde{\lambda}$. The following are equivalent:*

  (i)   *$S$ and $S'$ have conjugate shape changes by some choice of jeu de taquin;*

  (ii)   *$S$ and $S'$ are conjugate shape equivalent;*

 (iii)   *$S$ and $S'$ L–R correspond by some standard tableau $T_{\circ}$ on $\lambda$;*

 (iv)   *$S$ and $S'$ are conjugate L–R equivalent;*

  (v)   *$\sigma_{\nu/\lambda}(Q(w(S)))$ and $Q(w(S')^*)$ are conjugate standard tableaux.*

In particular, this gives a bijection between the Littlewood–Richardson skew tableaux on a given shape and those on its conjugate. Such a correspondence was given by Hanlon and Sundaram (1992); condition (v) can easily be seen to be equivalent to the condition in their correspondence, so their result follows from this proposition.

**Exercise 17** For the skew shape of the above example, find the three Littlewood–Richardson tableaux of content $\mu = (4,3,2)$ on $\nu/\lambda$, and the three corresponding Littlewood–Richardson tableaux of content $\tilde{\mu} = (3,3,2,1)$ on $\tilde{\nu}/\tilde{\lambda}$.

**Proof of Corollary 2** The equivalence of (i), (ii), (iii), and (iv) follows from the theorem. In addition, to prove the equivalence of these with (v), it suffices to consider skew tableaux whose rectifications are standard tableaux. In this case the correspondence given by (i)–(iv) is just conjugation. We must therefore show that for such a skew tableau $S$,

$$Q(w(S^\tau)^*)^\tau \;=\; \sigma_{\nu/\lambda}(Q(w(S))).$$

By Exercise 6 of §A.2 the left side is $Q(w(S^\tau)^{\text{rev}})$. Since $w(S^\tau)^{\text{rev}} = w_{\text{col}}(S)$, we are reduced to proving that

$$Q(w_{\text{col}}(S)) \;=\; \sigma_{\nu/\lambda}(Q(w(S))).$$

By the definition of $\sigma = \sigma_{\nu/\lambda}$, if $w_{\text{col}}(S) = v_1 \ldots v_m$, then $w(S) = v_{\sigma(1)} \ldots v_{\sigma(m)}$. We showed in §2.3 that $w_{\text{col}}(S)$ is $K'$-equivalent to $w(S)$. Hence the required equation follows from the following lemma.  $\square$

**Lemma 3** *Let $w = v_1 \ldots v_m$ be a permutation, and suppose $w' = v_{\sigma(1)} \ldots v_{\sigma(m)}$ for some $\sigma \in S_m$. Suppose $w'$ is $K'$-equivalent to $w$. Then $Q(w) = \sigma(Q(w'))$.*

**Proof** Since $K'$-equivalence is generated by elementary $K'$-transformations, it suffices to prove the lemma when $w'$ is obtained from $w = v_1 \ldots v_m$ by interchanging $v_i$ and $v_{i+1}$, when $v_i < v_{i-1} \le v_{i+1}$. For this it suffices to look at the row bumping in each case. Let $P = P(v_1 \ldots v_{i-1})$, and consider the row bumpings in

$$(P \leftarrow v_i) \leftarrow v_{i+1} \quad \text{and} \quad (P \leftarrow v_{i+1}) \leftarrow v_i.$$

In each case the same two boxes are added, since the words $v_1 \ldots v_{i-1} \cdot v_i \cdot v_{i+1}$ and $v_1 \ldots v_{i-1} \cdot v_{i+1} \cdot v_i$ are Knuth equivalent. However, these two boxes must be added in the opposite order, by the Row Bumping Lemma. Since the rest of the construction is the same for each, this shows that $Q(w)$ and $Q(w')$ are obtained from each other by applying the transposition that interchanges $i$ and $i+1$.  $\square$

## A.4  Variations on the R–S–K correspondence

Given a matrix, or a two-rowed array, there are several possibilities involving row or column bumping that one can use, in the spirit of the R–S–K correspondence, to make pairs of tableaux. In fact, all of these variations have "matrix-ball" constructions, using an appropriate orientation of the matrix, and appropriate orderings of the entries.

### A.4.1 The Burge correspondence

In Chapter 4 we saw three realizations of the R–S–K correspondence between matrices $A$ with nonnegative entries and pairs $(P, Q)$ of tableaux of the same shape. If $\begin{pmatrix} u_1 \, u_2 \, \ldots \, u_r \\ v_1 \, v_2 \, \ldots \, v_r \end{pmatrix}$ is the array corresponding to $A$, arranged in lexicographic order, the following three procedures give $(P, Q)$:

(1a) *Row bump* $v_1 \leftarrow v_2 \leftarrow \ldots \leftarrow v_r$, *and place in* $u_1, \ldots, u_r$.

(2a) *Column bump* $v_1 \rightarrow \ldots \rightarrow v_{r-1} \rightarrow v_r$, *and slide in* $u_r, \ldots, u_1$.

(3a) *Matrix-ball construction, with the northwest ordering.*

This **northwest** (nw) ordering refers to the fact that, when the balls in a matrix are numbered, a ball is numbered with the next highest number than the maximum number of all balls that are northwest (weakly above and left) of it.

If one takes the same array and tries to combine row bumping with sliding, or column bumping with placing, the second of the pair may not be a tableau. However, if one chooses another ordering for the pairs in the array, these procedures will give correspondences between arrays and pairs of tableaux of the same shape. Moreover, these two new procedures will give the same result, and in fact can be given by another matrix-ball construction.

The ordering on the array that works for this is what may be called the **antilexicographic** ordering, by which we mean that

$$u_i > u_{i+1}, \quad \text{or} \quad u_i = u_{i+1} \quad \text{and} \quad v_i \le v_{i+1}.$$

With this ordering, we will see that each of the following procedures gives a pair of tableaux $(R, S)$ of the same shape, and that this pair is the same for the two procedures. This correspondence is called the **Burge correspondence**. The two procedures are:

(1b) *Column bump* $v_1 \rightarrow \ldots \rightarrow v_r$, *and place in* $u_r, \ldots, u_1$.

(2b) *Row bump* $v_1 \leftarrow \ldots \leftarrow v_r$, *and slide in* $u_1, \ldots, u_r$.

For example, the array $\begin{pmatrix} 3\,3\,3\,2\,2\,1\,1\,1 \\ 1\,2\,3\,2\,3\,1\,2\,2 \end{pmatrix}$ leads by either of these procedures to the pair

$$R = \begin{array}{|c|c|c|c|c|} \hline 1 & 1 & 2 & 2 & 2 \\ \hline 2 & 3 \\ \cline{1-2} 3 \\ \cline{1-1} \end{array} \qquad S = \begin{array}{|c|c|c|c|c|} \hline 1 & 1 & 1 & 3 & 3 \\ \hline 2 & 2 \\ \cline{1-2} 3 \\ \cline{1-1} \end{array}$$

Moreover, if $A$ is the matrix corresponding to the array, we can define a

(3b) *Matrix-ball construction, using the NorthWest ordering.*

For this construction, replace each entry $t$ of the matrix with $t$ balls, arranged diagonally from southwest to northeast in the corresponding box of the matrix. Number the balls of this matrix from northwest to southeast, but numbering a ball one more than the maximum of the balls lying NorthWest of it, i.e., lying in a row strictly above it and a column strictly left of it, giving a matrix $A^{(1)}$. This may also be called the **strong** ordering. (In this context, the ordering in §4.2 will be called the **weak** ordering.) For example,

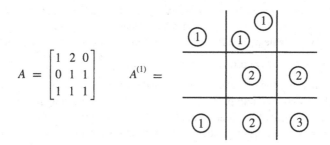

$$A = \begin{bmatrix} 1 & 2 & 0 \\ 0 & 1 & 1 \\ 1 & 1 & 1 \end{bmatrix} \qquad A^{(1)} =$$

The first columns of $R$ and $S$ are read from $A^{(1)}$, the $k$ th entry in the first column of $R$ (resp. $S$) being the number of the left-most column (resp. row) containing a ball numbered $k$. The balls with the same number $k$ in $A^{(1)}$ are arranged in a (weak) southwest to northeast order. For each successive pair, a ball is put in the next matrix, in the same row as the southwest member of the pair, and the same column as the northeast member, as in the matrix-ball construction of Chapter 4. These balls are numbered by the same rule as for $A^{(1)}$, getting a matrix of numbered balls $A^{(2)}$, from which the second columns of $R$ and $S$ are read; the process is repeated as before. In the example, the succeeding matrices of balls are

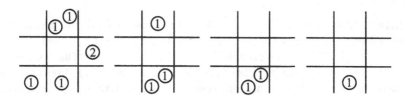

from which we read off the tableau-pair

$$R = \begin{array}{|c|c|c|c|c|} \hline 1 & 1 & 2 & 2 & 2 \\ \hline 2 & 3 \\ \cline{1-2} 3 \\ \cline{1-1} \end{array} \qquad S = \begin{array}{|c|c|c|c|c|} \hline 1 & 1 & 1 & 3 & 3 \\ \hline 2 & 2 \\ \cline{1-2} 3 \\ \cline{1-1} \end{array}$$

**Proposition 2** *The procedures* (1b), (2b), *and* (3b) *yield the same tableau pair from a given matrix, and this sets up a one-to-one correspondence between matrices (or arrays) and pairs of tableaux of the same shape.*

The analogues of conditions (i), (ii), and (iii) of the R–S–K Theorem are also valid for this Burge correspondence.

**Proof** For (1b) and (2b), proofs can either be given that are analogous to those of (1a) and (2a), or they can be reduced to these results as follows: If the given array is in antilexicographic order, then the array $\begin{pmatrix} u_r & \cdots & u_1 \\ v_r^* & \cdots & v_1^* \end{pmatrix}$ is in lexicographic order. When one does the row bumping $v_r^* \leftarrow \cdots \leftarrow v_1^*$, the new boxes one gets are the same, in the same order, as the new boxes one gets when one does the column bumping $v_1 \rightarrow \cdots \rightarrow v_r$; cf. Exercise 5 of §A.2. Hence placing $u_r, \ldots, u_1$ in these new boxes gives a tableau, and if $(R,S)$ is the pair arising from the construction (1b), then $(R^*,S)$ is the array corresponding to the array $\begin{pmatrix} u_r & \cdots & u_1 \\ v_r^* & \cdots & v_1^* \end{pmatrix}$ by the R–S–K correspondence. By the equivalence of (1a) and (2a) one knows one can get this same pair $(R^*,S)$ by column bumping $v_r^* \rightarrow \cdots \rightarrow v_1^*$ and sliding in $u_1, \ldots, u_r$; by the same reasoning, this means that (2b) also gives the pair $(R,S)$.

Next we show that the new matrix-ball construction gives the same pair. The proof is quite similar to the corresponding result in Chapter 4, so we only indicate the changes. This time the *last* position of $A$ is the position $(x,y)$ of the first nonzero entry in the last row that is nonzero. The pair $\binom{x}{y}$ is then the first pair $\binom{u_1}{v_1}$ of the antilexicographic array. Let $A_\circ$ be the matrix obtained from $A$ by subtracting 1 from this $(x,y)$ entry. Let $A^\flat$ denote the matrix whose $(i,j)$ entry is the number of balls in the matrix $A^{(1)}$ obtained from $A$ by the matrix-ball construction (3b). Let $R(A)$ and $S(A)$ denote the tableaux constructed from a matrix $A$ by this construction. As in §4.2, it suffices to prove the

**Claim** $R(A) = y \rightarrow R(A_\circ)$, *and* $S(A)$ *is obtained from* $S(A_\circ)$ *by placing* $x$ *in the new box.*

Let the ball that is in $A^{(1)}$ but not in $A_\circ^{(1)}$ be numbered $k$. Again, there are two cases. If there is no other ball in $A^{(1)}$ numbered $k$, then $A^\flat = (A_\circ)^\flat$, and $u_1 > u_2$ and $v_1 > v_2$. It follows that the column bumping puts $y$ at the end of the first column of $R(A_\circ)$, and $S(A)$ puts $x$ at the end of the first column of $S(A_\circ)$, from which the claim is evident. If there are other balls numbered with $k$, take $x' \leq x$ maximum, and $y' \geq y$ minimum, for which there is a ball in the $(x',y')$ place numbered $k$. In this case $y \rightarrow R(A_\circ)$ bumps $y'$ from the first column. The last position of $A^\flat$ is $(x,y')$, and $(A^\flat)_\circ = (A_\circ)^\flat$, from which it follows that $R(A^\flat) = y' \rightarrow R((A_\circ)^\flat)$, and the new box is the box in $S(A^\flat)$ that is not in $S((A_\circ)^\flat)$. The proof concludes by induction as before. □

The matrix-ball construction is clearly symmetric: if $A$ corresponds to a pair $(R,S)$, then the transpose $A^\tau$ corresponds to $(S,R)$. This proves the

**Symmetry Theorem (b)** *If an array* $\begin{pmatrix} u_1 & u_2 & \cdots & u_r \\ v_1 & v_2 & \cdots & v_r \end{pmatrix}$, *made antilexicographic, corresponds via the Burge correspondence to the pair of tableaux* $(R,S)$, *then the array* $\begin{pmatrix} v_1 & v_2 & \cdots & v_r \\ u_1 & u_2 & \cdots & u_r \end{pmatrix}$, *made antilexicographic, corresponds to the pair* $(S,R)$.

Symmetric matrices therefore correspond to pairs $(R,S)$ with $S = R$, so to tableaux $R$.

**Exercise 18** Show that if a symmetric matrix $A$ corresponds by this correspondence to a tableau $R$, then the number of odd diagonal entries of $A$ plus the number of odd diagonal entries of $A^{\flat}$ is the length of the first column of $R$. Deduce that the number of odd diagonal entries of $A$ is the number of rows of odd length in the Young diagram of $R$.

This correspondence also satisfied the expected duality:

**Duality Theorem (b)** *If an array* $\begin{pmatrix} u_1\ u_2\ \cdots\ u_r \\ v_1\ v_2\ \cdots\ v_r \end{pmatrix}$ *corresponds to a tableau pair*
$(R,S)$, *then the array* $\begin{pmatrix} u_r^{*}\ \cdots\ u_1^{*} \\ v_r^{*}\ \cdots\ v_1^{*} \end{pmatrix}$ *corresponds to the tableau pair* $(R^{*},S^{*})$.

**Proof** Using construction (1b), this array corresponds to a pair $(R^{*},X)$ for some $X$, and then the Symmetry Theorem (b) implies that $X = S^{*}$.  □

### A.4.2 The other corners of a matrix

We have two matrix-ball constructions, each using an ordering from the upper left corner of the matrix. It is natural to ask what happens when one does the same two constructions from the other three corners of the matrix. For example, one can number the balls from southwest to northeast, using the weak ordering, and putting in the first row of the first tableau the column numbers of the left-most column having balls labelled $1, 2, \ldots$ , and putting in the first row of the second tableau the dual numbers of the numbers of the lowest column with balls labelled $1, 2, \ldots$ . (Duals are used for measuring from the bottom or from the right). For example, with the same matrix $A$ as above, we have

from which one reads the tableau pair

Equivalently, one can put the array $\begin{pmatrix} u_1^* & \cdots & u_r^* \\ v_1 & \cdots & v_r \end{pmatrix}$ in lexicographic order, and construct the corresponding pair by the R–S–K correspondence. Note that the first of this tableau pair is the $R$ constructed by the Burge correspondence, and the second is the dual $S^*$ of the second. In fact, we will see that all eight possibilities for choices of corners and weak or strong orderings give tableaux consisting of the matrices $P$ or $R$ or their duals as the first in the pair, and $Q$ and $S$ or their duals in the second, and that all such pairs, with $P$ and $Q$ together, and $R$ and $S$ together, arise this way. The following theorem gives all the possible correspondences.

**Theorem 1** *The four tableau pairs arising from the lexicographic orderings of arrays and weak orderings in matrices are*

$$\begin{pmatrix} u \\ v \end{pmatrix} \quad \text{nw} \searrow (P,Q); \qquad \begin{pmatrix} u^* \\ v \end{pmatrix} \quad \text{sw} \nearrow (R,S^*);$$

$$\begin{pmatrix} u^* \\ v^* \end{pmatrix} \quad \text{se} \nwarrow (P^*,Q^*); \qquad \begin{pmatrix} u \\ v^* \end{pmatrix} \quad \text{ne} \swarrow (R^*,S).$$

*The four tableau pairs arising from the antilexicographic orderings of arrays and strong orderings of matrices are*

$$\begin{pmatrix} u \\ v \end{pmatrix} \quad \text{NW} \searrow (R,S); \qquad \begin{pmatrix} u^* \\ v \end{pmatrix} \quad \text{SW} \nearrow (P,Q^*);$$

$$\begin{pmatrix} u^* \\ v^* \end{pmatrix} \quad \text{SE} \nwarrow (R^*,S^*); \qquad \begin{pmatrix} u \\ v^* \end{pmatrix} \quad \text{NE} \swarrow (P^*,Q).$$

Notice that the pairs for the two cases are related by interchanging $P$ and $R$ and $Q$ and $S$. These eight correspondences all use orderings in which the rows of the matrix (or tops of the arrays) take precedence. Another eight correspondences, where the columns (or bottoms) take precedence, are obtained by reflecting in the arrows, and interchanging the two tableaux in each pair, as follows from the symmetry theorems.

**Proof** Let $\begin{pmatrix} u_1 & u_2 & \cdots & u_r \\ v_1 & v_2 & \cdots & v_r \end{pmatrix}$ be the lexicographic array corresponding to the matrix $A$, and let $(P,Q)$ be the tableau pair corresponding to this by the R–S–K correspondence. Working from the lower right corner of the matrix amounts to doing the R–S–K correspondence on the lexicographic array $\begin{pmatrix} u_r^* & \cdots & u_1^* \\ v_r^* & \cdots & v_1^* \end{pmatrix}$, and the Duality Theorem states that this gives the dual pair $(P^*,Q^*)$. The array $\begin{pmatrix} u_r & \cdots & u_1 \\ v_r^* & \cdots & v_1^* \end{pmatrix}$ is in antilexicographic order, and, as in the proof of Proposition 2 in §A.4.1, the second construction for this array leads to the pair $(P^*,Q)$. By the definition, this is the tableau pair of the last row of the theorem for the direction $\swarrow$. Similarly, $\begin{pmatrix} u_1^* & \cdots & u_r^* \\ v_1 & \cdots & v_r \end{pmatrix}$ is in antilexicographic order, and it corresponds to the tableau pair $(P,Q^*)$ by the Duality Theorem (b). The other four cases are proved by the same arguments. $\square$

These two correspondences determine a bijective transformation $(P,Q) \mapsto (R,S)$ from tableau pairs to tableau pairs, with each entry occurring the same number of times in $R$ as in $P$, and in $S$ as in $Q$. It should be interesting to study this transformation, or the corresponding transformation on matrices.

### A.4.3 Matrices of 0's and 1's

If $\begin{pmatrix} u_1 & u_2 & \dots & u_r \\ v_1 & v_2 & \dots & v_r \end{pmatrix}$ is a lexicographic array, what happens when one performs the column bumping $v_r \to \ \dots \ \to v_1$ and places in the entries $u_1, \dots, u_r$? The numbering of the Young diagram by these $u_i$ will be weakly increasing in rows and columns, but will not in general be a tableau. For example,

$$\begin{pmatrix} 1 & 1 \\ 1 & 2 \end{pmatrix} \longleftrightarrow \boxed{\begin{array}{c} 1 \\ 2 \end{array}} \ \boxed{1} \qquad \begin{pmatrix} 1 & 1 & 1 \\ 1 & 2 & 2 \end{pmatrix} \longleftrightarrow \boxed{\begin{array}{cc} 1 & 2 \\ 2 & \end{array}} \ \boxed{\begin{array}{cc} 1 & 1 \end{array}}$$

However, if each pair $\binom{x}{y}$ occurring in this array occurs no more than once, this second numbering will in fact be strictly increasing in the *rows*. By replacing this numbering by its conjugate, one arrives at a pair $\{\widetilde{P},\widetilde{Q}\}$ of tableaux, with the shapes of $\widetilde{P}$ and $\widetilde{Q}$ *conjugate* shapes. Equivalently, one gets the pair $\{\widetilde{P},\widetilde{Q}\}$ from the array by column bumping $v_r \to \ \dots \ \to v_1$ and successively placing $u_1, \dots, u_r$ in the conjugates of the new boxes. Let us call this procedure *conjugate placing* $u_1, \dots, u_r$. In brief:

(1) *Column bump* $v_r \to \ \dots \ \to v_1$; *conjugate place* $u_1, \ \dots \ u_r$.

For example, the array $\begin{pmatrix} 1 & 1 & 2 & 2 & 3 & 3 \\ 2 & 3 & 1 & 3 & 1 & 3 \end{pmatrix}$ corresponds by this procedure to the pair

$$\widetilde{P} = \boxed{\begin{array}{ccc} 1 & 1 & 2 \\ 3 & 3 & 3 \end{array}} \qquad \widetilde{Q} = \boxed{\begin{array}{cc} 1 & 1 \\ 2 & 2 \\ 3 & 3 \end{array}}$$

Arrays with no repeated pair correspond to matrices $A = (a(i,j))$ whose only entries are zeros and ones, where the entry $a(i,j)$ is 1 exactly when $\binom{i}{j}$ occurs in the array.

**Proposition 3** (Knuth) *The above procedure sets up a one-to-one correspondence between matrices $A$ whose entries are zeros and ones, or two-rowed arrays without repeated pairs, and pairs $\{\widetilde{P},\widetilde{Q}\}$ of tableaux with conjugate shapes.*

**Proof** The proof is entirely like that of the R–S–K correspondence in Chapter 4, using of course the Column Bumping Lemma in place of the Row Bumping Lemma. For example, if $i < k$ and $u_i = u_k$, then $v_i < \ \dots \ < v_k$. The Column Bumping

Lemma then says that the $k^{th}$ new box is strictly below and weakly left of the $i^{th}$ new box; in the conjugate shape $u_k$ is therefore placed strictly to the right of $u_i$, so there are no equal entries in any column.    □

As before, the $k^{th}$ row sum of $A$ is the number of times $k$ appears in $\widetilde{Q}$, and the $k^{th}$ column sum of $A$ is the number of times appears in $\widetilde{P}$. As in Chapter 4, this proves the following identity:

**Corollary** (Littlewood) $\prod_{i=1}^{n} \prod_{j=1}^{m} (1+x_i y_j) = \sum_\lambda s_\lambda(x_1, \ldots, x_n) s_{\tilde{\lambda}}(y_1, \ldots, y_m)$.
*Only those $\lambda$ with at most $n$ rows and $m$ columns occur on the right.*

**Exercise 19** Show that the number of $m \times n$ matrices of 0's and 1's with row sums $\lambda_1, \ldots, \lambda_m$ and column sums $\mu_1, \ldots, \mu_n$ is $\sum_\nu K_{\nu\lambda} K_{\tilde{\nu}\mu}$.

**Exercise 20** Deduce the following theorem of Gale and Ryser: For partitions $\lambda$ and $\mu$ of the same integer, the following are equivalent: (a) there is a matrix of 0's and 1's with row sums $\lambda_1, \ldots, \lambda_m$ and column sums $\mu_1, \ldots, \mu_n$; (b) $\lambda \trianglelefteq \tilde{\mu}$; (c) $\mu \trianglelefteq \tilde{\lambda}$. .

As with the R–S–K correspondence, there are other ways to make this correspondence, variations giving a conjugate pair $\{\widetilde{R}, \widetilde{S}\}$, and symmetry theorems, duality theorems, and matrix-ball constructions. We will state the results, but give only brief indications of proofs, since they are minor variations of those we have just seen. First, given the same lexicographic array, one can also

(2c) *Row bump $v_r \leftarrow \ldots \leftarrow v_1$, and conjugate slide in $u_r, \ldots, u_1$.*

This means that at each stage of the row bumping, which produces a new box, the conjugate of this new box is used to slide in the next entry of the second tableau.

There is also a matrix-ball construction, using a "northWest" ordering. For this, put $a(i, j)$ balls in the $(i, j)$ position of the matrix as before, and number the balls from the upper left corner, but numbering each ball one more than the maximum of a ball that lies northWest (weakly above and strictly left) of it. This gives a matrix $A^{(1)}$ of numbered balls, and the $k^{th}$ entry of the first column of $\widetilde{P}$ is the left-most column of $A^{(1)}$ containing a ball numbered $k$, and the $k^{th}$ entry of the first row of $\widetilde{Q}$ is the top-most row of $A^{(1)}$ containing a ball numbered $k$. For example, for the matrix $A$ corresponding to the array $\begin{pmatrix} 1\,1\,2\,2\,3\,3 \\ 2\,3\,1\,3\,1\,3 \end{pmatrix}$, we have

$$A = \begin{bmatrix} 0 & 1 & 1 \\ 1 & 0 & 1 \\ 1 & 0 & 1 \end{bmatrix} \qquad A^{(1)} = $$

which gives the first column of $\widetilde{P}$ to be $\begin{array}{|c|}\hline 1 \\ \hline 3 \\ \hline\end{array}$ and the first row of $\widetilde{Q}$ to be

$\begin{array}{|c|c|}\hline 1 & 1 \\ \hline\end{array}$ . Then one forms a new matrix $A^{\flat}$ by the same prescription as before, with a ball for each successive pair in $A^{(1)}$ that have the same number. This matrix $A^{\flat}$ has at most one ball in any place, and one can continue, constructing a sequence $A^{(2)}, A^{(3)}, \ldots,$ from which one reads the rest of $\widetilde{P}$ and $\widetilde{Q}$. In this example the successive constructions give

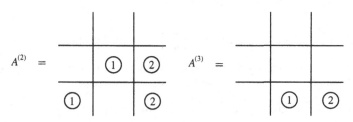

from which one finds the same $\widetilde{P}$ and $\widetilde{Q}$ as before. We denote this by

(3c) *Matrix-ball construction, using the northWest ordering.*

**Proposition 4** *The constructions* (1c), (2c), *and* (3c) *all lead to the same pair* $\{\widetilde{P}, \widetilde{Q}\}$ *of conjugate tableaux.*

There are three other constructions that give another correspondence between matrices $A$ of zeros and ones and pairs $\{\widetilde{R}, \widetilde{S}\}$ of conjugate tableaux. To describe them, take the corresponding array $\begin{pmatrix} u_1 \ u_2 \ \cdots \ u_r \\ v_1 \ v_2 \ \cdots \ v_r \end{pmatrix}$ in antilexicographic order: $u_i \geq u_{i+1}$, and $v_i < v_{i+1}$ if $u_i = u_{i+1}$. These constructions are:

(1d) Row bump $v_r \leftarrow \ldots \leftarrow v_1$ and conjugate place $u_r, \ldots, u_1$.

(2d) Column bump $v_r \rightarrow \ldots \rightarrow v_1$ and conjugate slide $u_1, \ldots, u_r$.

(3d) Matrix-ball construction, using the Northwest ordering.

This matrix-ball construction is obtained from the previous one by interchanging the roles of rows and columns; the numberings of the balls are used to label the rows of $\widetilde{R}$ and the columns of $\widetilde{S}$.

**Proposition 5** *The constructions* (1d), (2d), *and* (3d) *all lead to the same pair* $\{\widetilde{R}, \widetilde{S}\}$ *of conjugate tableaux, and each sets up a one-to-one correspondence between matrices of zeros and ones and pairs of conjugate tableaux.*

As before, one deduces a

**Symmetry Theorem** *If* $\{\widetilde{P}, \widetilde{Q}\}$ *and* $\{\widetilde{R}, \widetilde{S}\}$ *are the two pairs of conjugate tableaux corresponding to a matrix* $A$ *by these procedures, then the two pairs of conjugate tableaux corresponding to the transpose matrix* $A^{\tau}$ *are* $\{\widetilde{S}, \widetilde{R}\}$ *and* $\{\widetilde{Q}, \widetilde{P}\}$ *respectively.*

Note that this symmetry theorem interchanges the two procedures, as well as interchanging the tableaux in the pairs: if an array $\binom{u}{v}$ corresponds to a pair $\{\widetilde{P},\widetilde{Q}\}$ by the first procedure and to $\{\widetilde{R},\widetilde{S}\}$ by the second, then $\binom{v}{u}$ corresponds to $\{\widetilde{S},\widetilde{R}\}$ by the first, and to $\{\widetilde{Q},\widetilde{P}\}$ by the second.

**Theorem 2**  *The four conjugate tableau pairs arising from the lexicographic orderings of arrays and strong column, weak row orderings in matrices, are*

$$\binom{u}{v} \quad \text{nW} \searrow \{\widetilde{P},\widetilde{Q}\}; \qquad \binom{u^*}{v} \quad \text{sW} \nearrow \{\widetilde{R},\widetilde{S}^*\};$$

$$\binom{u^*}{v^*} \quad \text{sE} \nwarrow \{\widetilde{P}^*,\widetilde{Q}^*\}; \qquad \binom{u}{v^*} \quad \text{nE} \swarrow \{\widetilde{R}^*,\widetilde{S}\}.$$

*The four conjugate tableau pairs arising from the antilexicographic orderings of arrays and the weak column and strong row orderings of matrices are:*

$$\binom{u}{v} \quad \text{Nw} \searrow \{\widetilde{R},\widetilde{S}\}; \qquad \binom{u^*}{v} \quad \text{Sw} \nearrow \{\widetilde{P},\widetilde{Q}^*\};$$

$$\binom{u^*}{v^*} \quad \text{Se} \nwarrow \{\widetilde{R}^*,\widetilde{S}^*\}; \qquad \binom{u}{v^*} \quad \text{Ne} \swarrow \{\widetilde{P}^*,\widetilde{Q}\}.$$

The relations among these eight are exactly the same as in Theorem 1. As before, the proof is by calculating the first entry of each pair directly, and using the symmetry theorem to deduce what the other must be. In particular, this contains the

**Duality Theorem**  *If a lexicographic array* $\begin{pmatrix} u_1\,u_2\,\cdots\,u_r \\ v_1\,v_2\,\cdots\,v_r \end{pmatrix}$ *corresponds to a conjugate tableau pair* $\{\widetilde{P},\widetilde{Q}\}$ *then the antilexicographic array* $\begin{pmatrix} u_r^*\,\cdots\,u_1^* \\ v_r^*\,\cdots\,v_1^* \end{pmatrix}$ *corresponds to the conjugate tableau pair* $\{\widetilde{P}^*,\widetilde{Q}^*\}$. *If a lexicographic array* $\begin{pmatrix} u_1\,u_2\,\cdots\,u_r \\ v_1\,v_2\,\cdots\,v_r \end{pmatrix}$ *corresponds to a conjugate tableau pair* $\{\widetilde{R},\widetilde{S}\}$, *then the anti-lexicographic array* $\begin{pmatrix} u_r^*\,\cdots\,u_1^* \\ v_r^*\,\cdots\,v_1^* \end{pmatrix}$ *corresponds to the conjugate tableau pair* $\{\widetilde{R}^*,\widetilde{S}\}$.

If the entries in the top row or the bottom row of an array $\begin{pmatrix} u_1\,u_2\,\cdots\,u_r \\ v_1\,v_2\,\cdots\,v_r \end{pmatrix}$ are distinct, one can apply all four constructions, getting tableau pairs $(P,Q)$ and $(R,S)$ and conjugate pairs $\{\widetilde{P},\widetilde{Q}\}$ and $\{\widetilde{R},\widetilde{S}\}$

**Proposition 6**  (1) *If* $u_1,\dots,u_r$ *are distinct, then* $\widetilde{P}=R$, $\widetilde{Q}=S^\tau$, $\widetilde{R}=P$, *and* $\widetilde{S}=Q^\tau$, *so the four pairs are*

$$(P,Q),\ (R,S),\ \{R,S^\tau\}, \quad and \quad \{P,Q^\tau\}.$$

(2) *If* $v_1,\dots,v_r$ *are distinct, the four pairs are*

$$(P,Q),\ (R,S),\ \{P^\tau,Q\}, \quad and \quad \{R^\tau,S\}.$$

(3) *If* $u_1, \ldots, u_r$ *are distinct and* $v_1, \ldots, v_r$ *are distinct, then the four pairs are*

$$(P,Q), \quad (P^\tau,Q^\tau), \quad \{P^\tau,Q\}, \quad and \quad \{P,Q^\tau\}.$$

**Proof** Part (1) is derived easily from the definitions, and then (2) follows by symmetry, and (3) follows from (1) and (2). □

For example, if $w = v_1 \ldots v_r$ is a word, corresponding to the array $\begin{pmatrix} 1 & \ldots & r \\ v_1 & \ldots & v_r \end{pmatrix}$, and hence to a pair $(P,Q)$ with $P = P(w)$ and $Q = Q(w)$, then $R = P(w^{\text{rev}})$, where $w^{\text{rev}} = v_r \ldots v_1$, and $S = Q((w^*)^{\text{rev}})$, and the other tableaux are determined by (1). If $w$ is a permutation, then $Q = P(w^{-1})$, and all the other tableaux are determined by (3).

**Exercise 21** With the assumptions and notation of Proposition 3, show that, if $T$ is any tableau, and one performs the column-insertions $v_r \to \ldots \to v_1 \to T$ and successively places $u_1, \ldots, u_r$ in the conjugates of the new boxes that arise, then one gets a skew tableau $X$ with $\text{Rect}(X) = \tilde{Q}$.

**Exercise 22** With the notation of Proposition 5, show that if $T$ is any tableau, and one row bumps $T \leftarrow v_r \leftarrow \ldots \leftarrow v_1$, and places $u_r, \ldots, u_1$ in conjugates of the new boxes, one obtains a skew tableau $Y$ with $\text{Rect}(Y) = \tilde{S}$.

**Exercise 23** (a) Suppose a lexicographic array $\begin{pmatrix} u_1 & u_2 & \ldots & u_r \\ v_1 & v_2 & \ldots & v_r \end{pmatrix}$ corresponds to a tableau pair $(P,Q)$ by the R–S–K correspondence, and $(T, T_\circ)$ is any tableau pair such that the entries of $T_\circ$ are greater than each $u_i$. Perform the column bumping $v_1 \to \ldots \to v_r \to T$, and slide in the entries $u_r, \ldots, u_1$ by starting with $T_\circ$, performing reverse slides using the new boxes created by the column bumping, and placing the entries $u_r, \ldots, u_1$ successively in the upper left corner. Show that the entries $u_1, \ldots, u_r$ in the result form the tableau $Q$. (b) State and prove analogous results for the other three sliding constructions of §A.4.

**Exercise 24** (a) With the hypotheses of the preceding exercise, perform the row bumping $T \leftarrow v_1 \leftarrow \ldots \leftarrow v_r$, and slide in the entries $u_1^*, \ldots, u_r^*$, starting with $T_\circ$ and performing reverse slides, using the new boxes found in the row bumping, and successively putting $u_1^*, \ldots, u_r^*$ in the upper left corner; during the reverse slides, regard the entries $u_i^*$ as smaller than all entries of $T_\circ$. Show that the result is a tableau whose $r$ smallest entries form the tableau $Q^*$. (b) State and prove analogous results for the other three sliding constructions of §A.4.

**Exercise 25** If $\begin{pmatrix} u_1 & u_2 & \ldots & u_r \\ v_1 & v_2 & \ldots & v_r \end{pmatrix}$ has distinct entries in rows and columns, and corresponds by R–S–K to a tableau-pair $(P,Q)$, show that the array $\begin{pmatrix} u_1^* & \ldots & u_r^* \\ v_1 & \ldots & v_r \end{pmatrix}$ corresponds by R–S–K to the tableau pair $(P^\tau, (Q^\tau)^*) = (P^\tau, (Q^*)^\tau)$.

Suppose now the alphabets of both rows are $[r]$, and identify the opposite alphabet $[r]^*$ with $[r]$ by identifying $a^*$ identified with $r+1-a$. Taking $u_i = i$ for $1 \le i \le r$, the array $\begin{pmatrix} u_1^* & \cdots & u_r^* \\ v_1 & \cdots & v_r \end{pmatrix}$ is the array corresponding to the reversed word $w^{\text{rev}}$. This gives the

**Corollary 3** *If* $w$ *is a permutation and* $P = P(w)$ *and* $Q = Q(w)$, *then*

$$P(w^{\text{rev}}) = P^\tau \quad and \quad Q(w^{\text{rev}})^\tau = (Q^*)^\tau = (Q^\tau)^*.$$

**Exercise 26** Show that, for two skew tableaux of the same shape, their row words are $Q$-equivalent if and only if their column words are $Q$-equivalent.

**Exercise 27** State and prove analogues of Exercise 4 of §A.2 for the correspondences (b), (c), and (d).

## A.5  Keys

Another application of the ideas of Chapter 5 is the construction of the left and right "keys" of a given tableau. This notion was introduced by Lascoux and Schützenberger (1990) to analyze the combinatorics of standard bases of sections of line bundles on flag manifolds (cf. Fulton and Lascoux [1994]). It is based on the following fact:

**Proposition 7** *Let* $T$ *be a tableau. Let* $v/\lambda$ *be a skew diagram with the same number of columns of each length as* $T$. *Then there is a unique skew tableau* $S$ *on* $v/\lambda$ *that rectifies to* $T$.

**Proof** By Corollary 1 in §5.1 the number of such tableaux depends only on the shape $\mu$ of $T$. So we may take $T = U(\mu)$. In this case the skew tableau $S$ is the obvious one: the $i^{\text{th}}$ entry in each column is $i$. In fact, since $S$ must have $\mu_1$ 1's, and $S$ has exactly $\mu_1$ columns, these 1's must go at the top of the columns in order for $S$ to be a skew tableau. Similarly, since $S$ has exactly $\mu_2$ columns of length at least 2, the $\mu_2$ 2's must go in the second place in each such column, and similarly for all the entries. (The fact that $v/\lambda$ is a skew shape guarantees that the entries in rows are weakly increasing, so $S$ is a skew tableau.)  □

In fact, this proof shows that the entries of $S$ depend only on the (ordered) lengths of its columns. For given lengths, the most compact form is obtained by requiring that each successive pair of columns is aligned either at the top (if the left column is longer) or at the bottom (if the right column is longer), or both if they have the same length; the $S$ of the proposition for any other skew shape with these column lengths is obtained by stretching these columns. For example with

$$T = \begin{array}{|c|c|c|c|}\hline 1 & 1 & 2 & 2 \\\hline 2 & 3 & 3 \\\cline{1-3} 4 \\\cline{1-1}\end{array} \,,$$

if we look for skew tableaux with columns of lengths $(2,3,2,1)$, the compact form $S$ and another $S'$ that rectify to $T$ with these column lengths are

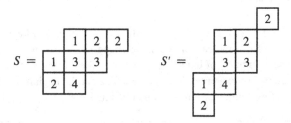

We will usually look for the compact form. When $T$ has two columns, it is easy to find $S$: one simply does reverse sliding, using the boxes at the bottom of the second column. For example,

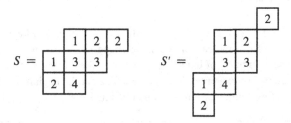

Let us call this process, or its inverse (sliding, using the boxes at the top of the first of two columns), when used on adjacent columns, an ***elementary move***. Elementary moves can be used successively on adjacent columns to find the skew tableau $S$ with given column lengths and given rectification. A skew tableau $S$ is called *frank* if its column lengths are a permutation of the column lengths of its rectification; if $T$ is the rectification of $S$, then $S$ is the unique skew tableau of its shape whose rectification is $T$. An elementary move applied to a frank skew tableau in compact form always produces a frank skew tableau.

For any permutation of the column lengths of $T$, the columns of the corresponding frank skew tableaux $S$ are uniquely determined. In fact, more is true:

**Corollary 1** *The left (resp. right) column of $S$ depends only on the length of that column.*

**Proof** All the other skew tableaux (in compact form) with given left (or right) column length can be found from a given one by performing elementary moves that do not involve that column. □

For a column length $c$ of $T$, let $\mathcal{L}_c$ (resp.$\mathcal{R}_c$) denote the set of elements in the left (resp. right) column of such a skew tableau.

**Corollary 2** *If $c < d$, then $\mathcal{L}_c \subset \mathcal{L}_d$ and $\mathcal{R}_c \subset \mathcal{R}_d$.*

**Proof** Use elementary moves to find $S$ with the two left-most columns of lengths $c$ and $d$, respectively. The elementary move with the first two columns, obtained by sliding by boxes at the top of the first column, adds some of the entries from the second column to those of the first. This shows that $\mathcal{L}_d$ contains $\mathcal{L}_c$. A similar argument works on the right side.     □

These nested sets $\{\mathcal{L}_c\}$ and $\{\mathcal{R}_c\}$ are called the left and right **keys** of $T$. Equivalently, one can form a tableau of the same shape as $T$, denoted $K_-(T)$ (resp. $K_+(T)$), whose column(s) of length $c$ consists of the elements in $\mathcal{L}_c$ (resp. $\mathcal{R}_c$) in increasing order. For example, if $T$ is the tableau at the beginning of this section, by performing a few elementary moves, one finds its keys:

$$
T = \begin{array}{|c|c|c|c|}
\hline 1 & 1 & 2 & 2 \\
\hline 2 & 3 & 3 \\
\cline{1-3}
4 \\
\cline{1-1}
\end{array}
\quad ; \quad
K_-(T) = \begin{array}{|c|c|c|c|}
\hline 1 & 1 & 1 & 1 \\
\hline 2 & 2 & 2 \\
\cline{1-3}
4 \\
\cline{1-1}
\end{array}
\quad , \quad
K_+(T) = \begin{array}{|c|c|c|c|}
\hline 2 & 2 & 2 & 2 \\
\hline 3 & 3 & 3 \\
\cline{1-3}
4 \\
\cline{1-1}
\end{array}
$$

**Exercise 28** If a frank skew tableau is split into several skew tableaux by vertical lines, show that each of the pieces is frank.

**Exercise 29** If $T$ and $U$ are tableaux, show that the following are equivalent: (1) $T * U$ is frank; (2) for any column $t$ of $K_+(T)$ and any column $u$ of $K_-(U)$, $t * u$ is frank; (3) $K_+(T) * K_-(U)$ is frank. Note that for columns $t$ and $u$, $t * u$ is frank exactly when they form a skew tableau when they are placed adjacent to each other in compact form, i.e., with tops or bottoms aligned.

**Exercise 30** For a column length $c$ of $T$, number the right boxes in the top $c$ rows of $T$ from bottom to top. Show that performing reverse row (resp. column) bumping $c$ times, using these boxes in this order, bumps out the entries of $\mathcal{R}_c$ (resp. $\mathcal{L}_c$).

Any chain $\mathcal{S}_1 \subset \mathcal{S}_2 \subset \ldots \subset \mathcal{S}_r$ of subsets of $[m]$ determines a permutation in $\mathcal{S}_m$. The word of this permutation is obtained by listing the elements of $\mathcal{S}_1$ in increasing order, followed by the elements of $\mathcal{S}_2 \setminus \mathcal{S}_1$ in increasing order, and so on, until finally one lists the elements of $[m] \setminus \mathcal{S}_r$ in increasing order. For the chains $\{\mathcal{L}_c\}$ and $\{\mathcal{R}_s\}$ coming from a tableau $T$, these permutations are denoted $w_-(T)$ and $w_+(T)$. For the above example, $w_-(T) = 1\ 2\ 4\ 3$ and $w_+(T) = 2\ 3\ 4\ 1$. These permutations play a role in standard monomial theory (see §10.5 and Fulton and Lascoux [1994]).

# On the topology of algebraic varieties

In this appendix we discuss the basic facts we have used about the cohomology and homology of complex algebraic varieties, including in particular the construction of the class of an algebraic subvariety. Although making constructions like this was one of the main motivating factors in the early development of topology, especially in the work of Poincaré and Lefschetz, it is remarkably difficult – nearly a century later – for a student to extract these basic facts from any algebraic topology text. The intuitive way to do this is to appeal to the fact that an algebraic variety can be triangulated, in such a way that its singular locus is a subcomplex; the sum of the top-dimensional simplices, properly oriented, is a cycle whose homology class is the desired class of the subvariety. Making this rigorous and proving the basic properties one needs from this can be done, but that require quite a bit of work.

An approach which avoids this difficulty, and has the desirable property of working also on a noncompact ambient space, is to use Borel–Moore homology. This is done in Borel and Haefliger (1961), and in detail in Iversen (1986). That approach, however, is based on sheaf cohomology and sheaf duality. In this appendix, we give an alternative but equivalent formulation which uses only standard facts about singular cohomology. (This is a simplified version of a general construction given in Fulton and MacPherson [1981]. Some of the techniques can be found in Dold [1980].) This approach requires basic properties of relative cohomology groups and the existence and basic properties of Thom classes of vector bundles, which can be found for example in Greenberg and Harper (1981), Dold (1980), and Spanier (1966). We also use some basic properties of differentiable manifolds, especially the existence of tubular neighborhoods, and partitions of unity, as in Guillemin and Pollack (1974) or Lang (1985).

In the first section we list the facts and properties that have been used in this text and are the goals of this appendix. The reader who is willing to accept the topology can use this section axiomatically. (For example, the same properties are valid if one uses Chow groups in place of cohomology groups; indeed, many of the proofs are easier in this context.) In sections B.2 and B.3 these facts and properties are proved; the last section discusses Chern classes and projective bundles.

In this appendix, a *variety* is assumed to be irreducible, and a *manifold* is assumed to be connected (and second countable), or at least a disjoint union of a finite number of connected manifolds of the same dimension.

## B.1 The basic facts

Any topological space $X$ has singular homology groups $H_i X$ and cohomology groups $H^i X$ (here taken always with integer coefficients). The cohomology $H^* X = \oplus H^i X$ has a ring structure, $H^i X \otimes H^j X \to H^{i+j} X$, sometimes written with a cup product, $\alpha \otimes \beta \mapsto \alpha \cup \beta$, or sometimes, as we will often do here, with a dot: $\alpha \cdot \beta$. This makes $H^* X$ into an associative, skew-commutative ring with unit $1 \in H^0 X$. The homology $H_* X = \oplus H_i X$ is a (left) module over the cohomology, by the cap product $H^i X \otimes H_j X \to H_{j-i} X$, $\alpha \otimes \beta \mapsto \alpha \cap \beta$.

Any continuous map $f: X \to Y$ determines *pullback* homomorphisms

(1)        $f^*: H^i Y \to H^i X$

and *pushforward* homomorphisms

(2)        $f_*: H_i X \to H_i Y$

for all $i$. Both of these are functorial: if also $g: Y \to Z$, then $(g \circ f)_* = g_* \circ f_*$ and $(g \circ f)^* = f^* \circ g^*$. The pullback $f^*$ is a homomorphism of rings, and there is the *projection formula*

(3)        $f_*(f^*(\alpha) \cap \beta) = \alpha \cap f_*(\beta)$    for    $\alpha \in H^i Y$, $\beta \in H_j X$.

Any projective nonsingular complex variety $X$ of dimension $n$ is a compact oriented $2n$-dimensional real manifold. This implies that its top homology group $H_{2n} X$ is canonically isomorphic to $\mathbb{Z}$, with a generator denoted $[X]$ and called the *fundamental class* of $X$. In addition, the *Poincaré duality map*

(4)        $H^i X \to H_{2n-i} X$, $\alpha \mapsto \alpha \cap [X]$,

given by capping with the fundamental class, is an isomorphism (see Greenberg and Harper [1981], Dold [1980], or Spanier [1966]). By means of this isomorphism, we can and will identify the homology groups with the cohomology groups of such a variety. In particular, if $f$ is a morphism from $X$ to $Y$, with $X$ and $Y$ nonsingular projective varieties of dimensions $n$ and $m$, respectively, then we get a pushforward homomorphism on cohomology, sometimes called a *Gysin* homomorphism:

(5)        $f_*: H^i X = H_{2n-i} X \to H_{2n-i} Y = H^{2m-2n+i} Y$.

In this notation, the projection formula (3) reads

(6)        $f_*(f^*(\alpha) \cdot \beta) = \alpha \cdot f_*(\beta)$    for    $\alpha \in H^i Y$, $\beta \in H^j X$.

Indeed,

$$f_*(f^*(\alpha) \cdot \beta) \cap [Y] = f_*((f^*(\alpha) \smile \beta) \smile [X]) = f_*(f^*(\alpha) \cap (\beta \cap [X]))$$

$$= \alpha \cap f_*(\beta \cap [X]) = (\alpha \cdot f_*(\beta)) \cap [Y].$$

One of the basic facts we will prove is that if $V$ is an irreducible closed $k$-dimensional subvariety of a nonsingular projective $n$-dimensional variety $X$, then $V$ determines a *fundamental class* $[V]$ in $H_{2k}X$, and therefore, by Poincaré duality, a cohomology class, also denoted $[V]$, in $H^{2n-2k}X = H^{2c}X$, where $c$ is the codimension of $V$ in $X$.

We list some of the basic facts about these classes. First, if $f$ is a morphism from $X$ to $Y$, it is a basic fact of algebraic geometry that $f(V)$ is an irreducible closed subvariety of $Y$ of dimension at most $k$, and if the dimension of $f(V)$ is $k$, then there is a Zariski open subset $U$ of $f(V)$ such that the mapping from $V$ to $f(V)$ determines a finite sheeted covering map from $f^{-1}(U) \cap V$ to $U$. The number of sheets of this covering is called the *degree* of $V$ over $f(V)$. Then

$$(7) \qquad f_*[V] = \begin{cases} 0 & \text{if } \dim(f(V)) < \dim(V) \\ d[f(V)] & \text{if } V \text{ has degree } d \text{ over } f(V). \end{cases}$$

In most situations, this fact from algebraic geometry, and the degree of $V$ over $f(V)$, will be evident; in the situations where this fact is used in this book, either $f$ maps $V$ onto a lower dimensional variety, or $f$ maps $V$ birationally onto $f(V)$, i.e., the degree of $V$ over $f(V)$ is 1. (See Shafarevich [1977], II.5, for the general proof.)

If $f$ is the constant map from $X$ to a point $Y$, the map $f_*$ from $H_0X$ to $H_0Y = \mathbb{Z}$ is an isomorphism, since $X$ is (path) connected. This map is called the *degree homomorphism*, and it takes the class $[x]$ of any point $x$ to 1.

Now suppose that $f$ is a smooth morphism from $X$ to $Y$. In our applications, $X$ will be a fiber bundle over $Y$; i.e., there will be a nonsingular projective variety $F$, and a covering of $Y$ by Zariski open sets $U_\alpha$, such that $f^{-1}(U_\alpha) \cong U_\alpha \times F$, with $f$ corresponding to the projection from $U_\alpha \times F$ to $U_\alpha$. In this case, if $V$ is a subvariety of $Y$, then $f^{-1}(V)$ is a subvariety of $X$ (locally defined by the pullbacks of the equations defining $V$ in $Y$). In this case, we have the formula

$$(8) \qquad f^*[V] = [f^{-1}(V)].$$

Suppose $V$ and $W$ are subvarieties of the projective nonsingular variety $X$, and suppose that the intersection $V \cap W$ is a union of subvarieties $Z_1, \ldots, Z_r$. Suppose that the intersection is *proper*, i.e., that the codimension of each $Z_i$ in $X$ is the sum of the codimension of $V$ and the codimension of $W$. Suppose, in addition, that $V$ and $W$ *meet transversally*, i.e., for points $z$ in a Zariski open set of each $Z_i$, the tangent spaces of $V$ and $W$ at $z$ intersect (transversally) in the tangent space of $Z_i$:

$$T_z Z_i = T_z V \cap T_z W \subset T_z X.$$

In this case we have the intersection formula

$$(9) \qquad [V] \cdot [W] = [Z_1] + \ldots + [Z_r].$$

If $V$ and $W$ have complementary dimensions then $[V] \cdot [W]$ is a class in $H_0 X$ whose degree is sometimes denoted $\langle [V], [W] \rangle$, and is called the **intersection number** of $V$ and $W$. If $V$ meets $W$ transversally in $r$ points, then $\langle [V], [W] \rangle = r$. Another notation for this intersection number is $([V] \cdot [W])$, or, by common abuse, simply $[V] \cdot [W]$.

In general, as long as the intersection is proper, one can assign (positive integer) intersection multiplicities $m_i$ to the components $Z_i$ so that $[V] \cdot [W] = \sum m_i [Z_i]$ (see (31) in §B.3) but we will not need this generality here.

Another important fact that we need is the following: if a projective nonsingular variety $X$ has a filtration $X = X_s \supset X_{s-1} \supset \ldots \supset X_0 = \emptyset$ by closed algebraic subsets, and $X_i \smallsetminus X_{i+1}$ is a disjoint union of varieties $U_{i,j}$ each isomorphic to an affine space $\mathbb{C}^{n(i,j)}$, then the classes $[\overline{U}_{i,j}]$ of the closures of these varieties give an additive basis for $H^*(X)$ over $\mathbb{Z}$.

Finally, we will need some basic facts about Chern classes of vector bundles. Here we need only consider algebraic vector bundles, which are fiber bundles $E \to X$ that are locally of the form $U_\alpha \times \mathbb{A}^e \to U_\alpha$, with the change of coordinates over $U_\alpha \cap U_\beta$ given by a morphism from $U_\alpha \cap U_\beta$ to the general linear group $GL_e \mathbb{C}$. First, a line bundle $L$ on a nonsingular projective variety $X$ has a **first Chern class** $c_1(L)$ in $H^2(X)$. Its basic properties are

(10)     $c_1(f^*(L)) = f^*(c_1(L))$   if   $f : Y \to X$.

(11)     $c_1(L \otimes M) = c_1(L) + c_1(M)$     for line bundles

$$L \text{ and } M \text{ on } X.$$

(12)     $c_1(L) = [D]$   if   $L \cong \mathcal{O}(D)$,     $D$ an irreducible

$$\text{hypersurface in } X.$$

Here, the equation $L = \mathcal{O}(D)$ means that $L$ has a section $s$ whose zeros cut out $D$ as a subvariety of $X$. More generally, any rational section $s$ of a line bundle $L$ determines a divisor $D = \sum n_i D_i$ where each $D_i$ is an irreducible hypersurface and $n_i$ is the order of vanishing of $s$ along $D_i$, and then $L \cong \mathcal{O}(D)$ and $c_1(L) = \sum n_i [D_i]$ in $H^2(X)$; we will not need this generality here. Each of (10), (11), or (12) implies that $c_1(L) = 0$ if $L$ is a trivial line bundle.

A vector bundle $E$ of rank $e$ has **Chern classes** $c_i(E)$ in $H^{2i}(E)$, with $c_0(E) = 1$, and $c_i(E) = 0$ if $i < 0$ or $i > e$. These satisfy the same functoriality as line bundles:

(13)     $c_i(f^*(E)) = f^*(c_i(E))$   if   $f : Y \to X$.

The other basic property is the **Whitney formula**: if $E'$ is a subbundle of $E$, with quotient bundle $E'' = E/E'$, then

(14)     $c_k(E) = \displaystyle\sum_{i+j=k} c_i(E') \cdot c_j(E'')$.

## B.2 Borel–Moore homology

We will use basic properties of the singular cohomology groups $H^i(X,Y)$, which are defined for any topological space $X$ and any subspace $Y$. In contrast to the typical situation in topology, we will use them only when $Y$ is an *open* subset of $X$. They are defined to be the cohomology groups of the complex $C^*(X,Y)$ of singular cochains (i.e., $\mathbb{Z}$-valued functions on singular chains) on $X$ that vanish on the chains in $Y$. From the definition one has natural long exact sequences, if $Z \subset Y \subset X$,

$$(15) \qquad \ldots \to H^i(X,Y) \to H^i(X,Z) \to H^i(Y,Z)$$
$$\to H^{i+1}(X,Y) \to H^{i+1}(X,Z) \to \ldots .$$

If $Y$ and $Z$ are open subsets of a space $X$, there is a cup product

$$(16) \qquad H^i(X,Y) \times H^j(X,Z) \to H^{i+j}(X,Y \cup Z),$$

with associativity and skew commutativity analogous to the absolute case. (This uses the Mayer–Vietoris property, to know that $H^*(X,Y \cup Z)$ is the cohomology of the complex of cochains on $X$ that vanish on chains that are in $Y$ or in $Z$.)

There is the standard *excision:* if $Y$ is open in $X$, and $A$ is a closed subset of $X$ that is contained in $Y$, then the natural map

$$(17) \qquad H^i(X,Y) \to H^i(X \smallsetminus A, Y \smallsetminus A) \quad \text{is an isomorphism.}$$

If $E$ is an oriented real vector bundle of rank $r$ on a topological space $X$, then there is a *Thom class* $\gamma_E$ in $H^r(E, E \smallsetminus \{0\})$, where $\{0\} \subset E$ denotes the image of the zero section. This has the property that, for any closed set $A$ in $X$, the map

$$(18) \qquad H^i(X, X \smallsetminus A) \to H^{i+r}(E, E \smallsetminus A), \quad \alpha \to \pi^*(\alpha) \cup \gamma_E$$

is an isomorphism; here $A$ is regarded as a subspace of $E$ by the embedding in the zero section, and $\pi$ is the projection from $E$ to $X$.

Suppose $M$ is a closed submanifold of a differentiable manifold $N$, of dimensions $m$ and $n$. We have an exact sequence of bundles on $M$:

$$0 \to T_M \to T_N|_M \to E \to 0,$$

where, by definition, $E$ is the normal bundle of $M$ in $N$. By this sequence, an orientation of $M$ and $N$ determines an orientation of the bundle $E$. If $E$ is oriented, we have a canonical isomorphism

$$(19) \qquad H^i(M, M \smallsetminus A) \cong H^{i+n-m}(N, N \smallsetminus A)$$

for any closed set $A$ in $M$. This can be defined as follows. Choose a tubular neighborhood $U$ of $M$ in $N$, with an isomorphism of the normal bundle $E$ with $U$ (cf. Guillemin and Pollack [1974], or Lang [1985]). Then

$$H^i(M, M \smallsetminus A) \cong H^{i+n-m}(E, E \smallsetminus A) \cong H^{i+n-m}(U, U \smallsetminus A)$$
$$\cong H^{i+n-m}(N, N \smallsetminus A),$$

the first isomorphism the Thom isomorphism, the second by the identification of $E$ with $U$, and the third by excision of $N \smallsetminus U$. This isomorphism is independent of choice of tubular neighborhood.

We define, for any topological space $X$ that can be embedded as a closed subspace of a Euclidean space $\mathbb{R}^n$, a version of **Borel–Moore homology groups**, denoted $\overline{H}_i X$, by the formula

(20) $\qquad \overline{H}_i X = H^{n-i}(\mathbb{R}^n, \mathbb{R}^n \smallsetminus X)$.

**Lemma 1** *The definition of $\overline{H}_i X$ is independent of choice of embedding in Euclidean space.*

**Proof** Let $\varphi: X \to \mathbb{R}^n$ and $\psi: X \to \mathbb{R}^m$ be two closed embeddings. We construct an isomorphism from $H^{n-i}(\mathbb{R}^n, \mathbb{R}^n \smallsetminus X_\varphi)$ to $H^{m-i}(\mathbb{R}^m, \mathbb{R}^m \smallsetminus X_\psi)$, where $X_\varphi$ and $X_\psi$ denote the images of $X$ by the embeddings $\varphi$ and $\psi$. The projection $\mathbb{R}^n \times \mathbb{R}^m \to \mathbb{R}^n$ is an oriented (trivial) vector bundle, so the Thom isomorphism (18) gives an isomorphism

(21) $\qquad H^{n-i}(\mathbb{R}^n, \mathbb{R}^n \smallsetminus X_\varphi) \cong H^{n+m-i}(\mathbb{R}^n \times \mathbb{R}^m, \mathbb{R}^n \times \mathbb{R}^m \smallsetminus X_{(\varphi,0)})$,

where $(\varphi,0): X \to \mathbb{R}^n \times \mathbb{R}^m$ is the closed embedding that maps $X$ to $\varphi(x) \times 0$.

By the Tietze extension theorem (cf. James [1987], 11.7) there is a continuous map $\widetilde{\psi}: \mathbb{R}^n \to \mathbb{R}^m$ such that $\widetilde{\psi} \circ \varphi = \psi$. Define a map $\vartheta: \mathbb{R}^n \times \mathbb{R}^m \to \mathbb{R}^n \times \mathbb{R}^m$ by the formula $\vartheta(v \times w) = v \times (w - \widetilde{\psi}(v))$ for $v \in \mathbb{R}^n$, $w \in \mathbb{R}^m$. Note that $\vartheta \circ (\varphi, \psi) = (\varphi, 0)$, where $(\varphi, \psi)$ is the closed embedding of $X$ in $\mathbb{R}^n \times \mathbb{R}^m$ that takes $x$ to $\varphi(x) \times \psi(x)$. Since $\vartheta$ is a homeomorphism mapping $X_{(\varphi,\psi)}$ onto $X_{(\varphi,0)}$, we have an isomorphism

(22) $\qquad \vartheta^*: H^{n+m-i}(\mathbb{R}^n \times \mathbb{R}^m, \mathbb{R}^n \times \mathbb{R}^m \smallsetminus X_{(\varphi,0)})$

$\qquad\qquad \overset{\cong}{\longrightarrow} H^{n+m-i}(\mathbb{R}^n \times \mathbb{R}^m, \mathbb{R}^n \times \mathbb{R}^m \smallsetminus X_{(\varphi,\psi)})$.

First we note that $\vartheta^*$ is independent of the choice of the extension $\widetilde{\psi}$ of $\psi$, for if $\widetilde{\psi}'$ were another, then $\vartheta_t(v \times w) = v \times (w - t \cdot \widetilde{\psi}(v) - (1-t) \cdot \widetilde{\psi}'(v))$ gives a homotopy from one to the other. The composite of (21) and (22) is an isomorphism

(23) $\qquad H^{n-i}(\mathbb{R}^n, \mathbb{R}^n \smallsetminus X_\varphi) \cong H^{n+m-i}(\mathbb{R}^n \times \mathbb{R}^m, \mathbb{R}^n \times \mathbb{R}^m \smallsetminus X_{(\varphi,\psi)})$.

By interchanging the roles of $\mathbb{R}^n$ and $\mathbb{R}^m$, we have similarly

(24) $\qquad H^{m-i}(\mathbb{R}^m, \mathbb{R}^m \smallsetminus X_\psi) \cong H^{m+n-i}(\mathbb{R}^m \times \mathbb{R}^n, \mathbb{R}^m \times \mathbb{R}^n \smallsetminus X_{(\psi,\varphi)})$.

To finish the proof, we must construct an isomorphism between the right sides of (23) and (24). Let $\tau: \mathbb{R}^m \times \mathbb{R}^n \to \mathbb{R}^n \times \mathbb{R}^m$ be the homeomorphism that reverses the factors: $\tau(v \times w) = w \times v$. Note that $\tau$ preserves or reverses the orientation according to whether $mn$ is even or odd, and that $\tau$ maps $X_{(\psi,\varphi)}$ onto $X_{(\varphi,\psi)}$.

Then $(-1)^{nm} \cdot \tau^*$ determines an isomorphism

$$(25) \qquad H^{n+m-i}(\mathbb{R}^n \times \mathbb{R}^m, \mathbb{R}^n \times \mathbb{R}^m \setminus X_{(\varphi,\psi)}) \xrightarrow{\cong}$$

$$H^{m+n-i}(\mathbb{R}^m \times \mathbb{R}^n, \mathbb{R}^m \times \mathbb{R}^n \setminus X_{(\psi,\varphi)}). \qquad \square$$

**Exercise 1** If $\psi = \varphi$, verify that the isomorphism constructed in this proof is the identity. Show that the isomorphisms constructed in the preceding proof are compatible, in the following sense: if $X \subset \mathbb{R}^p$ is a third closed embedding, then the diagram

$$H^{n-i}(\mathbb{R}^n, \mathbb{R}^n \setminus X) \longrightarrow H^{m-i}(\mathbb{R}^m, \mathbb{R}^m \setminus X)$$
$$\searrow \qquad \qquad \swarrow$$
$$H^{p-i}(\mathbb{R}^p, \mathbb{R}^p \setminus X)$$

commutes.

In fact, more is true:

**Lemma 2** *If a space $X$ is embedded as a closed subspace of an oriented differentiable manifold $M$, there is a canonical isomorphism*

$$\overline{H}_i X \cong H^{m-i}(M, M \setminus X), \quad \text{where} \quad m = \dim(M).$$

**Proof** Any manifold $M$ can be embedded as a closed submanifold of some Euclidean space $\mathbb{R}^n$. Then the isomorphism (19) gives an isomorphism

$$H^{m-i}(M, M \setminus X) \cong H^{n-i}(\mathbb{R}^n, \mathbb{R}^n \setminus X) = \overline{H}_i X. \qquad \square$$

**Exercise 2** Show that if $X$ is embedded as a closed subset of another manifold $N$ of dimension $n$, then there is a canonical isomorphism $H^{m-i}(M, M \setminus X) \cong H^{n-i}(N, N \setminus X)$, and that these isomorphisms are compatible in the sense of Exercise 1.

For an oriented $n$-dimensional manifold $M$, its Borel–Moore homology groups can be computed by embedding $M$ in itself:

$$(26) \qquad \overline{H}_i M = H^{n-i}(M, M \setminus M) = H^{n-i} M.$$

In particular, we see that $\overline{H}_i M = 0$ for $i > n$, and that $\overline{H}_n M = H^0 M$ is a free $\mathbb{Z}$-module, with one generator for each connected component of $M$. If $M$ is compact, the usual Poincaré duality $H^{n-i} M \cong H_i M$ shows that Borel–Moore homology is equal to ordinary homology for $M$. More generally, if $X$ is compact and locally contractible, and $X$ is embedded in an oriented manifold $M$, then Alexander–Lefschetz duality $H^{n-i}(M, M \setminus X) \cong H_i X$ (see Spanier [1966], Lemma 6.10.14) shows that Borel–Moore and singular homology agree for $X$; we won't need this generalization.

Unlike the ordinary singular homology groups, the Borel–Moore homology groups $\overline{H}_i$ are not covariant for arbitrary continuous maps. However, if $f: X \to Y$ is a

***proper*** continuous map (of spaces that admit closed embeddings in Euclidean spaces), i.e., the inverse image of any compact subset of $Y$ is compact in $X$, then there is a pushforward map $f_*: \overline{H}_i X \to \overline{H}_i Y$. This can be constructed as follows: Since $f$ is proper, there is a morphism $\varphi: X \to I^n \subset \mathbb{R}^n$, where $I \subset \mathbb{R}$ is a closed interval containing $0$ in its interior, such that the resulting map $(f, \varphi): X \to Y \times I^n$ is a closed embedding. Choose any closed embedding of $Y$ in $\mathbb{R}^m$, which determines a closed embedding $X \subset Y \times I^n \subset \mathbb{R}^m \times \mathbb{R}^n$. We must construct a homomorphism from $\overline{H}_i X = H^{m+n-i}(\mathbb{R}^m \times \mathbb{R}^n, \mathbb{R}^m \times \mathbb{R}^n \smallsetminus X)$ to $\overline{H}_i Y = H^{m-i}(\mathbb{R}^m, \mathbb{R}^m \smallsetminus Y)$. This is the composite of the restriction map

$$(27) \qquad H^{m+n-i}(\mathbb{R}^m \times \mathbb{R}^n, \mathbb{R}^m \times \mathbb{R}^n \smallsetminus X) \to$$

$$H^{m+n-i}(\mathbb{R}^m \times \mathbb{R}^n, \mathbb{R}^m \times \mathbb{R}^n \smallsetminus Y \times I^n)$$

followed by the inverses of the following two isomorphisms: the restriction

$$(28) \qquad H^{m+n-i}(\mathbb{R}^m \times \mathbb{R}^n, \mathbb{R}^m \times \mathbb{R}^n \smallsetminus Y \times \{0\})$$

$$\to H^{m+n-i}(\mathbb{R}^m \times \mathbb{R}^n, \mathbb{R}^m \times \mathbb{R}^n \smallsetminus Y \times I^n)$$

and the Thom isomorphism (18)

$$(29) \qquad H^{m-i}(\mathbb{R}^m, \mathbb{R}^m \smallsetminus Y) \to H^{m+n-i}(\mathbb{R}^m \times \mathbb{R}^n, \mathbb{R}^m \times \mathbb{R}^n \smallsetminus Y \times \{0\})$$

for the trivial bundle $\mathbb{R}^m \times \mathbb{R}^n \to \mathbb{R}^m$. That (28) is an isomorphism follows from the general fact that if $B$ is open in a space $A$, and $U' \subset U \subset A$ are open, with $U' \subset U$ and $U' \cap B \subset U \cap B$ deformation retracts, then the restriction maps $H^j(A, B \cup U) \to H^j(A, B \cup U')$ are isomorphisms. This fact is an easy consequence of the Mayer–Vietoris exact sequence, and it implies (28) by setting $B = \mathbb{R}^m \times \mathbb{R}^n \smallsetminus Y \times \mathbb{R}^n$, $A = \mathbb{R}^m \times \mathbb{R}^n$, $U' = \mathbb{R}^m \times \mathbb{R}^n \smallsetminus \mathbb{R}^m \times I^n$, and $U = \mathbb{R}^m \times \mathbb{R}^n \smallsetminus \mathbb{R}^m \times \{0\}$.

**Exercise 3** Show that $f_*$ is independent of the choices made in its construction. Show that these pushforward maps are functorial: if $g: Y \to Z$ is also proper, then $(g \circ f)_* = g_* \circ f_*$.

Let $U$ be an open subset of a space $X$ that admits a closed embedding in a Euclidean space (or a manifold). Then $U$ also admits such an embedding. Indeed, if $X$ is a closed subset of an oriented $n$-manifold $M$, then $U$ is closed in the oriented manifold $M^\circ = M \smallsetminus Y$, where $Y$ is the complement of $U$ in $X$. From this it follows that there is a canonical **restriction map** from $\overline{H}_i X$ to $\overline{H}_i U$. Indeed, this comes from the restriction map in cohomology:

$$(30) \qquad \overline{H}_i X = H^{n-i}(M, M \smallsetminus X) \to H^{n-i}(M^\circ, M^\circ \smallsetminus U) = \overline{H}_i U.$$

**Exercise 4** Show that this restriction map is independent of the choices made in its construction, and show that it is functorial: if $U' \subset U \subset X$ are open, the restriction from $\overline{H}_i X$ to $\overline{H}_i U'$ is the composite of that from $\overline{H}_i X$ to $\overline{H}_i U$ and that from $\overline{H}_i U$ to $\overline{H}_i U'$.

**Lemma 3** *If  U  is open in  X,  and  Y  is the complement of  U  in  X,  then there is a long exact sequence*

$$\ldots \to \overline{H}_i Y \to \overline{H}_i X \to \overline{H}_i U \to \overline{H}_{i-1} Y \to \overline{H}_{i-1} X$$

$$\to \overline{H}_{i-1} U \to \ \ldots .$$

**Proof** Choosing  $M$  as in the preceding construction, this is the long exact cohomology sequence (15) of the triple  $M \smallsetminus X \subset M \smallsetminus Y \subset M$.      □

**Exercise 5** Show that the maps in this sequence are independent of the choice of the embedding in  $M$.  Show that, if  $f: X' \to X$  is a proper map, and  $U' = f^{-1}(U)$,  $Y' = f^{-1}(Y)$,  then the diagram

$$\begin{array}{ccccccccc} \ldots \to & \overline{H}_i Y' & \to & \overline{H}_i X' & \to & \overline{H}_i U' & \to & \overline{H}_{i-1} Y' & \to & \overline{H}_{i-1} X' & \to & \ldots \\ & \downarrow & & \downarrow & & \downarrow & & \downarrow & & \downarrow & & \\ \ldots \to & \overline{H}_i Y & \to & \overline{H}_i X & \to & \overline{H}_i U & \to & \overline{H}_{i-1} Y & \to & \overline{H}_{i-1} X & \to & \ldots \end{array}$$

commutes, where the vertical maps are the proper pushforward maps.

**Exercise 6** If  $X$  is a disjoint union of a finite number of open subspaces  $X_\alpha$,  show that  $\overline{H}_i X$  is a direct sum of the  $\overline{H}_i(X_\alpha)$.

### B.3  Class of a subvariety

**Lemma 4** *Let  V  be an algebraic subset of a nonsingular algebraic variety, and let  k  be the dimension of  V.  Then  $\overline{H}_i V = 0$  for  $i > 2k$,  and  $\overline{H}_{2k} V$  is a free abelian group with a generator for each k-dimensional irreducible component of  V.*

**Proof** Consider first the case where  $V$  is nonsingular and purely  $k$-dimensional. In this case  $V$  is an oriented real  $2k$-manifold, and the conclusion follows from (26) and Exercise 6. Note that, for  $V$  nonsingular, its connected components are the same as its irreducible components; this follows from the fact that an irreducible variety is connected, and the fact that a point in the intersection of two or more components must be a singular point.

The proof of the lemma in the general case is by induction on  $k$.  There is a closed algebraic subset  $Z$  of  $V$  of dimension less than  $k$,  such that  $V \smallsetminus Z$  is nonsingular and purely  $k$-dimensional; indeed, one can take  $Z$  to be the union of all irreducible components of dimension less than  $k$,  together with the singular locus of  $V$.  Then  $\overline{H}_i(Z) = 0$  for  $i > 2k-2$  by induction, and  $\overline{H}_i(V \smallsetminus Z) = 0$  for  $i > 2k$  by the nonsingular case just discussed. The exact sequence of Lemma 3 then implies that  $\overline{H}_i V = 0$  for  $i > 2k$,  and gives an exact sequence

$$0 = \overline{H}_{2k} Z \to \overline{H}_{2k} V \to \overline{H}_{2k}(V \smallsetminus Z) \to \overline{H}_{2k-1} Z = 0.$$

Therefore  $\overline{H}_{2k} V \cong \overline{H}_{2k}(V \smallsetminus Z)$,  which is free on generators corresponding to irreducible components of  $V \smallsetminus Z$,  which are exactly the restrictions of the  $k$-dimensional components of  $V$.      □

Now suppose $V$ is an irreducible closed subvariety of a nonsingular projective (or compact) variety $X$. If $k$ is the dimension of $V$, then $\overline{H}_{2k}V = \mathbb{Z}$ has a canonical generator, and the closed embedding of $V$ in $X$ determines a pushforward homomorphism

$$\overline{H}_{2k}V \;\to\; \overline{H}_{2k}X \;=\; \overline{H}^{2n-2k}X \;=\; \overline{H}^{2c}X,$$

where $n$ is the dimension of $X$ and $c = n-k$ the codimension of $V$. The image in $H^{2c}X$ of the generator of $\overline{H}_{2k}V$ is called the **fundamental class** of $V$ in $X$, and is denoted $[V]$.

We next prove formula (7). We have a commutative diagram

$$\begin{array}{ccc}
\overline{H}_{2k}V & \to & \overline{H}_{2k}X = H^{2n-2k}X \\
\downarrow & & \downarrow \qquad \downarrow f_* \\
\overline{H}_{2k}f(V) & \to & \overline{H}_{2k}Y = H^{2m-2k}Y
\end{array}$$

where $k = \dim(V)$, $n = \dim(X)$, $m = \dim(Y)$. The fact that $f_*[V] = 0$ if $\dim(f(V)) < k$ follows from Lemma 4. Suppose $\dim(f(V)) = k$, and $U$ is an open set in $f(V)$ such that $f^{-1}(U) \cap V \to U$ is a $d$-sheeted covering space. Replacing $U$ by a small open ball contained in the nonsingular locus of $f(V)$, we may assume the covering is trivial, so $f^{-1}(U) \cap V$ is a disjoint union of $d$ open sets $U_\alpha$, each mapped isomorphically onto $U$. We have a commutative diagram (see Exercise 5)

$$\begin{array}{ccc}
\overline{H}_{2k}V & \to & \overline{H}_{2k}(f^{-1}(U)\cap V) = \oplus \overline{H}_{2k}(U_\alpha) \\
\downarrow & & \downarrow \\
\overline{H}_{2k}f(V) & \to & \overline{H}_{2k}U.
\end{array}$$

Now each restriction $\overline{H}_{2k}V \to \overline{H}_{2k}(U_\alpha)$ takes the generator to the generator, as does the restriction $\overline{H}_{2k}f(V) \to \overline{H}_{2k}U$, as well as each isomorphism $\overline{H}_{2k}(U_\alpha) \to \overline{H}_{2k}U$. From this it follows that the generator of $\overline{H}_{2k}V$ maps to $d$ times the generator of $\overline{H}_{2k}f(V)$, and this completes the proof of (7).

In order to prove (8) and (9), we again take advantage of the possibility of localizing to small open sets allowed by the restriction mapping for Borel–Moore homology. To carry this out, we note that an irreducible closed subvariety $V$ of any nonsingular variety $X$ has a **refined class**, denoted $\eta_V$, in the relative cohomology group $H^{2n-2k}(X, X \smallsetminus V) = H^{2c}(X, X \smallsetminus V)$, where $k = \dim(V)$, $n = \dim(X)$, and $c = n-k$ is the codimension. This class is the image of the canonical generator of $\overline{H}_{2k}V$ by the isomorphism of Lemma 2:

$$\overline{H}_{2k}V \;\cong\; H^{2n-2k}(X, X \smallsetminus V) \;=\; H^{2c}(X, X \smallsetminus V).$$

This has the property that, when $X$ is compact, the image of $\eta_V$ in $H^{2n-2k}(X)$ is the class $[V]$ of $V$. An important property, that follows directly from the definition, is that, if $X^\circ$ is any open subset of $X$ that meets $V$, and $V^\circ = V \cap X^\circ$, then the refined class $\eta_V$ restricts to $\eta_{V^\circ}$ by the restriction map from $H^{2c}(X, X \smallsetminus V)$ to $H^{2c}(X^\circ, X^\circ \smallsetminus V^\circ)$. Note that each of these relative cohomology groups is isomorphic to $\mathbb{Z}$, and these restriction maps are isomorphisms, so no information is lost by such restriction. It also follows from the definition that if $X = E$ is a complex vector bundle of rank $c$ over $V$, and $V$ is embedded as the zero section in $X$, then $\eta_V$ is the Thom class $\gamma_E$ of this vector bundle.

Now we prove (8). We prove something stronger:

**Lemma 5** *Let* $f: X \to Y$ *be a morphism of nonsingular varieties, and let* $V$ *be an irreducible subvariety of* $Y$ *of codimension* $c$, *such that* $W = f^{-1}(V)$ *is an irreducible subvariety of* $X$ *of codimension* $c$, *and the following condition holds: there is a neighborhood* $U$ *of a nonsingular point of* $V$ *on which* $V \cap U$ *is the submanifold defined by equations* $h_1, \ldots, h_c$, *such that* $W \cap f^{-1}(U)$ *is the submanifold of* $f^{-1}(U)$ *defined by the equations* $h_1 \circ f, \ldots, h_c \circ f$. *Then* $f^*(\eta_V) = \eta_W$, *where* $f^*$ *is the pullback from* $H^{2c}(Y, Y \smallsetminus V)$ *to* $H^{2c}(X, X \smallsetminus W)$.

**Proof** Since these relative cohomology groups are generated by $\eta_V$ and $\eta_W$, respectively, we must have $f^*(\eta_V) = d\eta_W$ for some integer $d$, and our object is to show that $d = 1$. By the discussion of the preceding paragraph, we may do this by replacing $Y$ by any open subset $Y^\circ$ of $Y$ that meets $V$, and replacing $X$ by any open subset $X^\circ$ of $f^{-1}(Y^\circ)$ that meets $W$. By taking $Y^\circ = U$ as in the hypotheses, we can reduce the situation to the case where $Y = E$ is a (trivial) vector bundle over $V$, with $V$ embedded as the zero section, and $X$ the pullback bundle $g^*E$, where $g$ is the morphism from $W$ to $V$ induced by $f$, and $W$ is the zero section of $g^*E$. The equation $\eta_W = f^*(\eta_V)$ is now the elementary fact that the Thom class of $g^*E$ is the pullback of the Thom class of $E$.  □

As the proof of this lemma shows, the functions $h_1, \ldots, h_c$ that locally cut out $V$ in $Y$ can be regular algebraic functions on a Zariski neighborhood $U$, or holomorphic functions on a classical neighborhood $U$.

Next we consider the relation between the intersection of two subvarieties $V$ and $W$ of a nonsingular compact variety $X$ and the product of their fundamental classes. Let $a = \dim(V)$, $b = \dim(W)$, and $n = \dim(X)$. We have the refined classes $\eta_V$ and $\eta_W$ in the relative cohomology groups $H^{2n-2a}(X, X \smallsetminus V)$ and $H^{2n-2b}(X, X \smallsetminus W)$. The cup product of these refined classes is an element of

$$H^{4n-2a-2b}(X, (X \smallsetminus V) \cup (X \smallsetminus W))$$
$$= H^{4n-2a-2b}(X, X \smallsetminus (V \cap W)) = \overline{H}_{2a+2b-2n}(V \cap W).$$

If the intersection is proper, i.e., each irreducible component $Z_i$ of $V \cap W$ has dimension $a + b - n$, then this group is free abelian with a generator $\eta_{Z_i}$ for each irreducible component $Z_i$ of $V \cap W$. Therefore

(31)     $\eta_V \cup \eta_W = m_1 \eta_{Z_1} + m_2 \eta_{Z_2} + \ldots + m_r \eta_{Z_r}$,

for some unique integers $m_1, \ldots, m_r$. We can take the coefficient $m_i$ as the definition of the **intersection multiplicity** of $Z_i$ in the intersection of $V$ and $W$ on $X$. It is a fact that these integers agree with those constructed in algebraic geometry (see Fulton

[1984], §19). What we need to prove to verify (9) is that, if the intersection along a component $Z = Z_i$ is generically transversal, then the coefficient $m_i$ is 1.

By our construction of these classes, they are compatible with restriction to an open subset $U$ of $X$; note that the compactness of $X$ was not needed for the construction of the refined classes. In particular, by restricting to a neighborhood of a point of $Z$ at which $V$ and $W$ meet transversally, we can take $U$ to be holomorphically isomorphic to a complex ball in $\mathbb{C}^n$, in such a way that $V \cap U$ and $W \cap U$ correspond to coordinate planes meeting transversally in $Z \cap U$. In particular, one is reduced to the case where $X$ is a direct sum of two vector bundles on the variety $Z$, with $V$ and $W$ the zero sections of these bundles. In this case the assertion amounts to the fact that the Thom class of a direct sum of bundles is the product of the Thom classes of the bundles.

**Lemma 6** *If $X = X_s \supset \ldots \supset X_0 = \emptyset$ is a sequence of closed algebraic subsets of an algebraic variety $X$, such that $X_i \smallsetminus X_{i+1}$ is a disjoint union of varieties $U_{i,j}$ each isomorphic to an affine space $\mathbb{C}^{n(i,j)}$, then the classes $[\overline{U}_{i,j}]$ of the closures of these varieties give an additive basis for the Borel–Moore homology groups $\overline{H}_*(X)$ over $\mathbb{Z}$.*

**Proof** Note first that $\overline{H}_i(\mathbb{C}^m) = \mathbb{Z}$ if $i = 2m$, and $\overline{H}_i(\mathbb{C}^m) = 0$ otherwise, as seen by the isomorphism $\overline{H}_i(\mathbb{C}^m) \cong H^{2m-i}(\mathbb{C}^m)$. We argue by induction on $p$ that the classes $[\overline{U}_{i,j}]$ for $i \leq p$ give a basis for $\overline{H}_*(X_p)$. Assuming the result for $p-1$, since all $\overline{H}_k(X_{p-1})$ and $\overline{H}_k(U_{i,j})$ vanish for $k$ odd, we deduce from Lemma 3 that $\overline{H}_k(X_p) = 0$ for $k$ odd, and we have exact sequences

$$0 \to \overline{H}_{2i}(X_{p-1}) \to \overline{H}_{2i}(X_p) \to \oplus \overline{H}_{2i}(U_{p,j}) \to 0.$$

The classes $[\overline{U}_{p,j}] \in \overline{H}_*(X_p)$ map to a basis of $\oplus\overline{H}_*(U_{p,j})$. From this it follows by induction that $\overline{H}_*(X_p)$ is free on the classes of the $[\overline{U}_{i,j}]$ for $i \leq p$.   $\square$

**Exercise 7** If a connected topological group $G$ acts on a space $X$, by a continuous map $G \times X \to X$, show that the induced actions on $H^i X$ and $H_i X$ are trivial. If $X$ is a nonsingular projective variety, deduce that $[g{\cdot}V] = g_*[V] = [V]$ for any $g$ in $G$ and any subvariety $V$ of $X$.

**Exercise 8** Let $s : \mathbb{P}^n \times \mathbb{P}^m \to \mathbb{P}^{nm+n+m}$ be the Segre embedding. Show that $s^*([H]) = [H_1 \times \mathbb{P}^m] + [\mathbb{P}^n \times H_2]$, where $H, H_1$, and $H_2$ are hyperplanes in $\mathbb{P}^{nm+n+m}$, $\mathbb{P}^n$, and $\mathbb{P}^m$.

## B.4 Chern classes

If $L$ is a complex line bundle on $X$, its Thom class $\gamma_L$ is a class in $H^2(L, L \smallsetminus X)$. The first Chern class $c_1(L)$ in $H^2 X$ can be defined to be the class whose pullback (by $\pi^*$, if $\pi$ is the projection from $L$ to $X$) to $H^2 L$ is the image of $\gamma_L$ in

$H^2 L$ (by the restriction map). Equation (10) then follows from the fact that the Thom classes are compatible with pullbacks.

To prove (12), we apply Lemma 5 to the morphism $s: X \to L$, and the subvariety $X \subset L$ embedded by the zero section. Then $D = s^{-1}(X)$, and the assumption that $s$ cuts out $D$ means that the hypotheses of Lemma 5 are satisfied. It follows that $s^*(\gamma_L) = \eta_D$, and from this it follows that $c_1(L) = s^*(\pi^*(c_1(L))) = [D]$.

A particular case of (12) is the assertion that if $\mathcal{O}(1)$ is the dual to the tautological line bundle on a projective space $\mathbb{P}(V)$, then $c_1(\mathcal{O}(1))$ is the class of a hyperplane. This follows from the fact that a nonzero vector in $V^*$ gives a section of $\mathcal{O}(1)$ that vanishes exactly on a hyperplane determined by vector. The class of a hyperplane is a universal first Chern class, in the following sense. For any line bundle $L$ on a manifold (or paracompact space) $X$, there is a continuous map $f: X \to \mathbb{P}^n$ such that $L \cong f^*(\mathcal{O}(1))$. This follows from the fact that $L$ has a finite number of continuous sections $s_0, \ldots, s_n$ such that at least one is nonvanishing at each point of $X$; the map $f$ is given by $f(x) = [s_0(x): \ldots : s_n(x)]$. More intrinsically, the sections give a surjection from the trivial bundle $\mathbb{C}^{n+1}{}_X$ to $L$, and $f$ is determined by the fact that this surjection is the pullback of the canonical surjection $\mathbb{C}^{n+1}{}_{\mathbb{P}^n} \to \mathcal{O}(1)$ on $\mathbb{P}^n$.

These considerations can be used to prove (11). Given line bundles $L$ and $M$ on $X$, let $f: X \to \mathbb{P}^n$ and $g: X \to \mathbb{P}^m$ be morphisms such that $L = f^*(\mathcal{O}(1))$ and $M = g^*(\mathcal{O}(1))$. Let $h: X \to \mathbb{P}^{nm+n+m}$ be the composite of $(f,g): X \to \mathbb{P}^n \times \mathbb{P}^m$ and the Segre embedding $s$. Then $L \otimes M$ is isomorphic to $h^*(\mathcal{O}(1))$, by Exercise 10 of §9.3. Using Exercise 8, we have

$$c_1(L \otimes M) = h^*([H]) = (f,g)^*([H_1 \times \mathbb{P}^m] + [\mathbb{P}^n \times H_2])$$

$$= f^*([H_1]) + g^*([H_2]) = c_1(L) + c_1(M).$$

To define Chern classes of a bundle $E$ of arbitrary rank $e$ on $X$, following Grothendieck, we can consider the associated projective bundle $p: \mathbb{P}(E) \to X$. On $\mathbb{P}(E)$ we have the tautological exact sequence

$$(32) \qquad 0 \to L \to p^*(E) \to Q \to 0,$$

where $L$ is the tautological line bundle, with $\mathcal{O}(1) = L^{\check{}}$ is its dual bundle. Let $\zeta = c_1(\mathcal{O}(1)) = -c_1(L)$ (by (11)). On any open set $U$ of $X$ over which $E$ is trivial, $\zeta$ restricts to the class of a hyperplane in $\mathbb{P}(E|_U) = U \times \mathbb{P}^{e-1}$. It follows by an easy Mayer–Vietoris argument that $H^*(\mathbb{P}(E))$ is free over $H^* X$ with basis $1, \zeta, \zeta^2, \ldots, \zeta^{e-1}$. Therefore there are unique classes $a_1, \ldots, a_e$ with $a_i$ in $H^{2i} X$, such that

$$(33) \qquad \zeta^e + p^*(a_1) \cdot \zeta^{e-1} + \ldots + p^*(a_{e-1}) \cdot \zeta + p^*(a_e)$$

$$= 0 \quad \text{in} \quad H^{2e}(\mathbb{P}(E)).$$

In other words, $H^*(\mathbb{P}(E)) \cong H^* X[T] / (T^e + a_1 \cdot T^{e-1} + \ldots + a_{e-1} \cdot T + a_e)$, with $T$ mapping to $\zeta$. Then we define $c_i(E)$ to the class $a_i$, and we define $c_0(E)$ to be 1, and $c_i(E) = 0$ if $i < 0$ or $i > e$. Note that when $e = 1$, $\mathbb{P}(E) = X$ and

$L = E$, so $\zeta = c_1(E^\vee) = -c_1(E)$, from which we see that the general definition agrees with the case of line bundles.

Property (13) follows easily from this definition, and the facts that $\mathbb{P}(f^*(E)) = Y \times_X \mathbb{P}(E)$ and the tautological line bundle on $\mathbb{P}(f^*(E))$ is the pullback of the tautological line bundle on $\mathbb{P}(E)$.

If the exact sequence (32) is tensored by the line bundle $\mathcal{O}(1) = L^\vee$, we see that the bundle $p^*(E) \otimes \mathcal{O}(1)$ has a trivial line subbundle, which is the same as saying that it has a nowhere zero section. We use this to prove the following lemma, which is the key to the Whitney formula (14).

**Lemma 7** *If a vector bundle $E$ is a direct sum of line bundles $L_1, \ldots, L_e$, then $c_i(E)$ is the $i^{\text{th}}$ elementary symmetric function in the variables $c_1(L_1), \ldots, c_1(L_e)$.*

**Proof** On $\mathbb{P}(E)$, the bundle $p^*(E) \otimes \mathcal{O}(1) = \oplus p^*(L_i) \otimes \mathcal{O}(1)$ has a nowhere vanishing section $s = \oplus s_i$, with $s_i$ a section of $p^*(L_i) \otimes \mathcal{O}(1)$. Let $U_i$ be the set where $s_i$ does not vanish. Since the restriction of $p^*(L_i) \otimes \mathcal{O}(1)$ to $U_i$ is a trivial line bundle, its first Chern class vanishes in $H^2(U_i)$. This means that there is some class $\alpha_i \in H^2(\mathbb{P}(E), U_i)$ whose image in $H^2(\mathbb{P}(E))$ is $c_1(p^*(L_i) \otimes \mathcal{O}(1)) = p^*(c_1(L_i) + \zeta$. The cup product $\alpha_1 \cup \ldots \cup \alpha_e$ is in $H^{2e}(\mathbb{P}(E), U_1 \cup \ldots \cup U_e)$, which vanishes since $U_1 \cup \ldots \cup U_e = \mathbb{P}(E)$. It follows that

$$(p^*(c_1(L_1) + \zeta) \cdot (p^*(c_1(L_2) + \zeta) \cdot \ldots \cdot (p^*(c_1(L_e) + \zeta) = 0.$$

Hence $\zeta^e + p^*(a_1) \cdot \zeta^{e-1} + \ldots + p^*(a_{e-1}) \cdot \zeta + p^*(a_e) = 0$, where $a_i$ is the $i^{\text{th}}$ elementary symmetric function in $c_1(L_1), \ldots, c_1(L_e)$. The assertion then follows from the definition of the Chern classes.   □

To use this, we need the ***splitting principle***:

**Lemma 8** *Given a vector bundle $E$ on $X$, there is a map $f: X' \to X$ such that $f^*: H^*(X) \to H^*(X')$ is injective and $f^*(E)$ is a direct sum of line bundles.*

**Proof** It follows from the description of $H^*(\mathbb{P}(E))$ as an algebra over $H^*X$ that the pullback $p^*: H^*X \to H^*(\mathbb{P}(E))$ is injective. On $\mathbb{P}(E)$ the pullback $p^*(E)$ has a line subbundle $L$. By choosing a Hermitian metric on $E$, we may take the perpendicular complement $E_1$ to $L$ in $p^*(E)$, so write $p^*(E) = L \oplus E_1$. By induction on the rank, we may construct $X' \to \mathbb{P}(E)$ to split $E_1$, with the map on cohomology injective, and then the composite $X' \to \mathbb{P}(E) \to X$ is the required map. In fact, one sees that one can take $X'$ to be the flag bundle of complete flags in $E$.   □

Now the Whitney sum formula (14) is easy to prove. Using a metric as above, one may assume $E = E' \oplus E''$. Using the splitting principle, one may assume each of $E'$ and $E''$ is a direct sum of line bundles. Then the conclusion is an immediate consequence of Lemma 7. Note in particular that $c_i(E) = 0$ for $i \neq 0$ if $E$ is a trivial bundle.

**Exercise 9** (i) For a bundle $E$ of rank $e$, show that $c_1(\wedge^e E) = c_1(E)$. (ii) If $E$ has rank $e$, and $L$ is a line bundle, show that

$$c_p(E \otimes L) = \sum_{i=0}^{p} \binom{e-i}{p-i} c_i(E) \cdot c_1(L)^{p-i}.$$

We have used one more fact in the text, in the case of bundles of rank 2:

**Lemma 9** *Let $E$ be a vector bundle of rank $e$ on a nonsingular projective variety $Y$. Let $X = \mathbb{P}(E)$, $p: X \to Y$ the projection. Let $L \subset p^*(E)$ be the tautological line bundle, and set $x = - c_1(L) \in H^2(X)$. Then*

$$p_*(x^{e-1}) = 1 \quad \text{in} \quad H^0(Y).$$

**Proof** The basic idea is to restrict to a fiber of $p$, which is a projective space – where we can compute everything – and then use formal properties of cohomology. Consider first the case when $Y$ is a point, so $X = \mathbb{P}^{e-1}$. In this case $x$ is the class of a hyperplane, as we saw earlier in this section. Therefore $x^{e-1}$ is the class of the intersection of $e-1$ general hyperplanes, which is the class of a point. Now take any point $y$ in $Y$, and let $F = p^{-1}(y)$ be the fiber over $y$. This fiber is a projective space, and $x$ restricts to the class of the tautological line bundle on $F$. By what we have just seen, together with the projection formula for the inclusion of $F$ in $X$, we have the formula $\langle x^{e-1}, [F] \rangle = 1$.

To prove the lemma, since $H^0(Y) = \mathbb{Z}$, we can write $p_*(x^{e-1}) = d \cdot 1$ for some integer $d$. Now

$$
\begin{aligned}
1 &= \langle x^{e-1}, [F] \rangle = \langle x^{e-1}, p^*([y]) \rangle \\
&= \langle p_*(x^{e-1}), [y] \rangle = d \cdot \langle 1, [y] \rangle = d,
\end{aligned}
$$

which shows that $d = 1$, as required.   □

The following exercise gives a simpler definition of Borel–Moore homology for an algebraic variety. Proving the equivalence of this definition with the one used here, or proving that this definition satisfies the required properties, however, requires more knowledge – e.g., that algebraic varieties can be triangulated.

**Exercise 10** If $X$ is an algebraic variety, and $X^+ = X \cup \{\cdot\}$ is its one-point compactification, show that

$$\overline{H}_i X \cong H_i(X^+, \{\cdot\}) = \widetilde{H}_i(X^+/\{\cdot\}),$$

where the homology groups in the middle are the singular homology groups of the pair, and those on the right are reduced singular homology groups.

# Answers and References

Included are references where the reader can find more about notions discussed in the text, with no attempt at completeness. The bibliographies of these references can be used for those who want more or are interested in tracking down original sources.

## Chapter 1

For the basic operations, see Schensted (1961), Knuth (1970; 1973), Schützenberger (1963; 1977), Lascoux and Schützenberger (1981), and Thomas (1976; 1978). Our use of skew tableaux has been mainly as a tool for studying ordinary tableaux, but much of the theory can be extended to skew tableaux in their own right; cf. Sagan and Stanley (1990) and the references contained there.

### *Exercises*

**1** The result is the same as that obtained earlier by row-inserting the entries of $U$ into $T$.

**2** Form the skew tableau $T * U * V$ as shown:

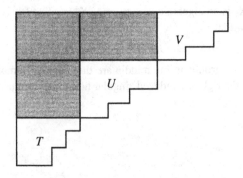

Sliding first with boxes in the rectangle above $T$ and left of $U$, one forms $(T \cdot U) * V$; then using the other boxes one forms $(T \cdot U) \cdot V$. Similarly, sliding first with boxes left of $V$ and above $U$, one forms $T * (U \cdot V)$, and then $T \cdot (U \cdot V)$.

## Chapter 2

The fact that the row bumping algorithm can be used to make the tableaux into a monoid was pointed out by Knuth (1970). This was developed by Lascoux and Schtüzenberger (1981), who coined the name "plactic monoid," and where one can find, sometimes in different form, all the results proved here.

### Exercises

**1** The answer for (5) is $\lambda_i + 1 \geq \mu_i \geq \lambda_i$ for all $i \geq 0$, together with the conditions $\mu_1 \geq \ldots \geq \mu_\ell \geq 0$ and $\sum \mu_i = \sum \lambda_i + p$.

**2** Consider the possible entries in the first $i$ rows.

**3** Dually, one may peel a hook off the upper right of $T$.

## Chapter 3

The basic reference for this is Schensted (1961). For a general analysis of the structure of increasing and decreasing sequences, see Greene (1980; 1991).

### Exercises

**1** Show that these numbers do not change by an elementary transformation, and examine the case of a word of a tableau.

**2** Compare the tableau with an $m \times n$ rectangle.

**3** $w = 4\ 1\ 2\ 5\ 6\ 3$.

## Chapter 4

A version of the correspondence, for permutations, was given by Robinson (1938), and later and independently by Schensted (1961), who also extended it to arbitrary words. The symmetry result for permutations can be found in Schützenberger (1977). Knuth (1970) made the generalization to arbitrary two-rowed arrays, or matrices; the general symmetry theorem is stated in Knuth 1970, although the proof there is a challenging "few moments' reflection." Knuth (1973) gives an algorithmic procedure for constructing the correspondence using tables. Exercises 12–15 are from Knuth (1973), §5.2.4. A construction similar to the matrix-ball construction given here has been found independently by Stanley and Fomin, and developed by Roby (1991). The proof of Proposition 2 is from Knuth (1970); to make this correspondence explicit, see Knuth (1970) and Bender and Knuth (1972). Combinatorial proofs of Stanley's formula (9) have been given by Remmel and Whitney (1983) and by Krattenthaler (1998).

Exercises

**1** A verification.

**2** A verification.

**3** $P = $

| 1 | 1 | 1 | 1 | 2 | 2 | 3 |
|---|---|---|---|---|---|---|
| 2 | 2 | 2 |   |   |   |   |
| 3 |   |   |   |   |   |   |

$Q = $

| 1 | 1 | 1 | 1 | 2 | 3 | 3 |
|---|---|---|---|---|---|---|
| 2 | 2 | 2 |   |   |   |   |
| 3 |   |   |   |   |   |   |

**4** If the balls labelled $k$ in $A^{(1)}$ include a ball on the diagonal, this contributes 1 to the trace of $A$ but gives no balls on the diagonal of $A^{(2)}$, while if the balls labelled $k$ include no balls on the diagonal, this contributes 0 to the trace of $A$ but puts one ball on the diagonal of $A^{(2)}$. By induction the trace of $A^b$ is the number of even columns of $P$, so the trace of $A$ is the number of odd columns.

**5** The matrix is the one displayed just before the exercise.

**6** Such an involution has the form $(a_1 \, b_1) \cdot (a_2 \, b_2) \cdot \ldots \cdot (a_k \, b_k)$, where the $a_i$ and $b_i$ are distinct in $[n]$, uniquely determined up to the order in each pair and the order of the pairs. The number of involutions with $k$ pairs is therefore $n!/(n - 2k)! \cdot 2^k \cdot k!$, which proves the formula.

In fact, $\sum_{k=0}^{[n/2]} \frac{n!}{(n-2k)! \cdot 2^k k!}$ is $n!$ times the coefficient of $z^n$ in the function $\exp(z + z^2/2)$. An asymptotic formula can be found in Knuth (1973), §5.1.4. Although we have simple formulas for the sum of the $f^\lambda$ and the sum of the $(f^\lambda)^2$, nothing similar is known for higher powers. P. Diaconis (1988) points out how even an asymptotic estimate for $\sum_{\lambda \vdash n} (f^\lambda)^3$ would be useful.

**7** To each subset $\{b_1 < b_2 < \ldots < b_{k-1}\}$ of $[k + r - 1]$, set $b_0 = 0$ and $b_k = k + r$, and construct $(a_1, \ldots, a_k)$ by letting $a_i = b_i - b_{i-1} - 1$.

**8** This is a direct consequence of the R–S–K correspondence and the definition of the Kostka numbers.

**9** It suffices by induction on the number of rows to see that the product of the hook lengths for boxes in the first row is

$$\frac{\ell_1!}{(\ell_1 - \ell_k) \cdot (\ell_1 - \ell_{k-1}) \cdot \ldots \cdot (\ell_1 - \ell_2)},$$

and this is straightforward: the first box has hook length $\ell_1$, the second $\ell_1 - 1$, and so on until the $\lambda_k^{\text{th}}$, which has hook length $\ell_1 - \lambda_k + 1$; the next has hook length $\ell_1 - \lambda_k - 1$, provided $\lambda_k < \lambda_{k-1}$, since the next column is one shorter, and this accounts for the missing $\ell_1 - \lambda_k$, which is the first term in the denominator. In general, if

$$\lambda_k = \lambda_{k-1} = \ldots = \lambda_{k-p} < \lambda_{k-p-1},$$

the next hook length is $\ell_1 - \lambda_k - p - 1$, which corresponds to the missing $(\ell_1 - \ell_k)\cdot\ldots\cdot(\ell_1 - \ell_{k-p})$ that appear in the denominator; and so on along the row, with terms in the denominator appearing each time the column length gets shorter.

10  To deduce the formula from the identity, set $x_i = \ell_i$ and $t = -1$, and note that

$$\ell_1 + \ldots + \ell_k - \binom{k}{2} = \lambda_1 + \ldots + \lambda_k = n.$$

To prove the identity, the fact that the left side is skew-symmetric in the variables $x_1, \ldots, x_k$ implies that it is divisible by $\Delta(x_1, \ldots, x_k)$, and by degree considerations, the ratio is a homogeneous linear polynomial in the $k+1$ variables. The two sides are clearly equal when $t = 0$, and to find the coefficient of $t$ it suffices to compute one case where both sides have nonzero value; for example one can take $t = 1$ and $x_i = k - i$. Reference: Knuth (1973), §5.1.4.

11  Apply the formula of Exercise 9, with $n = k$, and use (4).

12  By the results of Chapter 3, such permutations correspond to pairs $(P, Q)$ of standard tableaux with shape $\lambda$ as prescribed. For the example, the answer is

$$(f^{(15,4,1,1)})^2 + (f^{(15,3,2,1)})^2 + (f^{(15,2,2,2)})^2,$$

which by the hook length formula is

$$361,760^2 + 587,860^2 + 186,200^2 = 511,120,117,200.$$

13  The corresponding pairs $(P, Q)$ have two rows of $m$ and $n-m$ boxes; there are $2m-n+1$ choices for $P$ and $f^{(m, n-m)}$ choices for $Q$. Reference: Schensted (1961).

14  On the left is the sum of products $\prod_{i=1}^{m} x_i^{a(i,i)} \prod_{i < j} (x_i x_j)^{a(i,j)}$, one for each symmetric $m \times m$ matrix $A$ with nonnegative integer entries; on the right is the sum of all $x^P$ for tableaux $P$ with entries in $[m]$; apply the correspondence between such $A$ and $(P, P)$. References: Bender and Knuth (1972), Stanley (1971; 1983).

15  Compare with the sum of all $t^{\Sigma i a(i,j)}$, over all symmetric matrices $A = (a(i,j))$ of nonnegative integers such that $a(i,j) = 0$ if $i$ or $j$ is not in $S$. Write the exponent as the sum of the diagonal terms $\sum_i i a(i,i)$ and the sum of the off-diagonal terms $\sum_{i < j} (i+j) a(i,j)$. For more formulas of this type, see Bender and Knuth (1972).

16  They correspond to the pairs $(T, U)$ with $U$ standard.

17  By the Row Bumping Lemma, the sign is $+$ if $i+1$ is weakly above and strictly right of $i$ in $Q$, and $-$ if it is strictly below and weakly left. Reference: Schützenberger (1977).

## Chapter 5

There are now many proofs of the Littlewood–Richardson rule, many of them closely related to each other and to the proof given here; cf. Remmel and Whitney (1984), Thomas (1978), White (1981), and Zelevinsky (1981). For other approaches see Fomin and Greene (1993), Macdonald (1979), Lascoux and Shützenberger (1985), Bergeron and Garsia (1990), and Van der Jeugt and Fack (1991), and for a generalization to products of more than two Schur polynomials, see Benkart, Sottile, and Stroomer (1996). P. Littelmann (1994) has a generalization to representations of general semisimple Lie algebras.

### *Exercises*

1 Look at $\mathcal{T}(\lambda,\mu,U(\nu))$ and use Lemma 2.

2 The only skew tableau on $\nu/\lambda$ rectifying to $U(\mu)$ has its $i^{\text{th}}$ row consisting entirely of $i$'s.

3 The equation $T{\cdot}U = U(\nu)$ forces $U$ to be $U(\mu)$.

4 A tableau on $\nu$ with entries $1 < \ldots < k < \bar{1} < \ldots < \bar{\ell}$ is the same as a tableau on some $\lambda \leq \nu$ with entries from $\{1,\ldots,k\}$ and a skew tableau on $\nu/\lambda$ with entries from $\{\bar{1},\ldots,\bar{\ell}\}$.

5 There are $c_{\lambda\,\mu}^{\nu}$ with each of the $f^{\mu}$ possible rectifications.

6 (a) Such permutations are exactly the words of skew tableaux of content $(1,\ldots,1)$ on $\nu(s)/\lambda(s)$; apply the preceding exercise. The sum in (b) is the number of pairs of skew tableaux on $\nu(s)/\lambda(s)$ and $\nu(t)/\lambda(t)$ whose rectifications are standard tableaux of the same shape, and such rectifications correspond by the Robinson correspondence to permutations of the desired form. Use the hook formula and the Littlewood–Richardson rule for (c). Reference: Foulkes (1979).

   MacMahon (see Stanley [1986], p. 69) proves that, if the descents are at $s_1,\ldots,s_k,\ 1 \leq s_1 < \ldots < s_k \leq n-1$, the number of such permutations is

$$n!{\cdot}\det\left[1/(s_{j+1} - s_i)!\right] \;=\; \det\left[\binom{n-s_i}{s_{j+1} - s_j}\right],$$

   where the determinants are taken over $0 \leq i,j \leq k$, with $s_0 = 0$ and $s_{k+1} = n$.

7 A calculation.

8 Use the correspondence between standard tableaux and reverse lattice words, together with the hook length formula. Reference: Knuth (1973).

9 See the preceding exercise for the count. Reference: Knuth (1973). Another correspondence $T \longleftrightarrow (P,Q)$ is obtained by taking $P$ and $S$ as in the exercise, but taking $Q = Q(w(S))$ the insertion tableau of $w(S)$.

10 For (a), there are $c_{\lambda\,\mu}^{\nu}$ such skew tableaux rectifying to each of the $K_{\mu r}$ tableaux with content $r$. For (b), consider all tableaux on $\nu$ with $r_1$ 1's, $\ldots, r_p$ $p$'s, and $s_1$ $\bar{1}$'s, $\ldots, s_q$ $\bar{q}$'s, with $1 < \ldots < p < \bar{1} < \ldots < \bar{q}$.

**11** Such words correspond to pairs $(P,Q)$ of tableaux on the shape $v$, with $Q$ standard, and $P$ consists of $P(u_o)$ in the shape $\lambda$ and a skew tableau on $v/\lambda$ whose rectification is $P(v_o)$. Reference: Lascoux (1980).

## Chapter 6

See Macdonald (1979) for a thorough treatment of the algebra of symmetric functions. For connections between symmetric functions and tableaux, see Stanley (1971). A tableau-theoretic proof of the formula in Exercise 4 can be found in Bender and Knuth (1972). A nice proof of a generalization of the Jacobi–Trudi formula that uses tableaux has been given by Wachs (1985). For another generalization of the Jacobi–Trudi formula, see Pragacz (1991) and Pragacz and Thorup (1992).

### *Exercises*

**1** A quick way is to use the generating function $E(t) = \prod(1+x_i t) = \sum e_p t^p$ and the equation

$$E'(t)/E(t) = \frac{d}{dt}\log(E(t)) = \sum \frac{d}{dt}\log(1 + x_i t).$$

Similarly for the second, using $H(t) = \prod(1 - x_i t)^{-1} = \sum h_p t^p$.

**2** Equation (8) implies that $h_p(x) = t_{(p)}$, and then that the functions $t_\lambda$ satisfy the same formula (6) of §2.2 as the $s_\lambda(x)$. By the invertibility of the matrix $(K_{\lambda\mu})$ of Kostka numbers, it follows that $s_\lambda(x) = t_\lambda$ for all $\lambda$. To prove (9), let $a^{(p)}(\ell_2, \ldots, \ell_m)$ be the determinant formed as above, but using only the $m-1$ exponents from $\ell_2$ to $\ell_m$ and the $m-1$ variables $x_1, \ldots, x_{p-1}, x_{p+1}, \ldots, x_m$. Expanding the determinant along the top row,

$$a(\ell_1, \ldots, \ell_m) = \sum_{p=1}^{m} (-1)^{p+1}(x_p)^{\ell_1} \cdot a^{(p)}(\ell_2, \ldots, \ell_m).$$

By induction on the number of variables, the left side of (9) is

$$\sum_{p=1}^{m} (-1)^{p+1}(x_p)^{\ell_1} \cdot (1 - x_p)^{-1} \cdot a^{(p)}(\ell_2, \ldots, \ell_m) \prod_{i \neq p}(1 - x_i)^{-1}$$

$$= \sum_{p=1}^{m} (-1)^{p+1} \sum (x_p)^{n_1} \cdot a^{(p)}(n_2, \ldots, n_m)$$

$$= \sum a(n_1, n_2, \ldots, n_m),$$

the sum over all $n_1 \geq \ell_1$ and $n_2 \geq \ell_2 > \ldots > n_m \geq \ell_m$. To finish, it suffices to show that the terms with $n_2 \geq \ell_1$ cancel, and this follows from the alternating property of determinants: $a(n,n,n_3, \ldots, n_m) = 0$ and

$$a(n_1,n_2,n_3, \ldots, n_m) + a(n_2,n_1,n_3, \ldots, n_m) = 0.$$

**3** Equation (5) is obtained by taking determinants in the last display of the exercise, and using (7). Note that, when $\lambda = (0, \ldots, 0)$, taking determinants shows that

$$\det\left[(-1)^{m-i} e_{m-i}^{(j)}\right]_{1 \le i,j \le m} = \det\left[(x_j)^{m-i}\right]_{1 \le i,j \le m}.$$

**4** The first statement follows from the expansion

$$\det\left[h_{\lambda_i+j-i}(x)\right]_{1 \le i,j \le m}$$
$$= \sum_{\sigma \in S_m} \mathrm{sgn}(\sigma) h_{\lambda_1+\sigma(1)-1}(x) \cdot \ldots \cdot h_{\lambda_m+\sigma(m)-m}(x)$$

and equation (8) of §2.2. Equation (6) is similarly equivalent to the same formula, applied to the conjugate of $\lambda$.

**5** The numerator of (7) becomes a Vandermonde determinant.

**6** Compute the limit as $x \to 1$ (or divide by the appropriate powers of $x-1$) in the formula of the preceding exercise, noting that $(x^p-1)/(x^q-1) \to p/q$. Use Exercise 9 of §4.3 to see that this formula for $d_\lambda(m) = s_\lambda(1, \ldots, 1)$ is the same as that in formula (9) of §4.3.

**7** See Macdonald (1979), §I.5.

**8** A proof can be found in Fulton (1984), Lemma A.9.2. For basic properties of super-symmetric polynomials, including a generalization by Sergeev and Pragacz of the Jacobi–Trudi identity which implies this formula, see Pragacz and Thorup (1992).

## Chapter 7

One of the many possible references for basic facts about complex representations of finite groups is Fulton and Harris (1991). References for representations of the symmetric group are James (1978), James and Kerber (1981) (with an extensive bibliography), Peel (1975), and Carter and Lusztig (1974), all of which emphasize extensions to positive characteristic; Sagan (1991), for an elementary text that brings out the relations with standard tableaux; and Robinson (1961). For a variety of applications, see Diaconis (1988). The module $\widetilde{M}^\lambda$ constructed in §7.4 is isomorphic, but not canonically, to a dual construction in James and Kerber (1981), p. 318, where column tabloids are not oriented.

### *Exercises*

**1** (a) follows from the fact that $R(T)$ and $C(T)$ are subgroups of $S_n$, together with the fact that $\mathrm{sgn}(q_1 \cdot q_2) = \mathrm{sgn}(q_1) \mathrm{sgn}(q_2)$. (b) follows from the fact that each element in a subgroup $G$ of $S_n$ can be written $\#G$ ways as a product of two elements of $G$.

**2** The number of ways to distribute $n$ integers into subsets with $m_r$ sets of size $r$, without regard to order, is $n!/\left(\prod(r!)^{m_r} \cdot \prod m_r!\right)$; choosing a cycle for a subset with $r$ elements multiplies by $r!/r$.

**3** Since $C(\sigma \cdot T) = \sigma \cdot C(T) \cdot \sigma^{-1}$ (see (1) of §7.1) and $\operatorname{sgn}(\sigma \cdot q \cdot \sigma^{-1}) = \operatorname{sgn}(q)$, $v_{\sigma \cdot T} = \sum_{q \in C(T)} \operatorname{sgn}(q) \{\sigma \cdot q \cdot T\} = \sigma \cdot v_T$.

**4** Note that $M^{(n)}$ is the trivial representation, and $M^{(1^n)} \cong A$ is the regular representation.

**5** Induct on $m$.

**6** Use the Jacobi–Trudi formula.

**7** The coefficient of $x_1^{\ell_1} \cdot \ldots \cdot x_k^{\ell_k}$ in

$$\prod_{\sigma \in S_k} \operatorname{sgn}(\sigma) x_k^{\sigma(1)-1} \cdot \ldots \cdot x_1^{\sigma(k)-k} \cdot (x_1 + \ldots + x_k)^n$$

is the product of $n! / \prod(\ell_i!)$ and the determinant of a matrix that can be column reduced to the Vandermonde determinant $\prod(\ell_i - \ell_j)$. See Fulton and Harris (1991), §4.1, for details.

**8** Write $\mathbb{U}_n = \mathbb{C} \cdot u$, choose a standard tableau $T$ on $\lambda$, and map $M^{\tilde{\lambda}}$ to $S^\lambda \otimes \mathbb{U}_n$ by sending $\{\sigma \cdot T^\tau\}$ to $\sigma \cdot (v_T \otimes u)$. Check that this is a well-defined map of representations, and that $v_{T^\tau}$ is mapped to $\#R(T) \cdot (v_T \otimes u)$, so its restriction to $S^\lambda$ is not zero. Compare James (1978).

**9** That $M^\lambda \cong A \cdot a_T$ follows from the fact that the $\sigma \cdot a_T$ are linearly independent, as $\sigma$ varies over representatives of $S_n / R(T)$. This isomorphism maps $v_T = b_T \cdot \{T\}$ to $b_T \cdot a_T = c_T$, so it maps $S^\lambda = A \cdot v_T$ onto $A \cdot c_T$.

**10** If $\sigma$ ranges over coset representatives for $S_n / R(T)$, and $p$ ranges over $R(T)$, an element $x = \sum_{\sigma, p} x_{\sigma, p} \sigma p$ is in the kernel of this map exactly when the sum $\sum_p x_{\sigma, p}$ vanishes for every $\sigma$, which means that $x = \sum_{\sigma, p} x_{\sigma, p} \sigma (p - 1)$. For the second assertion, write $p = p_1 \cdot \ldots \cdot p_r$ as a product of transpositions in $R(T)$, and write $p - 1$ as the sum $\sum(p_1 \cdot \ldots \cdot p_{i-1})(p_i - 1)$.

**11** For (a), note that $M^{(n-1,1)} = \mathbb{C}^n$, and $M^{(n-1,1)} = S^{(n-1,1)} \oplus \mathbb{1}_n$ by Young's rule. For (b), use the fact that $V_n$ restricts to the direct sum of $V_{n-1}$ and the trivial representation of $S_{n-1}$, so $\wedge^p(V_n)$ restricts to $\wedge^p(V_{n-1}) \oplus \wedge^{p-1}(V_{n-1})$.

**12** The proofs are entirely analogous to those in §7.2.

**13** For (b), use the isomorphism of $\Lambda$ with $R$ and equation (4) of §6.1. For (c), take coset representatives $\sigma$ for $S_n / C(T)$, and note that an element $x = \sum \operatorname{sgn}(q) x_{\sigma, q} \sigma q$ is in the kernel exactly when $\sum_q x_{\sigma, q} = 0$ for all $\sigma$; therefore $x = \sum \operatorname{sgn}(q) x_{\sigma, q} \sigma (q - \operatorname{sgn}(q) \cdot 1)$. And if $q = q_1 \cdot \ldots \cdot q_r$ is a product of transpositions in $C(T)$, then $q - \operatorname{sgn}(q) \cdot 1$ can be written in the form

$$\sum_{i=1}^{r} (-1)^{r-i} (q_1 \cdot \ldots \cdot q_{i-1})(q_i + 1).$$

**14** For a nonempty subset $Y$ of the $(j+1)^{\text{st}}$ row, define $\tilde{\gamma}_Y(T) = \sum \{S\}$, the sum over $S$ obtained from $T$ by interchanging a subset $Z$ of $Y$ with a subset of the $j^{\text{th}}$ row. Show that $\tilde{\gamma}_Y(T)$ is in the kernel of $\beta$, using the fact that if $t$ transposes two elements in a column of a numbering $pT$, then $t \cdot [pT] = -[pT]$.

Show that

$$\{T\} + (-1)^k \, \tilde{\pi}_{j,k}(T) = \sum_Y (-1)^{\#Y} \, \tilde{\gamma}_Y(T),$$

the sum over subsets $Y$ of the first $k$ elements in the $(j+1)^{\text{st}}$ row of $T$.

**15** Compute the $2 \times 2$ matrix of a possible isomorphism between them.

**16** Part (a) is a direct calculation, looking at the terms that appear on each side; there is no cancellation. From (a) we have

$$k \cdot ([T] - \pi_{j,k}(T)) = (k-1) \cdot ([T] - \pi_{j,k-1}(T))$$
$$+ \sum_{i=1}^{m} ([T_{i,k}] - \pi_{j,k-1}(T_{i,k})) + \left([T] - \sum_{i=1}^{m} [T_{i,k}]\right);$$

this implies that $k \cdot N_k \subset N_{k-1} + N_{k-1} + N_1$, which implies (b). Reference: Towber (1979). See also Carter and Lusztig (1974), §3.2.

**17** By (a), the map $[T] \mapsto F_T$ gives a homomorphism of $S_n$-modules from $\widetilde{M}^\lambda$ onto the subspace they span, so it suffices to verify that generators for $Q^\lambda$ map to zero. By expanding the Vandermonde determinants, one sees that $F_T = \sum_{q \in C(T)} \operatorname{sgn}(q) G_{qT}$, where $G_S = \prod_{(i,j) \in \lambda} (x_{S(i,j)})^{i-1}$. Since $t \cdot G_S = G_S$ for $t$ a transposition of two elements in the same row of $S$, the same argument as in the proof of Claim 1 works here. Reference: Peel (1975).

**18** (a) The equation $c_T \cdot c_T = n_\lambda c_T$ is equivalent to the equation $c_T \cdot v_T = n_\lambda v_T$, which was seen in the proof of Lemma 5. (b) If $t$ transposes the two elements, then $c_{T'} \cdot c_T = c_{T'} \cdot t \cdot t \cdot c_T = -c_{T'} \cdot c_T$, since $a_{T'} \cdot t = a_{T'}$ and $t \cdot b_T = -b_T$. If $c_{T'} \cdot c_T = 0$, and there is no such pair of entries, then $p' \cdot T' = q \cdot T$ for some $p' \in R(T')$, $q \in C(T)$ by Lemma 1 of §7.1. Then $c_{p' \cdot T'} = p' \cdot c_{T'}$ and $c_{q \cdot T} = \pm c_T \cdot q^{-1}$, so $c_{p' \cdot T'} \cdot c_{q \cdot T} = 0$, contradicting (a). (c) To see that the sum of the ideals $A \cdot c_T$ is direct, suppose $\sum x_T \cdot c_T = 0$, $x_T \in A$, the sum over standard $T$. Multiply on the right by the $c_{T_\circ}$ with $T_\circ$ the minimal $T$ such that $x_T \neq 0$; use the corollary in §7.1 to see that $0 = n_\lambda \cdot x_{T_\circ} \cdot c_{T_\circ}$, a contradiction. Since the dimension $n!$ of $A$ is the sum $\sum (f^\lambda)^2$ of the dimensions of the $A \cdot c_T$, the sum of these ideals must be all of $A$. (d) The smallest example has $\lambda = (3, 2)$, with

$$T = \begin{array}{|c|c|c|} \hline 1 & 2 & 3 \\ \hline 4 & 5 \\ \cline{1-2} \end{array} \quad \text{and} \quad T' = \begin{array}{|c|c|c|} \hline 1 & 3 & 5 \\ \hline 2 & 4 \\ \cline{1-2} \end{array}$$

Compare James and Kerber (1981), p. 109.

**19** For any numbering $T$, $c_T^2 = n_\lambda c_T$. Consider the $S_n$-endomorphism of $A = \mathbb{C}[S_n]$ given by right multiplication by $c_T$, where $T$ is standard on $\lambda$

and $T$ minimal in the ordering defined in §7.1. This vanishes on all $A \cdot c_U$ for all standard $U \neq T$ as in the preceding exercise, and it is multiplication by $n_\lambda$ on $A \cdot c_T$. The trace of the endomorphism is seen to be $n!$ by using the basis of $A$ consisting of permutations, and it is $n_\lambda \cdot \dim(A \cdot c_T) = n_\lambda f^\lambda$ by the preceding observation.

**20** Right multiplications by $b_T$ and $a_T$ give the isomorphisms. Use Exercise 18(a).

**21** (a) For the compatibility with $\omega$, use Corollary 2 to Proposition 2 of §5.1. (b) The map $R_n \to R_p \otimes R_q$ takes the class $[V]$ of a representation $V$ of $S_n$ to the sum of $[U] \otimes [W]$, where the restriction of $V$ to $S_p \times S_q$ is a direct sum of tensor products $U \otimes W$ of representations $U$ of $S_p$ and $W$ of $S_q$. Reference: Liulevicius (1980).

## Chapter 8

The modules we call "Schur modules" $E^\lambda$ were defined in a different but equivalent form by Towber (1977; 1979), where they are denoted $\bigwedge_R^\mu E$, with $\mu = \tilde{\lambda}$. When $E$ is free, Akin, Buchsbaum, and Weyman (1982) constructed "Schur functors," denoted $L_\mu E$; it follows from their theorem that $E^\lambda$ is isomorphic to $L_\mu E$ in case $E$ is free. Carter and Lusztig (1974) construct from a free module $\overline{V}$ a "Weyl module" $\overline{V}^\mu$ as a submodule of the tensor product $\overline{V}^{\otimes n}$ satisfying properties dual to our properties (1)–(3). When $E$ is free, if $\overline{V}$ denotes the dual module, our $E^\lambda$ is the dual of $\overline{V}^\mu$.

Green (1980) develops the representation theory of $GL_m$ algebraically from scratch, without use of Lie groups or symmetric groups, and uses results about $GL_m$-representations to prove facts about representations of symmetric groups. For more about these constructions and the relations among them – as well as $q$-analogues – see Martin (1993).

Similarly, the algebras in §8.4 have been studied under a variety of names and guises, such as "shape algebras" (Towber [1977; 1979]) and "bracket rings" (Désarménien, Kung, and Rota [1978]; Sturmfels and White [1989]; DeConcini, Eisenbud, and Procesi [1980]), besides their appearance as homogeneous coordinate rings of Grassmannians and flag varieties (Hodge and Pedoe [1952]). For more on Deruyts' construction, see Green (1991). The "quadratic relations" were prominent in the nineteenth century, with nearly every algebraic geometer and invariant theorist making contributions; they were developed by Young (1928) and continue to flourish under names like "straightening laws."

The original approach of Schur was to consider the subalgebras of the algebra of $\mathrm{End}(E^{\otimes n})$ generated by $\mathrm{End}(V)$ and by $\mathbb{C}[S_n]$, showing that each is the commutator algebra of the other. In this way, the fact that polynomial representations of $GL(E)$ are direct sums of irreducible representations is deduced from the semisimplicity of $\mathbb{C}[S_n]$.

P. Magyar (1998) has recently extended many of the ideas described here about representations and their characters from the case of Young diagrams to a much larger class of diagrams.

We have limited our attention to the most "classical" case, which in representation theory corresponds to the general (or special) linear group. For a sketch of the role in the representation theory of the other classical groups, see Sundaram (1990). For more on the Lie group–Lie algebra story, as well as more about representations of $GL_m\mathbb{C}$, see Fulton and Harris (1991).

### *Exercises*

**1** This can be done by explicit formulas. A simpler way is the following: let $'E^\lambda$ be the module constructed with the restricted relations, so we have a canonical surjection $'E^\lambda \twoheadrightarrow E^\lambda$. When $E$ is finitely generated and free, the same proof as for $E^\lambda$ shows that the canonical map from $'E^\lambda$ to $R[Z]$ maps $'E^\lambda$ isomorphically onto $D^\lambda$, so the map from $'E^\lambda$ to $E^\lambda$ must be an isomorphism as well. The conclusion for general $E$ follows from the free case by mapping a free module to $E$, with basis mapping to the entries of any given $\mathbf{v}$ in $E^{\times\lambda}$.

**2** Since the functor $E \mapsto E^\lambda$ is compatible with base change, one may reduce to the case where $R$ is finitely generated over $\mathbb{Z}$, so Noetherian. Let $I$ (resp. $I_\lambda$) be the ideal generated by $m \times m$ (resp. $d_\lambda(m) \times d_\lambda(m)$) minors of a matrix for $\varphi$ (resp. $\varphi^\lambda$) with respect to some bases. Now $\varphi$ is a monomorphism $\iff I$ contains a nonzerodivisor $\iff I$ is not contained in any associated prime of $R$, and similarly for $\varphi^\lambda$. It suffices to show that $I^\lambda \subset \mathfrak{p} \Rightarrow I \subset \mathfrak{p}$ for a prime ideal $\mathfrak{p}$, and conversely if $\lambda$ has at most $m$ rows. Localize at $\mathfrak{p}$ and take the base extension to $R/\mathfrak{p}$ to reduce to the case where $R$ is a field, which is easy.

　　P. Murthy points out that (i) $\Rightarrow$ (ii) can also be proved as follows: For $R$ Noetherian, if the depth of $R$ is 0, the image of $E$ in $F$ must be a direct summand, since the cokernel must be free. In general, if a prime $\mathfrak{p}$ is associated to the kernel of $E^\lambda \to F^\lambda$, localize at $\mathfrak{p}$ and induct on the dimension of $R$.

**3** This follows from the defining equation $g{\cdot}e_j = \sum g_{i,j}e_i$ by multilinearity. Note that this calculation takes place in $E^{\otimes\lambda}$, but the conclusion therefore follows in the image $E^\lambda$.

**4** This follows from the multilinearity of determinants in the row vectors.

**5** Since $e_T$ maps to $D_T$, this follows from Exercises 3 and 4.

**6** The proofs are entirely similar.

**7** Look at the kernel and image of a map $L: V \to W$, and of $L - \lambda I$ when $V = W$ and $\lambda$ is an eigenvalue of $L$.

**8** If $\varphi$ is surjective, with kernel $K$, then $K$ has a complementary submodule $N'$ so $K \oplus N' = M$, $N' \xrightarrow{\cong} N$. Then $E(K) \oplus E(N') = E(M)$ and $E(N') \xrightarrow{\cong} E(N)$.

**9** Since $\mathbb{C}[S_n \times S_m] = \mathbb{C}[S_n] \otimes_\mathbb{C} \mathbb{C}[S_m]$,

$$N \circ M = \mathbb{C}[S_{n+m}] \otimes_{\mathbb{C}[S_n] \otimes_\mathbb{C} \mathbb{C}[S_m]} (N \otimes_\mathbb{C} M),$$

so

$$E(N \circ M) = E^{\otimes(n+m)} \otimes {}_{\mathbb{C}[S_{n+m}]}(\mathbb{C}[S_{n+m}] \otimes {}_{\mathbb{C}[S_n]} \otimes {}_{\mathbb{C}\mathbb{C}[S_m]}(N \otimes {}_{\mathbb{C}}M))$$

$$= (E^{\otimes n} \otimes {}_{\mathbb{C}}E^{\otimes m}) \otimes {}_{\mathbb{C}[S_n]} \otimes {}_{\mathbb{C}\mathbb{C}[S_m]}(N \otimes {}_{\mathbb{C}}M)$$

$$= (E^{\otimes n} \otimes {}_{\mathbb{C}[S_n]}N) \otimes {}_{\mathbb{C}}(E^{\otimes m} \otimes {}_{\mathbb{C}[S_m]}M)$$

$$= E(N) \otimes E(M).$$

**10** The proof is exactly like the proof of the proposition.

**11** This is similar to the version for columns, but a little trickier and uses characteristic zero. Order the $T$'s using the last entry of the last row in which they differ. Suppose $T$ has weakly increasing rows, and the $k^{\text{th}}$ entry in the $j^{\text{th}}$ row is the first entry in that row that is greater than or equal to the entry below it. Suppose the entries in these two rows are $x_1, \ldots, x_p$ and $y_1, \ldots, y_q$, with $p = \lambda_j$, $q = \lambda_{j+1}$. Let $S$ be the filling of $\lambda$ that has $j^{\text{th}}$ row $y_1, \ldots, y_{k-1}, x_k, \ldots, x_p$; $(j+1)^{\text{st}}$ row $x_1, \ldots, x_{k-1}, y_k, \ldots, y_p$; and other rows the same as $T$. Apply the "quadratic" relations $\tilde{\pi}_{j,k}$ to $S$. This gives an equation $e_S = (-1)^k c \cdot e_T + \sum m_{T'} e_{T'}$, a sum over $T' > T$, where $c$ is a positive number. Applying $\tilde{\pi}_{j,k-1}$ to $S$ gives an equation $e_S = (-1)^{k-1} d \cdot e_T + \sum n_{T'} e_{T'}$, with $d$ positive. Subtracting one from the other gives an equation that writes $(c+d)e_T$ as a linear combination of $e_{T'}$ for $T' > T$.

**12** By Pieri, $\mathrm{Sym}^p E \otimes \mathrm{Sym}^q E \cong (\mathrm{Sym}^{p+1} E \otimes \mathrm{Sym}^{q-1} E) \oplus E^{(p,q)}$ and $\wedge^p E \otimes \wedge^q E \cong (\wedge^{p+1} E \otimes \wedge^{q-1} E) \oplus E^{(2^q \, 1^{p-q})}$.

**13** Use the preceding exercise. See also Exercise 16 of §7.3.

**14** It is enough to check this for $V = S^\lambda$, for $\lambda \vdash n$, where the weight space has basis $e_T$ for $T$ the standard tableaux on $\lambda$.

**15** The character of $\bigoplus_k \mathrm{Sym}^k(E \oplus \wedge^2 E)$ is

$$\prod_{i=1}^m (1 - x_i)^{-1} \cdot \prod_{1 \le i < j \le m} (1 - x_i x_j)^{-1}.$$

Apply Exercise 14 of Chapter 4. Writing this as a sum of monomials corresponding to symmetric matrices $A = (a(i,j))$, the degree $k$ is $\sum_{i<j} a(i,j) + \sum_i a(i,i)$; the number of boxes is $2 \sum_{i<j} a(i,j) + \sum_i a(i,i)$, and the number of odd columns is $\sum_i a(i,i)$ by Exercise 4 of Chapter 4. Reference: Stanley (1983).

**16** Reference: Knutson (1973).

**17** Show that $V_1 + \ldots + V_i = V_1 \oplus \ldots \oplus V_i$ by induction on $i$. If $V_{i+1} \cap (V_1 + \ldots + V_i) \neq 0$, then $V_{i+1} \subset V_1 \oplus \ldots \oplus V_i$, but $\mathrm{Hom}(V_{i+1}, V_j) = 0$ for $j \le i$ by Schur's Lemma.

## Chapter 9

There is a vast literature on invariant theory and its connections with representation theory; for a start, see Weyl (1939), Howe (1987), Désarmenien, Kung, and Rota (1978), DeConcini, Eisenbud, and Procesi (1980), and Fulton and Harris (1991). For a discussion of Schubert calculus, see Kleiman and Laksov (1972) and Stanley (1977). Other approaches to and applications of Schubert calculus can be found in Griffiths and Harris (1978), and Fulton (1984), §14. The basic facts we have used from algebraic geometry can be found in many texts, such as Harris (1992), Shafarevich (1977), or Hartshorne (1977).

### *Exercises*

1  Apply (1) to the two sequences $j_2, \ldots, j_{d+1}$ and $j_1, i_1, \ldots, i_d$.

2  With $(i_1, i_2) = (1, 2)$, the matrix $A$ is $\begin{pmatrix} 1 & 0 & -1 & -2 \\ 0 & 1 & 2 & 1 \end{pmatrix}$, and the subspace is the kernel of the corresponding linear map from $\mathbb{C}^4$ to $\mathbb{C}^2$.

3  The image in (i) is defined by linear equations, and the images in (ii) and (iii) are again defined by quadratic equations. Reference: Harris (1992).

4  As in Proposition 2, the span of the $D_{i_1, \ldots, i_p}$ for $p \in \{d_1, \ldots, d_s\}$ in the polynomials of degree $a$ has dimension $\sum d_\lambda(m)$, the sum over partitions $\lambda$ of $a$ that have columns of lengths in $\{d_1, \ldots, d_s\}$; and $\mathrm{Sym}^a(V^{\oplus m}) \cong \bigoplus (V^\lambda)^{\oplus d_\lambda(m)}$, the sum over all partitions $\lambda$ of $a$ in at most $n$ parts. To prove (a) it suffices to show that, with $G = G(d_1, \ldots, d_s)$, $(V^\lambda)^G$ is one-dimensional when $\lambda$ has columns of lengths in $\{d_1, \ldots, d_s\}$, and zero otherwise. This is seen by looking at the action of matrices in $G$ on the basis of $V^\lambda$ corresponding to tableaux on $\lambda$; the unique fixed basis vector, when $\lambda$ has its column lengths in $\{d_1, \ldots, d_s\}$, corresponds to the tableau $U(\lambda)$ with all $i$'s in the $i^{\text{th}}$ row. For (b), following the proof of the corollary, it suffices to verify that $G$ is connected with no nontrivial characters, a fact that can be proved by induction on $s$, noting that a nilpotent group such as the group of lower triangular matrices with 1's on the diagonal has no nontrivial characters. Reference: Kraft (1984), §II.3.

5  Consider the localizations of the homogeneous coordinate ring of $X$, of the form $\{F/T^m : F$ homogeneous of degree $m\}$, where $T$ is a linear form on $\mathbb{P}^n$ not vanishing on $X$. These are unique factorization domains. As $T$ varies over a basis of linear forms, these localizations correspond to an open covering of $X$. If $\mathfrak{p}$ is the homogeneous ideal corresponding to a subvariety of codimension one in $X$, find an $F$ that generates the corresponding localizations of $\mathfrak{p}$ for such a covering.

6  The incidence variety $I_Z \subset \mathbb{P}(E) \times Gr^n E$ consisting of $(P, F)$ with $P$ in $Z \cap \mathbb{P}(F)$ is an irreducible subvariety of dimension one less than the dimension $n(m-n)$ of $Gr^n E$, and the map from $I_Z$ to $H_Z$ is birational. For (b), fix a general pair $A \subset B \subset E$ of linear subspaces of codimensions $n+1$ and $n-1$, and consider the line

$$\ell = \{F \in Gr^n E : A \subset F \subset B\}.$$

Then $\mathbb{P}(B)$ meets $Z$ in $d = \deg(Z)$ points, which gives $d$ points on the line that are in $H_Z$. This shows, once transversality is checked, that the two degrees are equal. See Harris (1992) for details.

**7** Consider the localizations of the multihomogeneous coordinate ring of $X$, of the form

$$\{F/T_1^{m_1} \cdot \ldots \cdot T_r^{m_r} : F \text{ is multihomogeneous of degree } (m_1, \ldots, m_r)\},$$

where $T_i$ is a linear form on $\mathbb{P}^{n_i}$ not vanishing on $X$. These are unique factorization domains, and correspond to open coverings of $X$. The fact that hyperplane sections generate the divisor class group follows as in the case of one factor, and their independence follows from the fact that the projections to the factors are nontrivial.

**8** Note that $E_{i,j} \cdot (e_{U(\lambda)}{}^*) = 0 \iff E_{j,i} \cdot (e_{U(\lambda)}) = 0$, and $E_{j,i} \cdot (e_{U(\lambda)})$ is the sum of all $e_T$, for all $T$ obtained by replacing an $i$ in $U(\lambda)$ by $j$.

**9** Since both subvarieties are orbits by the action of $G$, it suffices to verify that the basic flag $F_1 \subset \ldots \subset F_s \subset E$ is sent to the point $[e_{U(\lambda)}{}^*]$ by the embedding in §9.1.

**10** This is straightforward from the definitions.

**11** This is similar to the standard proof that $V$ is the space of sections of $\mathcal{O}_{\mathbb{P}^*(V)}(1)$. For $X \subset \mathbb{P}^{m-1}$, with homogeneous coordinate ring

$$A = \mathbb{C}[X_1, \ldots, X_m]/I(X) = \mathbb{C}[x_1, \ldots, x_m]$$

a unique factorization domain, a section of $\mathcal{O}_X(n)$ is given by a collection of elements $s_i$ in the localizations $A_{x_i}$ that are homogeneous of degree zero, and satisfy the transition equations $(x_i/x_j)^n \cdot s_i = s_j$ in $A_{x_i x_j}$. We must show that there is a homogeneous element $f$ of degree $n$ in $A$ such that $s_i = f/x_i^n$ for all $i$. The elements $x_i \cdot s_i$ have the same image, say $f$, in $A_{x_1 \ldots x_m}$. This $f$ can be written uniquely in the form $f = \prod_{i=1}^{m} x_i^{p_i} \cdot g$, with $p_i$ integers and $g$ a homogeneous element of $A$ that is not divisible by any $x_i$. The fact that $f$ is in $A_{x_i}$ means that each $p_j$ for $j \neq i$ is nonnegative, and this implies that $f$ is in $A$, which is the required conclusion. A similar proof works in the multihomogeneous case. In fact, only the normality of the multihomogeneous coordinate ring is needed; cf. Hartshorne (1977), II, Ex. 5.14.

**12** Fix any nonzero vector $y$ in the fiber of $L$ over the fixed point $x$ of $P$, and map $L(\chi)$ to $L$ by the formula $g \times z \mapsto g \cdot zy = z(g \cdot y)$. Check that this is a well-defined isomorphism. The second assertion is immediate from the definition.

**13** The closure of $\Omega_\lambda^\circ$ consists of those subspaces whose echelon matrix has its 1's at or to the right of the spots specified by $\lambda$. For (d), construct a filtration $Gr^n(\mathbb{C}^m) = Y_0 \supset Y_1 \supset \ldots$ as in (iv), where $Y_p$ is the union of all $\Omega_\lambda$ with $|\lambda| \geq p$.

**14** This follows immediately from the definitions, i.e., that $A_i$ is spanned by the first $n + i - \lambda_i$ of the $m = n + r$ basis vectors, and $B_{r+1-i}$ is spanned by the last $n + (r+1-i) - \mu_{r+1-i}$ basis vectors.

**15** If $v_i$ is a general vector in $A_i \cap B_{r+1-i}$, then the space spanned by $v_1, \ldots, v_r$ is in the intersection of $\Omega_\lambda$ and $\tilde{\Omega}_\mu$.

**16** For (a), note that a basis vector $e_p$ of $\mathbb{C}^m$ is in $C$ exactly when it is in $C_j$ for some $1 \leq j \leq r$, which means that

$$\text{(i)} \qquad j + \mu_{r+1-j} \leq p \leq n + j - \lambda_j \quad \text{for some} \quad 1 \leq j \leq r;$$

and $e_p$ is in $\bigcap_{i=0}^{r}(A_i + B_{r-i})$ when, setting $\lambda_0 = \mu_0 = n$,

$$\text{(ii)} \qquad p \leq n + i - \lambda_i \quad \text{or} \quad p > i + \mu_{r-i} \quad \text{for all} \quad 0 \leq i \leq r.$$

To show that (i) $\Rightarrow$ (ii), suppose (i) holds for $j$; if $i < j$, then $i + \mu_{r-i} < j + \mu_{r+1-j} \leq p$, and if $i \geq j$, then $p \leq n + j - \lambda_j \leq n + i - \lambda_i$. For (ii) $\Rightarrow$ (i), look at the smallest $j$ such that $p \leq n+j-\lambda_j$; the fact that the first condition in (ii) fails for $j - 1$ means that $p > (j-1) + \mu_{r-(j-1)}$, which is (i).

**17** With $n = m - r$, this is the $(r\cdot n)$-fold intersection of the special Schubert class $\sigma_1$, which, by repeated application of Pieri's formula, is the number of standard tableaux on the shape $(n^r)$. Use the hook-length formula.

**18** If $\lambda_{i-1} = \lambda_i$, the condition for $\lambda_i$ implies that for $\lambda_{i-1}$. If any of these conditions is omitted, one has the conditions for a partition of a smaller integer, so of a larger Schubert variety.

**19** With what we have proved, this follows formally from the Pieri formula (7) of §2.2.

**20** Use Exercise 18 for the last assertion.

**21** The $\sigma_k$ correspond to the special Schubert classes, which are the images of $h_k \in \Lambda$. The kernel of $\Lambda \to H^*(Gr^n(\mathbb{C}^m))$ is generated by all $h_i$ for $i > n$ and all $e_p$ for $p > r$. We know that $e_p = s_{(1^p)} = \det(h_{1+j-i})_{1 \leq i,j \leq p}$. From the equations $e_p - h_1 e_{p-1} + \ldots + (-1)^p h_p = 0$ it follows that only those $e_p$ for $p \leq m$ are needed. This presentation of the cohomology of the Grassmannian was given by Borel. For an interesting variation of these ideas, see Gepner (1991).

## Chapter 10

The formulas for the Schubert varieties in flag manifolds $G/B$ for a general semisimple group were given independently by Bernstein, Gelfand, and Gelfand (1973) and by Demazure (1974). For some background to this story, as well as the origins of the Bruhat order, see Chevalley (1994). The explicit representatives as Schubert *polynomials* were found and studied by Lascoux and Schützenberger. Our treatment follows the plan of Billey and Haiman (1995), where they construct analogues of these polynomials for the other classical groups; we thank S. Billey for discussions about this.

A purely algebraic construction of these polynomials is also possible, using a little more knowledge about the symmetric group (see Macdonald [1991a; 1991b]). See

Billey, Jockusch, and Stanley (1993), Fomin and Kirillov (1993), Fomin and Stanley (1994), Reiner and Shimozono (1995) and Sottile (1996) for some recent work on Schubert polynomials. For more about the Bruhat order, see Deodhar (1977).

Proposition 3 is a special case of a theorem of Borel (1953).

### *Exercises*

**1** (a) It suffices to do this when $V = E^\lambda$, where $E = \mathbb{C}^m$. Choose any total ordering of the tableaux $T$ on $\lambda$ with entries in $[m]$, such that $T < T'$ whenever the sum of the entries in $T$ is less than the sum of the entries in $T'$. Let $V_k$ be the subspace of $V$ spanned by the first $k$ basis elements $e_T$, using this ordering on the $T$'s. Note that $E_{i,j}(V_k) \subset V_{k-1}$ if $i < j$, so $V_k$ is mapped to itself by $B$.

(b) Consider the chain $Z \cap \mathbb{P}(V_1) \subset Z \cap \mathbb{P}(V_2) \subset \ldots \subset Z \cap \mathbb{P}(V_r) = Z$, each of which is invariant by $B$. Since an algebraic subset of projective space that is not finite must meet any hyperplane, it follows that one of these sets $Z \cap \mathbb{P}(V_k)$ must be finite and nonempty. Since $B$ is connected, each of its points must be fixed by $B$.

**2** The set $U_w$ is open, since it is defined by the condition that certain minors are nonzero: those using the first $p$ rows and the columns numbered $w(1), \ldots, w(p)$, for $1 \le p \le m$. Note that the minor obtained from the first $p$ rows and the columns numbered $w(1), \ldots, w(p-1), q$ is, up to sign, the entry in the $(p,q)^{\text{th}}$ position of the matrix. It follows that, under the mapping $\mathbb{C}^n \to F\ell(m) \subset \mathbb{P}^r$, each of the coordinates of a point in $\mathbb{C}^n$ corresponding to a star appears, up to sign, as one of the homogeneous coordinates in the image point in $\mathbb{P}^r$; similarly, one of the homogeneous coordinates is 1. This implies that the map is an embedding.

**3** The first assertion follows directly from our definition of the length of a permutation. The second assertion follows from the first.

**4** (a) The second statement follows from the definition of $X_w^\circ$, and the first follows from the second. (b) The assertions about the diagram are translations of the descriptions given in the text.

**5** A calculation, as in the preceding.

**6** We know it if $w = d\ d-1 \ldots 2\ 1$; use descending induction on $\ell(w)$. Write $w = w(1) \ldots w(d-p)\ p\ p-1 \ldots 2\ 1\ p+1\ w(d+2) \ldots$. Let

$$w' = w(1) \ldots w(d-p)\ p+1\ p\ p-1 \ldots 2\ 1\ w(d+2) \ldots$$

$$= w \cdot s_d \cdot s_{d-1} \cdot \ldots \cdot s_{d-p+1}.$$

Then $\mathfrak{S}_{w'} = X_1{}^{w(1)-1} X_2{}^{w(2)-1} \cdot \ldots \cdot X_{d-p}{}^{w(d-p)-1} X_{d-p+1}{}^p \cdot \ldots \cdot X_d$ by induction, and

$$\mathfrak{S}_w = \partial_{d-p+1} \circ \ldots \circ \partial_d (\mathfrak{S}_{w'})$$

$$= X_1{}^{w(1)-1} X_2{}^{w(2)-1} \cdot \ldots \cdot X_{d-p}{}^{w(d-p)-1} X_{d-p+1}^{p-1} \cdot \ldots \cdot X_{d-1}.$$

**7** The proof is similar to that of Proposition 6.

**8** This is another way of saying that $r_u(p,q) \geq r_v(p,q)$ for all $q$.

**9** (a) Note first that if $w(k) < w(k+1)$, and $w^* = w \cdot s_k$, then

$$r_{w^*}(p,q) = \begin{cases} r_w(p,q) - 1 & \text{if } p = k,\ w(k) \leq q < w(k+1) \\ r_w(p,q) & \text{otherwise.} \end{cases}$$

Suppose $u \leq v$, but $r_{u^*}(p,q) < r_{v^*}(p,q)$ for some $p,q$. We must have $r_u(p,q) = r_v(p,q)$, $r_{u^*}(p,q) = r_u(p,q) - 1$, $r_{v^*}(p,q) = r_v(p,q)$, and $p = k$ and $q \notin [v(k), v(k+1))$. If $q < v(k)$, then

$$r_v(k,q) = r_v(k-1,q) \leq r_u(k-1,q) = r_u(k,q) - 1,$$

a contradiction. If $q \geq v(k+1)$, then

$$r_v(k,q) = r_v(k+1,q) - 1 \leq r_u(k+1,q) - 1 = r_u(k,q) - 1,$$

a contradiction. The converse is similar.

(b) Let $v = t_1 \cdot \ldots \cdot t_\ell$, with $\ell(v) = \ell$, and each $t_i \in \{s_1, \ldots, s_{m-1}\}$. If $u \leq v$, by Lemma 11, we may assume $u = v \cdot (j,k)$, with $j < k$, $v(j) > v(k)$, and $v(i) \notin (v(k), v(j))$ for $i \in (j,k)$. Let $t_\ell = s_p$, so $v(p) > v(p+1)$. If $u(p) > u(p+1)$, then by (a), $u \cdot s_p \leq v \cdot s_p = t_1 \cdot \ldots \cdot t_{\ell-1}$, and, by induction on $\ell$, $u \cdot s_p$ is obtained from $t_1 \cdot \ldots \cdot t_{\ell-1}$ by removing one of the $t_i$'s; therefore $u$ is obtained from $v$ by removing one of the $t_i$'s, for $i \leq \ell-1$. Otherwise $u(p) < u(p+1)$, while $v(p) > v(p + 1)$. One sees easily that there is no such $p$ unless $k = j+1$, with $p = j$, in which case $u = t_1 \cdot \ldots \cdot t_{\ell-1}$.

Conversely, suppose $u$ is obtained from $v$ by removing $t_i$ from a reduced expression for $v$, with $\ell(u) = \ell-1$. If $i = \ell$, that $u \leq v$ is clear from the definition. If $i < \ell$, then $\ell(u \cdot s_p) \leq \ell-2 < \ell(u)$, so $u(p) > u(p+1)$. Then $u \cdot s_p < v \cdot s_p$ by induction on $\ell$, and one concludes by (a).

**10** Suppose $w(p) < w(p+1)$. If $w(p) \leq q$, let $b$ be the smallest integer in $T \cup \{m\}$ that is larger than $p$. Then $r_w(b,q) = r_w(p,q) + (b - q)$, and the result follows from the fact that $\dim(E_p \cap F_q) \geq \dim(E_b \cap F_q) - (b - q)$.

In case $w(q) > p$, let $a$ be the largest integer in $T \cup \{0\}$ that is less than $p$. If $a = 0$, then $r_w(p,q) = 0$. Otherwise, $r_w(p,q) = r_w(a,q)$ and the result follows from the fact that $\dim(E_p \cap F_q) \geq \dim(E_a \cap F_q)$.

A minimal set of $(p,q)$ for which the condition must be checked is given in Fulton (1992), §3.

**11** (a) amounts to the preceding corollary. For (b), note that the quadratic generators of the ideal of the flag variety map to zero, as do the generators of $J_w$ described above. The image of the homomorphism is a subring whose quotient field is $\mathbb{C}(\{A_{i,j}\})$, so the image is a domain of dimension equal to $\dim(X_w) + m$. Since this is the dimension of the multihomogeneous coordinate ring of $X_w$, the map from $\mathbb{C}[\{X_I\}]/I(X_w)$ to $\mathbb{C}[\{A_{ij}\}]$ must be injective.

**12** The coefficient $a_w$ is the intersection number of $[Z]$ with the dual Schubert variety $[X_w]$. This is a nonnegative integer, since one can find an element $g$ of $GL_n\mathbb{C}$ such that $Z$ and $g \cdot X_w$ meet transversally in a finite number of points. Similarly, one can replace $\Omega_v$ by a translate by an element of the group, so that $\Omega_u$ meets $h \cdot \Omega_v$ properly. The coefficient $c_{u,v}^{w}$ is the number of points in $\Omega_u \cap h \cdot \Omega_v \cap g \cdot X_w$ for some general elements $g$ and $h$ in $GL_m\mathbb{C}$. These facts are proved more generally by Kleiman, see Hartshorne (1977), III, Thm. 10.8.

**13** Note that $\dim(E_p \cap F_q) = p - \mathrm{rank}(E_p \to (F'_{m-q})^*)$ and

$$\mathrm{rank}(E_p \to (F'_{m-q})^*) = \mathrm{rank}(F'_{m-q} \to (E_p)^*)$$
$$= (m - q) - \dim(E'_{m-p} \cap F'_{m-q}).$$

The fact that $X_w$ corresponds to $X_{w_o \cdot w \cdot w_o}$ follows by counting. The corresponding statement for $\Omega_w$ follows by replacing $w$ by $w_o \cdot w$.

## Appendix A

Our treatment of duality is derived from Lascoux and Schützenberger (1981). Column bumping can be found in most of the references cited for Chapter 1. The relation between Littlewood–Richardson correspondences and shape changes in the jeu de taquin that we include in §A.3 was first noticed by M. Haiman, see Haiman (1992) and Sagan (1991); most results in these sources are for tableaux with distinct entries. Many of the correspondences discussed in §A.4 can be found in Knuth (1970; 1973) and Bender and Knuth (1972). E. Gansner, in his MIT thesis in the late 1970s, looked at the symmetries of a matrix and proved Theorem 1; see Vo and Whitney (1983) and Burge (1974) for developments of this idea. The various "matrix-ball" constructions are new here; in particular, they prove the corresponding symmetry theorems, which have seldom received much justification in the literature. Exercise 30 was provided by A. Buch.

### *Exercises*

**1** Look directly at the bumping that occurs when row-inserting $v_n{}^*, \ldots, v_1{}^*$.

**2** See the proof of Proposition 1 in §2.1.

**3** See the Row Bumping Lemma in Chapter 1.

**4** Induct on the number of pairs in the array.

**5** At each stage the two tableaux are dual, so have the same shape by the Duality Theorem.

**6** The new boxes for row bumping $v_r \leftarrow \ldots \leftarrow v_1$ are conjugate to those for column bumping $v_1 \to \ldots \to v_r$, which are the same as those for row bumping $v_r{}^* \leftarrow \ldots \leftarrow v_1{}^*$. See §A.4 for more on this.

**7** Both of these follow readily from the definitions. Note that $S(Q_{\text{row}}(\lambda))$ inserts $n$, $n-1$, . . . , $n-\lambda_1+1$ in the bottoms of the columns of $\lambda$, then the next $\lambda_2$ integers in the bottoms of the unfilled columns, and so on until $\lambda$ is filled.

**8** This follows from the definition of the dual tableau using sliding.

**9** Look at the corresponding equation $U^* \cdot T^* = V_o^*$. Fix $U_o$ on $\mu$ with alphabet ordered before that of $U(\lambda)$. Define two lexicographic arrays by

$$(U^*, U_o) \longleftrightarrow \begin{pmatrix} y_1 & \cdots & y_m \\ v_r^* & \cdots & v_{n+1}^* \end{pmatrix}, \quad (T^*, U(\lambda)) \longleftrightarrow \begin{pmatrix} x_1 & \cdots & x_n \\ v_n^* & \cdots & v_1^* \end{pmatrix}$$

and note by Proposition 1 of §5.1 that (using the notation of Proposition 2 of §5.1)

$$(V_o^*, (U_o)s) \longleftrightarrow \begin{pmatrix} y_1 & \cdots & y_m & x_1 & \cdots & x_n \\ v_r^* & & \cdots & & & v_1^* \end{pmatrix}.$$

Successive column-insertion of $v_r$, . . . , $v_{n+1}$ gives $U$, and then successive column-insertion of $v_n$, . . . , $v_1$ gives $V_o$; numbering the new boxes $x_1$, . . . , $x_n$ gives $S$, cf. Exercise 5. Exercise 1 shows that $x_1$, . . . , $x_n$ is the sequence $1, 1, . . . , k$.

**10** If $w$ is a reverse lattice word, the fact that $U(w) = Q(w^*)$ follows from the definition of row bumping. From the Duality Theorem of §A.1 one sees that $Q(w^*)$ determines $Q(w)$, from which the uniqueness follows.

**11** By the theorem, it is enough to do this when $S = S^\natural$ is a Littlewood–Richardson skew tableau, when it is clear.

**12** Use Exercise 11.

**13** Both assertions follow from the definitions. For example, consider the construction of $P(w)$ by row bumping. When the $v_i$ is bumped in, the fact that $i$ is maximal with $I(i) = J(k)$ assures that it does not bump a $k$, so $v_i - 1$ goes to the same place. If any later $v_j = k - 1$ bumps this $v_i$, there will be a $k$ to bump in the other sequence. For the second assertion, note that $v_i$ cannot be preceded by a $k$ in its row, or have a $k - 1$ above it in its column.

**14** The independence of choice follows immediately from the equivalence of (iii) and (iv) in the Shape Change Theorem. If there are such words, they correspond to $(U, Q)$ and $(U', Q)$ for the same $Q$. Inserting $1, . . . , m$ in the new boxes will produce the same skew tableau, so $[T, U]$ corresponds to $[T', U']$ by $S(\nu/\lambda, Q)$.

**15** Write out the lexicographic arrays corresponding to $(T, T_o)$, $(U, U_o)$, $(T \cdot U, (T_o)s)$, $(T', T_o)$, $(U', U_o)$, and $(T' \cdot U', (T_o)s)$, and apply the duality operator.

**16** This follows directly from the definitions, once it is known that the correspondences are independent of choices.

**17** The row words of the former are

$$3\,1\,2\,3\,1\,2\,2\,1\,1, \quad 2\,1\,3\,3\,1\,2\,2\,1\,1, \quad \text{and} \quad 1\,2\,3\,3\,1\,2\,2\,1\,1,$$

and the row words of the corresponding latter are

$$3\,4\,2\,2\,3\,1\,2\,1\,1,\quad 2\,4\,1\,3\,3\,2\,2\,1\,1,\quad \text{and}\quad 2\,3\,1\,2\,4\,1\,3\,2\,1.$$

**18** See Exercise 4 in §4.2.

**19** This follows from Proposition 3.

**20** For (b) $\Rightarrow$ (c), note that (b) is equivalent to the assertion that, for all $k$, $\lambda$ has at least as many boxes below row $k$ as $\tilde{\mu}$ does; if $p$ is minimal with $\mu_1 + \ldots + \mu_p > \tilde{\lambda}_1 + \ldots + \tilde{\lambda}_p$, this is contradicted with $k = \tilde{\lambda}_p$. From the preceding exercise, (a) is equivalent to the existence of $\nu$ with $\lambda \trianglelefteq \nu$ and $\mu \trianglelefteq \tilde{\nu}$, or to $\nu$ with $\lambda \trianglelefteq \nu \trianglelefteq \tilde{\mu}$, which is equivalent to (b).

**21** Take a tableau $U$ with conjugate shape to $T$, with entries smaller than the $u_i$, and let the pair $(T, U)$ correspond to a lexicographic array $\begin{pmatrix} s_1 & \ldots & s_n \\ t_1 & \ldots & t_n \end{pmatrix}$.

The lexicographic array $\begin{pmatrix} s_1 & \ldots & s_n & u_1 & \ldots & u_r \\ t_1 & \ldots & t_n & v_n & \ldots & v_r \end{pmatrix}$ corresponds to a conjugate pair $\{\widetilde{P \cdot T}, (U)_X\}$. Turning this array upside down and putting it in antilexicographic order, the symmetry theorem implies that the corresponding conjugate pair has the form $\{(U)_X, \widetilde{P \cdot T}\}$. Remove the $n$ smallest entries, and appeal to Lemma 3 of §3.2.

**22** See Exercise 21.

**23** If $(T, T_\circ)$ corresponds to a lexicographic array $\begin{pmatrix} x_1 & \ldots & x_n \\ y_1 & \ldots & y_n \end{pmatrix}$, consider the array $\begin{pmatrix} u_1 & \ldots & u_r & x_1 & \ldots & x_n \\ v_1 & \ldots & v_r & y_1 & \ldots & y_n \end{pmatrix}$.

**24** See Exercise 23.

**25** Use Theorem 1 and Proposition 6(3).

**26** Suppose the row words are $Q$-equivalent. As in the proof of the Shape Change Theorem in §A.3, one may assume the skew tableaux have distinct entries, since $w_{\mathrm{col}}(S^\#) = w_{\mathrm{col}}(S)^\#$. $Q$-equivalence of the row words is the same as shape equivalence, which is preserved for such tableaux by taking transposes. This means that the reverses of their column words are $Q$-equivalent. Apply Exercise 6.

**27** In each case one can start at any place in the middle and use (1) and (2) in any order to work to the ends.

**28** This follows from the fact that one can perform arbitrary elementary moves on a frank skew tableau, always obtaining a frank skew tableau. Using only adjacent columns in one piece, one transforms it to a tableau whose column lengths are a permutation of its column lengths.

**29** If $T$ and $U$ are frank skew tableaux (e.g., tableaux), one can perform elementary moves within $T$ and $U$, preserving frankness. In order to permute among all columns, one needs exactly to be able to permute a last column of $T'$ with a first column of $U'$, where $T'$ and $U'$ are obtained from $T$ and $U$ by elementary moves, and this is precisely the condition that one can permute $t$ and $u$ as in (2).

**30** The reverse row (resp. column) bumping bumps out a column tableau $C$ of length $c$, leaving another tableau $U$. The skew tableau $C * U$ (resp. $U * C$) is a frank skew tableau whose rectification is $T$.

## Appendix B

For the required topology, a knowledge of Greenberg and Harper (1981) should suffice, with an occasional reference to Spanier (1966), Dold (1980), or Husemoller (1994). For facts about differentiable manifolds, see Guillemin and Pollack (1974) or Lang (1985). We also assume some basic facts about algebraic varieties, for which Shafarevich (1977) is a good reference.

Atiyah and Hirzebruch (1962) give another way to define the fundamental class of a subvariety of a complex manifold.

### *Exercises*

**1–6** These involve a lot of verification, but no new ideas beyond those in the text. For example, to prove that the diagram in Exercise 5 commutes, by factoring $f$ through $X' \to X \times I_n \to X$, it suffices to do two cases: where $f$ is a closed embedding, and where $f$ is a projection from $X \times I^n$ to $X$. For the first, a closed embedding of $X$ in an oriented manifold determines embeddings of the other spaces, and the vertical maps in the diagram are induced by restriction maps, which are compatible with long exact sequences. For the second, note that the induced maps $\overline{H}_i(X \times I^n) \to \overline{H}_i X$ are isomorphisms: for a closed embedding of $X$ in an oriented $m$-manifold $M$, the inverse is the composite

$$\overline{H}_i X = H^{m-i}(M, M \smallsetminus X) \to H^{m+n-i}(M \times \mathbb{R}^n, M \times \mathbb{R}^n \smallsetminus X \times \{0\})$$

$$\to H^{m+n-i}(M \times \mathbb{R}^n, M \times \mathbb{R}^n \smallsetminus X \times I^n) = \overline{H}_i(X \times I^n),$$

where the first map is the Thom isomorphism for the trivial bundle, and the second is a restriction map. Both of these maps commute with the maps in the long exact sequences.

**7** A path from the identity element of $G$ to $g$ gives a homotopy from the identity map on $X$ to left multiplication by $g$. That $g_*[V] = [g \cdot V]$ follows from the construction of $[V]$.

**8** By the Künneth formula, $s^*([H]) = a[H_1 \times \mathbb{P}^m] + b[\mathbb{P}^n \times H_2]$ for some integers $a$ and $b$. Then $a$ is the degree of the intersection of $s_*([H])$ with $[\ell_1 \times \mathbb{P}^m]$, where $\ell_1$ is a line in $\mathbb{P}^n$. By the projection formula, this is the same as the degree of the intersection of $s_*[\ell_1 \times \mathbb{P}^m] = [s(\ell_1 \times \mathbb{P}^m)]$ with $H$, and for general $\ell_1$ and $H$, these varieties meet transversally in one point.

**9** Use the splitting principle.

**10** If $X$ is not compact, embed $X$ as an open subvariety of a compact variety $Y$, and let $Z = Y \smallsetminus X$. The fact that the pair $(Y, Z)$ can be triangulated implies that

$X^+ = Y/Z$ is compact and locally contractible. Therefore for any embedding of $X^+$ in an oriented $n$-manifold $M$,

$$H_i(X^+) \cong H^{n-i}(M, M \smallsetminus X^+)$$

(by Spanier [1966], 6.10.14), i.e., $H_i(X^+) \cong \bar{H}_i X^+$. Apply Lemma 3 to the open subspace $X$ of $X^+$ to finish. (Alternatively, embed $Y$ in an oriented $n$-manifold $M$, and use the duality isomorphism

$$H_i(Y, Z) \cong H^{n-i}(M \smallsetminus Z, M \smallsetminus Y),$$

which is valid since $(Y, Z)$ is a Euclidean neighborhood retract.)

# Bibliography

S. Abeasis, "On the Plücker relations for the Grassmann varieties," Advances in Math. **36** (1980), 277–282.

K. Akin, D. A. Buchsbaum and J. Weyman, "Schur functors and Schur complexes," Advances in Math. **44** (1982), 207–278.

M. F. Atiyah and F. Hirzebruch, "Analytic cycles and complex manifolds," Topology **1** (1962), 25–45.

E. A. Bender and D. E. Knuth, "Enumeration of plane partitions," J. of Combin. Theory, Ser. A **13** (1972), 40–54.

G. Benkart, F. Sottile and J. Stroomer, "Tableau switching: algorithms and applications," J. of Combin. Theory, Ser. A **76** (1996), 11–43.

N. Bergeron and A. M. Garsia, "Sergeev's formula and the Littlewood–Richardson rule," Linear and Multilinear Algebra **27** (1990), 79–100.

I. N. Bernstein, I. M. Gelfand and S. I. Gelfand, "Schubert cells and cohomology of the spaces G/P," Russian Math. Surveys **28:3** (1973), 1–26.

S. Billey and M. Haiman, "Schubert polynomials for the classical groups," J. Amer. Math. Soc. **8** (1995), 443–482.

S. C. Billey, W. Jockusch and R. P. Stanley, "Some combinatorial properties of Schubert polynomials," J. Algebraic Combinatorics **2** (1993), 345–374.

A. Borel, "Sur la cohomologie des espaces fibrés principaux et des espaces homogènes des groupes de Lie compacts," Annals of Math. **57** (1953), 115–207.

A. Borel and A. Haefliger, "La classe d'homologie fondamentale d'un espace analytique," Bull. Soc. Math. France **89** (1961), 461–513.

W. H. Burge, "Four correspondences between graphs and generalized Young tableaux," J. of Combin. Theory, Ser. A **17** (1974), 12–30.

R. W. Carter and G. Lusztig, "On the modular representations of the general linear and symmetric groups," Math. Zeit. **136** (1974), 193–242.

Y. M. Chen, A. M. Garsia and J. Remmel, "Algorithms for plethysm," in *Combinatorics and Algebra*, Contemporary Math. **34** (1984), 109–153.

C. Chevalley, "Sur les décompositions cellulaires des espaces G/B," Proc. Symp. Pure Math. **56**, Part 1 (1994), 1–23.

C. DeConcini, D. Eisenbud and C. Procesi, "Young diagrams and determinantal varieties," Invent. Math. **56** (1980), 129–165.

M. Demazure, "Désingularization des variétés de Schubert généralisées," Ann. Scuola Norm. Sup. Pisa Cl. Sci. (4) **7** (1974), 53–88.

V. V. Deodhar, "Some characterizations of Bruhat ordering on a Coxeter group and determination of the relative Möbius function," Invent. Math. **39** (1977), 187–198.

J. Désarménien, J. Kung and G.-C. Rota, "Invariant theory, Young bitableaux, and combinatorics," Advances in Math. **27** (1978), 63–92.

P. Diaconis, *Group Representations in Probability and Statistics*, Institute of Mathematical Statistics, Hayward, CA, 1988.

A. Dold, *Lectures on Algebraic Topology*, Springer-Verlag, 1980.

D. Foata, "A matrix-analog for Viennot's construction of the Robinson correspondence," Linear and Multilinear Algebra **7** (1979), 281–298.

S. Fomin and C. Greene, "A Littlewood–Richardson miscellany," Europ. J. Combinatorics **14** (1993), 191–212.

S. Fomin and A. Kirillov, "The Yang–Baxter equation, symmetric functions, and Schubert polynomials," in *Proceedings of the 5th International Conference on Formal Power Series and Algebraic Combinatorics,* Firenze (1993), 215–229; to appear in Discrete Math.

S. Fomin and R. Stanley, "Schubert polynomials and the nilCoxeter algebra," Advances in Math. **103** (1994), 196–207.

H. O. Foulkes, "Enumeration of permutations with prescribed up–down and inversion sequences," Discrete Math. **15** (1976), 235–252.

W. Fulton, *Intersection Theory*, Springer-Verlag, 1984, 1998.

W. Fulton, "Flags, Schubert polynomials, degeneracy loci, and determinantal formulas," Duke Math. J. **65** (1992), 381–420.

W. Fulton and J. Harris, *Representation Theory: A First Course*, Springer-Verlag, 1991.

W. Fulton and A. Lascoux, "A Pieri formula in the Grothendieck ring of a flag bundle," Duke Math. J. **76** (1994), 711–729.

W. Fulton and R. MacPherson, *Categorical Framework for the Study of Singular Spaces*, Memoirs Amer. Math. Soc. **243**, 1981.

D. Gepner, "Fusion rings and geometry," Commun. Math. Phys. **141** (1991), 381–411.

J. A. Green, *Polynomial Representations of $GL_n$* , Lecture Notes in Math. **830**, Springer-Verlag, 1980.

J. A. Green, "Classical invariants and the general linear group," in *Representation Theory of Finite Groups and Finite-Dimensional Algebras*, Progress in Math. **95**, Birkhäuser, (1991), 247–272.

M. J. Greenberg and J. R. Harper, *Algebraic Topology: A First Course*, Benjamin/Cummings, 1981.

C. Greene, "An extension of Schensted's theorem," Advances in Math. **14** (1974), 254–265.

C. Greene, "Some partitions associated with a partially ordered set," J. of Combin. Theory, Ser. A **20** (1976), 69–79.

C. Greene, A. Nijenhuis and H. S. Wilf, "A probabilistic proof of a formula for the number of Young tableaux of a given shape," Advances in Math. **31** (1979), 104–109.

P. Griffiths and J. Harris, *Principles of Algebraic Geometry*, Wiley, 1978.

V. Guillemin and A. Pollack, *Differential Topology*, Prentice-Hall, 1974.

M. D. Haiman, "Dual equivalence with applications, including a conjecture of Proctor," Discrete Math. **99** (1992), 79–113.

P. Hanlon and S. Sundaram, "On a bijection between Littlewood–Richardson fillings of conjugate shape," J. of Combin. Theory, Ser. A **60** (1992), 1–18.

J. Harris, *Algebraic Geometry: A First Course*, Springer-Verlag, 1992.

R. Hartshorne, *Algebraic Geometry*, Springer-Verlag, 1977.

W. V. D. Hodge and D. Pedoe, *Methods of Algebraic Geometry*, Vols. 1, 2, and 3, Cambridge University Press, 1947, 1952, 1954.

R. Howe, "$(GL_n, GL_m)$-duality and symmetric plethysm," Proc. Indian Acad. Sci. (Math. Sci.) **97** (1987), 85–109.

D. Husemoller, *Fibre Bundles*, 3rd edition, Springer-Verlag, 1994.

B. Iversen, *Cohomology of Sheaves*, Springer-Verlag, 1986.

G. D. James, *The Representation Theory of the Symmetric Groups*, Lecture Notes in Math. **682**, Springer-Verlag, 1978.

G. James and A. Kerber, *The Representation Theory of the Symmetric Group*, Encyclopedia of Mathematics and Its Applications, vol. 16, Addison-Wesley, 1981.

I. M. James, *Topological and Uniform Spaces*, Springer-Verlag, 1987.

S. L. Kleiman and D. Laksov, "Schubert calculus," Amer. Math. Monthly **79** (1972), 1061–1082.

D. E. Knuth, "Permutations, matrices and generalized Young tableaux," Pacific J. Math. **34** (1970), 709–727.

D. E. Knuth, *The Art of Computer Programming III*, Addison-Wesley, 1973.

D. Knutson, *$\lambda$-Rings and the Representation Theory of the Symmetric Group*, Lecture Notes in Math. **308**, Springer-Verlag, 1973.

H. Kraft, *Geometrische Methoden in der Invariantentheorie*, Fried. Vieweg & Sohn, Braunschweig, 1984.

C. Krattenthaler, "An involution principle-free bijective proof of Stanley's hook-content formula," Discrete Math. Theor. Comput. Sci. (http://dmtcs.loria.fr) **3** (1998), 11–32.

V. Lakshmibai and C. S. Seshadri, "Geometry of G/P – $\underline{V}$ ," J. of Algebra **100** (1986), 462–557.

S. Lang, *Differentiable Manifolds*, Addison-Wesley, 1971, Springer-Verlag, 1985.

A. Lascoux, "Produit de Kronecker des représentations du groupe symétrique," in *Séminaire Dubreil-Malliavin 1978–1979*, Lecture Notes in Math. **795** (1980), Springer-Verlag, 319–329.

A. Lascoux and M. P. Schützenberger, "Le monoïde plaxique," in *Non-Commutative Structures in Algebra and Geometric Combinatorics*, Quaderni de "La ricerca scientifica," n. 109, Roma, CNR (1981), 129–156.

A. Lascoux and M. P. Schützenberger, "Schubert polynomials and the Littlewood–Richardson rule," Letters in Math. Physics **10** (1985), 111–124.

A. Lascoux and M. P. Schützenberger, "Keys and standard bases," in *Invariant Theory and Tableaux*, D. Stanton, ed., Springer-Verlag (1990), 125–144.

P. Littelmann, "A Littlewood–Richardson rule for symmetrizable Kac–Moody algebra," Invent. Math. **116** (1994), 329–346.

A. Liulevicius, "Arrows, symmetries and representation rings," J. Pure App. Algebra **19** (1980), 259–273.

I. G. Macdonald, *Symmetric Functions and Hall Polynomials*, Clarendon Press, Oxford, 1979, 1995.

I. G. Macdonald, *Notes on Schubert Polynomials*, Département de mathématiques et d'informatique, Université du Québec, Montréal, 1991a.

I. G. Macdonald, "Schubert polynomials," in *Surveys in Combinatorics*, Cambridge University Press, (1991b) 73–99.

P. Magyar, "Borel–Weil theorem for configuration varieties and Schur modules," Adv. Math. **134** (1998), 328–366.

S. Martin, *Schur Algebras and Representation Theory*, Cambridge University Press, 1993.

D. Monk, "The geometry of flag manifolds," Proc. London Math. Soc. **9** (1959), 253–286.

M. H. Peel, "Specht modules and symmetric groups," J. of Algebra **36** (1975), 88–97.

P. Pragacz, "Algebro-geometric applications of Schur S- and Q-polynomials," in *Séminaire d'Algèbre Dubreil-Malliavin 1989-1990*, Lecture Notes in Math. **1478** (1991), Springer-Verlag, 130–191.

P. Pragacz and A. Thorup, "On a Jacobi–Trudi identity for supersymmetric polynomials," Advances in Math. **95** (1992), 8–17.

R. A. Proctor, "Equivalence of the combinatorial and the classical definitions of Schur functions," J. of Combin. Theory, Ser. A **51** (1989), 135–137.

A. Ramanathan, "Equations defining Schubert varieties, and Frobenius splitting of diagonals," Publ. Math. I.H.E.S. **65** (1987), 61–90.

V. Reiner and M. Shimozono, "Plactification," J. of Algebraic Combinatorics **4** (1995), 331–351.

J. B. Remmel and R. Whitney, "A bijective proof of the hook formula for the number of column-strict tableaux with bounded entries," European J. Combin. **4** (1983), 45–63.

J. B. Remmel and R. Whitney, "Multiplying Schur functions," J. of Algorithms **5** (1984), 471–487.

G. de B. Robinson, "On the representations of the symmetric group," Amer. J. Math. **60** (1938), 745–760.

G. de B. Robinson, *Representation Theory of the Symmetric Group*, University of Toronto Press, 1961.

T. W. Roby, "Applications and extensions of Fomin's generalization of the Robinson–Schensted correspondences to differential posets," MIT PhD Thesis, 1991.

B. E. Sagan, *The Symmetric Group*, Wadsworth, 1991.

B. E. Sagan and R. P. Stanley, "Robinson–Schensted algorithms for skew tableaux," J. of Combin. Theory, Ser. A **55** (1990), 161–193.

C. Schensted, "Longest increasing and decreasing subsequences," Canad. J. Math. **13** (1961), 179–191.

M. P. Schützenberger, "Quelques remarques sur une construction de Schensted," Math. Scand. **12** (1963), 117–128.

M. P. Schützenberger, "La correspondance de Robinson," in *Combinatoire et Représentation du Groupe Symétrique*, Lecture Notes in Math. **579** (1977), Springer-Verlag, 59–135.

I. Shafarevich, *Basic Algebraic Geometry*, Springer-Verlag, 1977.

F. Sottile, "Pieri's rule for flag manifolds and Schubert polynomials," Annales Fourier **46** (1996), 89–110.

E. H. Spanier, *Algebraic Topology*, McGraw-Hill, 1966.

R. P. Stanley, "Theory and applications of plane partitions," Studies in Appl. Math. **1** (1971), 167–187 and 259–279.

R. P. Stanley, "Some combinatorial aspects of the Schubert calculus," in *Combinatoire et Représentation du Groupe Symétrique*, Lecture Notes in Math. **579** (1977), Springer-Verlag, 225–259.

R. P. Stanley, "GL(n, $\mathbb{C}$) for combinatorialists," in *Surveys in Combinatorics*, E. K. Lloyd (ed.), Cambridge University Press, 1983.

R. P. Stanley, *Enumerative Combinatorics*, Vol. I, Wadsworth and Brooks/Cole, 1986.

R. Steinberg, "An occurrence of the Robinson–Schensted correspondence," J. of Algebra **113** (1988), 523–528.

J. R. Stembridge, "Rational tableaux and the tensor algebra of $gl_n$," J. of Combin. Theory, Ser. A **46** (1987), 79–120.

B. Sturmfels and N. White, "Gröbner bases and invariant theory," Advances in Math. **76** (1989), 245–259.

S. Sundaram, "Tableaux in the representation theory of the classical Lie groups," in *Invariant Theory and Tableaux*, D. Stanton (ed.), Springer-Verlag, 1990, 191–225.

G. P. Thomas, "A generalization of a construction due to Robinson," Canad. J. Math. **28** (1976), 665–672.

G. P. Thomas, "On Schensted's construction and the multiplication of Schur-functions," Advances in Math. **30** (1978), 8–32.

J. Towber, "Two new functors from modules to algebras," J. of Algebra **47** (1977), 80–104.

J. Towber, "Young symmetry, the flag manifold, and representations of GL(n)," J. of Algebra **61** (1979), 414–462.

J. Van der Jeugt and V. Fack, "The Pragacz identity and a new algorithm for Littlewood–Richardson coefficients," Computers Math. Appl. **21** (1991), 39–47.

G. Viennot, "Une forme géométrique de la correspondance de Robinson–Schensted," in *Combinatoire et Représentation du Groupe Symétrique*, Lecture Notes in Math. **579** (1977), Springer-Verlag, 29–58.

K.-P. Vo and R. Whitney, "Tableaux and matrix correspondences," J. of Combin. Theory, Ser. A **35** (1983), 328–359.

M. L. Wachs, "Flagged Schur functions, Schubert polynomials, and symmetrizing operators," J. of Combin. Theory, Ser. A **40** (1985), 276–289.

H. Weyl, *The Classical Groups. Their Invariants and Representations*, Princeton University Press, 1939.

D. E. White, "Some connections between the Littlewood–Richardson rule and the construction of Schensted," J. of Combin. Theory, Ser. A **30** (1981), 237–247.

A. Young, "On quantitative substitutional analysis III," Proc. London Math. Soc. (2) **28** (1928), 255–292.

A. V. Zelevinsky, "A generalization of the Littlewood–Richardson rule and the Robinson–Schensted–Knuth correspondence," J. of Algebra **69** (1981), 82–94.

# Index of Notation

# General Index

algebraic subset, 128–9
alphabet, 2
alternating representation, 80
antilexicographic ordering, 198

base change, 107
binary tree, 70
birational, 168, 213
Borel–Moore homology, 215–9, 225
Borel's fixed point theorem, 155–6
branching rule, 93
Bruhat order, 173–7
bumping route, 9, 187
Burge correspondence, 198–201

canonical construction of $P(w)$, 22
Cauchy–Littlewood formula, 52, 121
character, 91, 120
Chern class, 161, 214, 222–5
Chow variety, 140
class of subvariety, 219–22
closed embedding, 129
cohomology, 212
   of flag manifold, 161–2, 181
   of Grassmannian, 152–3
column bumping, 186–9
Column Bumping Lemma, 187
column group, 84
column-insertion, 186–9
column tabloid, 95
column word, 27, 187
complete flag, 145, 154

complete symmetric polynomial, 3, 72, 77
conjugate diagram, 2
conjugate L–R equivalence, 196
conjugate placing, 203
conjugate shape equivalence, 196
content of a tableau, 25, 64

decreasing sequences, 34–5, 56, 71
degree, 213
degree homomorphism, 213
Deruyts' construction, 104, 111, 126
determinant representation, 112
determinantal formulas, 75, 77, 146
diagram of a permutation, 158
difference operators, 165–6, 173
dimension of variety, 130
dominance order, dominate, 26
dual class, 150
dual flag variety, 182
dual Schubert cell, 148, 158
dual Knuth equivalence, 191
dual tableau, 184
duality isomorphism, 152, 182
duality theorems, 149, 160, 201, 206

elementary Knuth transformation, 19
   dual Knuth transformation, 191
elementary move, 209
elementary symmetric polynomial, 4, 72, 77
equivariant line bundle, 143
Erdös–Szekeres theorem, 34
evacuation, 184
exchange, 81, 98, 102, 105